# TEMPERATURE AND ENVIRONMENTAL EFFECTS ON THE TESTIS

# ADVANCES IN EXPERIMENTAL MEDICINE AND BIOLOGY

---

## Recent Volumes in this Series

---

A Continuation Order Plan is available for this series. A continuation order will bring delivery of each new volume
immediately upon publication. Volumes are billed only upon actual shipment. For further information please contact
the publisher.

# TEMPERATURE AND ENVIRONMENTAL EFFECTS ON THE TESTIS

Edited by

## Adrian W. Zorgniotti

New York University School of Medicine
New York, New York

PLENUM PRESS • NEW YORK AND LONDON

Library of Congress Cataloging in Publication Data

Conference on Temperature and Environmental Factors and the Testis (1989: New
York University School of Medicine.
 Temperature and environmental effects on the testis / edited by Adrian W.
Zorgniotti.
     p.     cm. — (Advances in experimental medicine and biology: v. 286)
 "Proceedings of a Conference on Temperature and Environmental Factors and the
Testis, held Dec. 8-9, 1989, at the New York University School of Medicine, New York,
New York" — T.p. verso
 Includes bibliographical references.
 Includes index.
 ISBN-13: 978-1-4684-5915-9
 1. Testis — Effect of heat on — Congresses. 2. Infertility, Male — Environmental
aspects — Congresses. 3. Testis — Thermography — Congresses. I. Zorgniotti, Adrian W.,
1925-    II. Title. III. Series.
 (DNLM: 1. Body Temperature — congresses. 2. Infertility, Male — physiopathology —
congresses. 3. Testis — congresses. 4. Thermography — congresses. W1 AD559 v.
286 / WJ 830 C748t 1989]
QP255.C653   1989
616.6'8 — dc20
DNLN/DLC                                                            90-14362
for Library of Congress                                                  CIP

Proceedings of a Conference on Temperature and Environmental Factors
and the Testis, held December 8-9, 1989, at the New York University
School of Medicine, New York, New York

ISBN-13: 978-1-4684-5915-9     e-ISBN-13: 978-1-4684-5913-5
DOI: 10.1007/ 978-1-4684-5913-5


© 1991 Plenum Press, New York
Softcover reprint of the hardcover 1st edition 1991
A Division of Plenum Publishing Corporation
233 Spring Street, New York, N.Y. 10013

For Diane-Marie

# PREFACE

> *It must be considered that there is nothing more difficult to carry out,*
> *nor more doubtful of success, nor more dangerous to handle, than to initiate*
> *a new order of things.*
>
> *Machiavelli: The Prince (1513)*

These are the Proceedings of a Conference on Temperature and Environmental Factors and the Testis which took place at New York University School of Medicine, December 8th and 9th, 1989. There is good reason to believe that this was the first of its kind to address, exclusively, the implications of temperature for this highly thermosensitive organ and its precious genetic cargo. The organizers of the Conference hoped to stimulate interest in this area which, paradoxically, has a considerable literature but which has received scant attention and sometimes outright opposition from clinicians expert in male infertility.

There have been studies of the relationship of temperature to reproduction starting in the mid-18th Century with observations of the relationship of water temperature to spawning of fish. There is also a vast literature on the deleterious effects of externally applied heat upon spermatogenesis but little study of the possibility that intrinsic heat may be an important etiologic factor in subfertile semen. Today, fertility research has largely ignored this in favor of research in areas which have not produced successes, in terms of live births, comparable to what can be obtained by varicocelectomy (when appropriate) or scrotal hypothermia: viz.

1. Concentration upon the endocrine aspects of testicular function and its relation to spermatogenesis.
2. Gamete manipulation to solve problems of male infertility.

The attitude of a large segment of the scientific community toward temperature and testis function is exemplified by an opinion expressed by leaders such as the Manns[1] with regard to application of cooling to the subfertile testis: "There are numerous (but not necessarily well-founded) statements concerning spectacular improvement in sperm output and fertility resulting from changing tight for loose clothing; by the same token, cold irrigations of the scrotal area is said to have a miraculous influence on the performance of subfertile men. Se non é vero, ben trovato."

It has been established that application of minimal heat to the scrotum can affect spermatogenesis and epididymal maturation. This observation becomes increasingly relevant when we observe a decline in fertility and confront imminent global climatologic changes.

---

[1]Mann, T. and Lutwak-Mann, C. 1981. Male reproductive function and semen. Springer-Verlag, Berlin, p. 89.

Some of the matters addressed by the Conferees were:

- Basic temperature physiology of the testis and scrotum.
- Testis thermometry.
- The heat exchanger function of the pampiniform plexus.
- The thermodynamics of heat loss by the testis and scrotum.
- Feedback thermoregulation in the testis.
- The effect of temperature on the biochemistry of the testis.
- Temperature effects on the epididymis.
- Evidence that environmental temperature elevation affects sperm output and fertility.
- Evidence that infertile men have significantly higher intrascrotal temperature than normals.
- Treatment of subfertile semen by altering temperature.

Hopefully, these Proceedings will persuade the reader that intrinsic and extrinsic temperature alterations do play a major role in testis physiology and male fertility and are worthy of study. Areas which still remain largely unexplored are the possible implications for genetic alteration, fetal wastage and possibly testicular cancer. Investigators who do not incorporate the study of temperature into their research run the risk of finding their efforts rendered naught if, indeed, temperature is central to testicular function and male fertility.

I am grateful to Maria S. Chan and Charles O. Chan for their editorial work on this book.

Adrian W. Zorgniotti
Conference Chairman

# CONTENTS

# VIEWS OF TESTICULAR FUNCTION FROM ANTIQUITY TO THE PRESENT

Adrian W. Zorgniotti

Department of Urology
New York University School of Medicine

## VARICOCELE

Interestingly enough, the earliest references to alteration of the testis have to do with varicocele. Celsus (Born ca. 25 AD) spoke of cirsocele (varicocele): "When the disease has spread also over the testicle and its cord, the testicle sinks a little lower and becomes smaller than its fellow inasmuch as its nutrition has become defective."

The earliest reference to the effect of varicocele on semen appears to be that of Fortunatus Fidelis (1602) who is quoted by Amann (1689): "Fidelis considers varicose men to be impotent, but one should distinguish that this is not true for those who have varices of the thighs, legs and feet. But varicose women can conceive and why shouldn't varicose men be fertile also? But when the testicles are infested with an infinity of things like knots or varices and the part becomes very heavy so that the production of prolific (fertile) semen is impeded and this disease is called cirsocele."

A major advance was the identification of spermatozoa by Leeuwenhoek and his pupil Hamm (1677) with his newly invented microscope. Many medical historians have said that Leeuwenhoek thought that what he saw were parasites in the semen. However, in a letter to the Royal Society (1685), he wrote: "-- when a man is unable to beget children, -- no living animalcules will be found in the seed of such a man, or that, should any living animalcules be found in it, they are too weakly to survive long enough in the womb." Unquestionably the first description of azoospermia and oligoasthenospermia. This is of additional interest in that infertile marriage was generally considered to be of female origin; here is clear proof of the male factor.

Gosselin (1853) wrote of a patient who was found to be azoospermic by microscopic examination. The patient had a left varicocele with considerable atrophy of the testicle. The man also had a right gonorrheal orchitis. This represents an early reference to the microscope specifically for the diagnosis of infertility.

While Tulloch (1951) is generally credited with the introduction of varicocelectomy for the reversal of subfertile semen, there were others ahead of him. Tulloch, performed surgery on an azoospermic man and this resulted in a count of 27 million active sperm and a pregnancy. This was fortuitous since we know that a good result is most unusual in the azoospermic patient. Barwell (1885) observed improvement in testis size and consistency and a reversal of the infertile state after varicocelectomy in one man who produced two children within seven years of the surgery. At that time, the operation was not performed to improve fertility. Macomber and Sanders, pioneers in semen analysis (1929), reported a planned varicocelectomy which resulted in an increase in

*Temperature and Environmental Effects on the Testis*
Edited by A. W. Zorgniotti, Plenum Press, New York, 1991

spermatozoa count from 20 to 73 million/ml and a pregnancy. Similarly, Wilhelm (1937) performed varicocelectomy for the subfertile state but it was not until Tulloch's publication that surgeons became interested and varicocelectomy became prevalent.


EXTRINSIC TEMPERATURE AND THE TESTIS

Understanding of the thermoregulatory function of the scrotum in most mammals and the implications of temperature for spermatogenesis did not begin until the closing years of the 19th Century, although Cooper (1830) had noted the effect of ambient temperature on the scrotum: "The scrotum varies greatly in its appearance and size, for under the influence of cold, it is small, contracted and wrinkled; under heat it is relaxed, smooth on its surface, and greatly extended."

That there was some understanding of temperature and fertility in the 17th Century can be found in the diary of Samuel Pepys (1664). After a dinner for a godchild, Pepys asked the ladies present for advice on how to get his wife, Elizabeth, pregnant since they were married for a number of years without issue. Most of their advice had to do with the frequency of intercourse and what beverages to drink but two recommendations are of interest. 1. "Wear cool Holland-drawers." 2. "To lie with our heads where our heels do, or at least to make the bed high at feet and low at head." Both can have the effect of decreasing testicular temperature.

The early work began with studies of natural and artificial cryptorchidism and it was not until the 1920's that application of extrinsic heat to the scrotum resulted in a clear understanding of the effects of heat on spermatogenesis. Felizet and Branca (1898, 1902) are generally credited with the original suggestion that elevated temperature affected spermatogenesis in the cryptorchid testis. However, Griffiths (1893) showed that spermatogenic activity of the dog would degenerate when replaced in the abdominal cavity. Crew (1922), who had a doctorate in philosophy as well as a degree in medicine and was a member of the Animal Breeding Research Department at the University of Edinburgh, published a paper which, while not based on experimental evidence, is the point of departure for research on the significance of scrotal temperature. His conclusions were:

1.    In mammals that move "impulsively", the testis migrates from its position in the region of the kidneys into the scrotum. That ultimately the structure of the inguinal canal developed in order to permit the descent of the testis. This Lamarckian explanation never prevailed.

2.    The scrotum is capable of thermoregulation and that there is an optimum temperature for spermatogenesis found within the scrotum and not within the abdominal cavity.

3.    The imperfectly descended testis is aspermatic because of the temperature of the abdominal position.

Crew's "suggestion", as he chose to call it, was timely in that it coincided with serious experimentation on the effect of extrinsic temperature on the testis. The leading figure was Carl R. Moore of the Hull Zoological Laboratory of the University of Chicago. Moore (1922) began with surgically produced cryptorchidism in rats and guinea pigs observing that the tubular degeneration left only a layer of Sertoli Cells on the basement membrane. He and his coworkers (1923, 1924) soon conceived the idea of studying the effects of heat application and insulation of the scrotum, showing conclusively that increase in temperature resulted in seminiferous degeneration. Moore's experiments involved the study of excised testicular tissue. Seven years later Phillips and McKenzie (1934) studied the semen of rams who had scrotal insulation, obtaining their specimens at the time of service. The mean increase in scrotal or testis temperature under their experimental conditions was 2.2°C demonstrating alterations in count, motility and morphology.

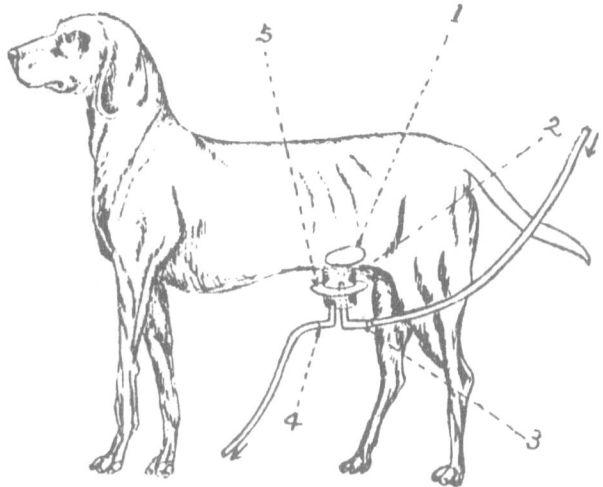

1. Experimentally retained testicle
2. Cooler
3. Inlet tube from tank 2m high
4. Outlet tube
5. Margin of fixing band

Figure 1. Fukui's classical experiment (1923) of cooling the surgically created
cryptorchid testis in the dog. The apparatus circulated cool water over the
underlying testis. In this way spermatogenic degeneration, which was expected to
occur owing to the higher intrabdominal temperature, was obviated.

Another investigator was Fukui of the Kyoto Imperial Medical School. Fukui
(1923a) conducted similar experiments exposing the testis of the rabbit to a variety of
sources of external heat (sunlight, warm water and warm air) and was able to
demonstrate seminiferous degeneration. He called these "heat testicles" a term which was
never adopted. Fukui (1923b) also performed a key experiment in which he applied
external cooling to a surgically cryptorchid testis in the dog, thereby preventing the
expected seminal degeneration (Figure 1). Fukui is quoted in all reviews but seems to
have been overshadowed, in the West at least, by Moore. These initial experiments
opened the door to scores of reports on the effect of heat application to the testes of a
wide variety of species by a surprising variety of thermal agencies, including radar and
hot paraffin, and at temperature levels which, at times, can only be considered lethal.
These were, for the most part, published from 1920 - 1960 although even today external
heat to the testes experiments are still being published without adequate review of the
prior literature. More or less, all came to the same conclusion: extrinsic heat application
is deleterious to spermatogenesis.

INTRINSIC TEMPERATURE AND SEMINIFEROUS DEGENERATION

The possibility that intrinsic elevation of scrotal temperature could be related to
seminiferous degeneration has, on the other hand, not caught the interest of the majority
of investigators and thus has not been studied as extensively as extrinsic temperature
elevation.

Davidson (1954a) stated that over half of its patients had a disturbance of testis
temperature regulation although he did not offer much in the way of experimental
evidence to support this. He went on to report (1954b) on twelve men who had been
instructed to immerse the scrotum in cold water before and after varicocelectomy with
good results. Unfortunately, we cannot be certain which of the two agencies was
actually the effective one.

Robinson and Rock (1967) were the first to compare temperature in normal men and oligospermic men using scrotal rectal (temperature) differentials (SRD). Their observation, not subjected to statistical evaluation, was that 21 euspermic men had a mean SRD of 1.75°C while 37 oligospermics had SRD of 1.9°C. In another study 36 normals (no reference to semen) had SRD of 2.4°C, this difference in means being 0.5°C. Their results were clearly inconclusive but this can be attributed to three causes: (1) The use of SRD is not as accurate as actual intrascrotal temperature. (2) The accuracy of their electronic thermometer was probably of the order of ±0.5°C, which would make it incapable of detecting small differences in temperature. (3) Data have to be gleaned from more than one experiment. They also noted that chronic lowering of the SRD by 1.0°C, using scrotal insulation, resulted in suppression of spermatogenesis. (A decrease in SRD indicates increase in testicular temperature assuming that rectal temperature does not change. Thus 1.0°C increase in temperature can be said to disrupt spermatogenesis in humans.) The authors do not seem to have pursued the matter further.

Agger (1971) noted small elevations in varicocele patients over normals and concluded that the presence of a temperature increase could not be excluded.

Hanley (1956) reported a decrease in testis temperature 1.0°C following varicocelectomy.

Zorgniotti and MacLeod (1973) showed significant intrascrotal temperature differences (0.6 - 0.7°C) between oligospermic varicocele patients and normals. In order to test the supposition that low level hyperthermia was related to infertility, Zorgniotti et al. (1980) constructed a testicular hypothermia device (THD) which lowered temperature 2.0°C, demonstrating that this improves semen and produces pregnancy.

Lynch et al. (1986) demonstrates improvement in semen parameters by avoidance of environmental hyperthermia such as tight briefs, hot baths, environmental occupational heat and so forth. This did not have an appreciable effect on pregnancies.

More recently, Mieusset (1987a,b) presented data on normal and infertile men. His observation was that a significant number of patients with infertile semen have elevated intrascrotal temperature when compared to fertile men, concluding that elevated temperature is "a causal or at least an aggravating factor" in subfertile semen. Zorgniotti and Sealfon (1987, 1988) compared temperatures of men with excellent semen and those with oligospermia and concluded that over 80% of such infertile men had elevated intrascrotal temperature. The two studies are comparable in that both investigators used the same method for measuring intrascrotal temperature. Mirone (1987) has proposed that elevated temperature, in the absence of varicocele, be called Idiopathic Testicular Hyperthermia.

REFERENCES

Agger, P. 1971. Scrotal and testicular temperature: Its relation to sperm count before and after varicocelectomy. Fertil. Steril., 22: 286.

Amann, P. 1689. "Irenicum Numae Pompilii", Sumptibus Autoris, Frankfurt and Leipzig, p. 142 (Available in NLM).

Barwell, R. 1875. Varicocele and its curative operation by subcutaneous loop. Lancet, I: 820.

Celsus, A.C. (b. ca. A.D. 25). "De Medicina", Loeb Classical Library, V. 3, p.388.

Cooper, A. Sir. 1830. "Observations on the structure and diseases of the testis". Longman, Rees, Orme, Browne and Greene, London.

Crew, F.A.E. 1922. A suggestion as the cause of the aspermatic condition of the imperfectly descended testis. J. Anat., 56: 98.

Davidson, H.A. 1954a. Disturbances of testicular temperature regulation. Proc. Roy. Soc. Med., 47: 710.

Davidson, H.A. 1954b. Treatment of male subfertility. Practitioner, 173: 704.

Felizet, G. and Branca, A. 1898. Histologie du testicule ectopique. J. Anat. Physiol. (Paris), 34: 589.

Felizet, G. and Branca, A. 1902. Recherches sur le testicule en ectopie. J. Anat. Physiol. (Paris), 38: 329.

Fukui, N. 1923a. On a hitherto unknown action of heat ray on the testicles. Japan Med. World, 3: 27.

Fukui, N. 1923b. Action of body temperature on the testicle. Japan Med. World, 3: 160.

Gosselin, M. 1853. Etudes sur l'obliteration des voies spermatiques. Arch. Gen. Med., 5ème Serie 2: 268.

Griffiths, J. 1893. Structural changes in the testicle of the dog when it is replaced in the abdominal cavity. J. Anat. Physiol., 27: 482.

Hanley, H.C. 1956. Surgical correction of errors of temperature regulation. In: Proceedings of Second World Congress of Fertility and Sterility. Amsterdam, World Fertility Association, p. 953.

Leeuwenhoek, A. van. 1677, 1685. "The Select Works". S. Hoole, ed. H. Fry, London, 1798-99.

Lynch, R. et al. 1986. Improved seminal characteristics in infertile men after a conservative treatment regimen based on the avoidance of testicular hyperthermia. Fertil. Steril., 46: 476.

Macomber, D. and Sanders, M.B. 1929. The spermatozoa count. New Engl. J. Med., 200: 981.

Mieusset, R. 1987. Températures scrotales et infertilité. In: Résumés du 5ème Forum Internationale d'Andrologie, Abstract #163, G. Arvis, ed., Hôpital Saint Antoine Paris.

Mieusset, R. et al. 1987. Association of scrotal hyperthermia with impaired spermatogenesis in infertile men. Fertil. Steril., 48: 1006.

Mirone, V. et al. 1987. La terapia ipotermica dell'ipertermia testicolare essenzialle e da varicocele: Il THD, Valutazione preliminare, in progressi. In: "Andrologia", G. D'Ottavio and D. Pozza (eds.), Acta Medica Edizioni e Congressi s.r.l. Rome, p. 331.

Moore, C.R. 1923. On the relationship of the germinal epithelium to the position of the testis. Anat. Record, 25: 142.

Moore, C.R. and Oslund, R. 1924. Experiments on the sheep testis - cryptorchidism, vasectomy and scrotal insulation. Am. J. Physiol., 67: 595.

Pepys, S. 1644. The diary of Samuel Pepys. Latham, R. and Matthews, W. (eds.), University of California Press, Berkeley, v. 5, p. 222.

Phillips, R.W. 1931. Observations on the spermatozoa of the ram and their applications to the determination of fertility. Master's thesis, University of Missouri.

Phillips, R.W. and McKenzie, F.F. 1934. In: "The thermoregulatory function and mechanism of the scrotum". University of Missouri College of Agriculture; Agricultural Research Station, Research Bulletin 217.

Rock, J. and Robinson, D. 1965. Effect of induced intrascrotal hyperthermia on testicular function in man. Am. J. Obstet. Gynecol., 93: 793.

Tulloch, W.S. 1951-2. Consideration of sterility factors in light of subsequent pregnancies: subfertility in the male, Trans. Edinburgh Obstet. Soc., 59: 29.

Wilhelm, S.F. 1937. Sterility in the male. In: Oxford Loose Leaf Surgery, 764 (298 17A).

Zorgniotti, A.W. and MacLeod, J. 1973. Studies in temperature, human semen quality and varicocele. Fertil. Steril., 24: 854-863.

Zorgniotti, A.W. et al. 1980. Chronic scrotal hypothermia as a treatment for poor semen quality. Lancet, I: 904-6.

Zorgniotti, A.W. and Sealfon, A.I. 1987. Elevated scrotal temperature in 300 consecutive infertile patients. In: Résumés du 5ème Forum Internationale d'Andrologie, Abstract #162, G. Arvis (ed.), Hôpital Saint Antoine, Paris.

Zorgniotti, A.W. and Sealfon, A.I. 1988. Measurement of intrascrotal temperature in normal and subfertile men. J. Reprod. Fert., 82: 563-6.

# SECTION 1

## TESTIS THERMOREGULATION: AN OVERVIEW

# THERMOREGULATION OF THE SCROTUM AND TESTIS:

## STUDIES IN ANIMALS AND SIGNIFICANCE FOR MAN

Geoffrey M.H. Waites

World Health Organization
Special Programme of Research, Development and Research Training
in Human Reproduction
1211 Geneva 27, Switzerland

ABSTRACT

This article reviews the extensive information available from experiments on animals concerning the thermal monitoring provided by the scrotum. Cutaneous temperature receptors initiate responses which follow unique pathways and undergo "switching" processing within the central nervous system. These pathways evoke reflex responses which are subject to control from receptors in other regions of the body, including the skin and temperature sensitive neurones in the brain and spinal cord.

The local thermoregulatory responses of the scrotum, e.g., sweating and vasomotor changes, clearly have a role to play in the protection of the testis against temperature elevation in both man and animals. It is as yet more difficult to propose how general reflex responses may be of benefit in the protection of spermatogenesis against heat damage. This meeting should provide the stimulus for further work. The need is urgent since it may also provide a means to apply this knowledge to a better understanding of male infertility and possibly to speed the development of male methods of fertility regulation.

## INTRODUCTION

In his fictional anticipation of the future, Aldous Huxley (1932) caused one of his characters to comment: "the male gametes ... have to be kept at thirty-five instead of thirty-seven. Full blood heat sterilizes. Rams wrapped in thermogene beget no lambs". Other papers from this meeting will trace the history of these perceptive remarks and will close the gap between the 1930's and today. My task is to review how knowledge of the neuro-physiological basis of thermo-regulation of the testis and its integration with the reflexes initiated from the scrotum in animals may help in the understanding of the medical conditions associated with the thermal vulnerability of spermatogenesis. In addition to the reviews in other articles in this publication, the animal studies have been reviewed by Hales and Hutchinson (1971), Hellon (1982; 1983), and Waites (1970; 1973; 1976).

## MATERIALS AND METHODS

The classical approach to understanding whole-body responses to thermal stress in animals has been to enclose the animals in climate rooms wherein air temperature and humidity can be controlled. The metabolic and other responses were then recorded by instruments of ever increasing sophistication external to the rooms. In this way respiratory and body gas exchanges, blood and skin temperature changes and, in a variety of tissues, e.g., brain and central nervous system, adaptations could be recorded often for weeks or months at a time. More recently, space exploration research has added refinements in the form of techniques with remote sensing options which have reduced the degree of interference with the animal (or man).

The methods used in the articles to be reviewed here have in general been more specific because they have sought to record differential regional responses (e.g., Waites and Voglmayr, 1963; Ingram and Legge, 1972), sometimes integrating these with the classical whole-body responses (e.g., Hales and Hutchinson, 1971). For example, to explore the local responses of the scrotum to temperature, a portable climate chamber was designed in which the temperature and humidity of the circulating air could be controlled together with the wall temperature so that the skin could be subjected to radiant as well as convective heat (Fig. 1). The more recent neuro-physiological results have been acquired with sensitive techniques to record from cutaneous temperature receptors (e.g., Iggo, 1969; Hellon et al., 1975) together with central recording of neuronal responses from the spinal cord and areas of the brain (e.g., Hellon and Misra, 1973; Hellon and Taylor, 1982). While these latter studies have required the use of general anaesthesia and were limited to studying neural pathways in rats, they have nevertheless been essential to the more complete understanding of the neural patterns underlying the general responses.

## RESULTS AND DISCUSSION

### Reflex Responses from Scrotal Warming in Animals

Some years ago it was reported that a series of powerful thermo- regulatory reflexes were initiated by altering the temperature of the scrotum of the ram (Waites, 1961, 1962,

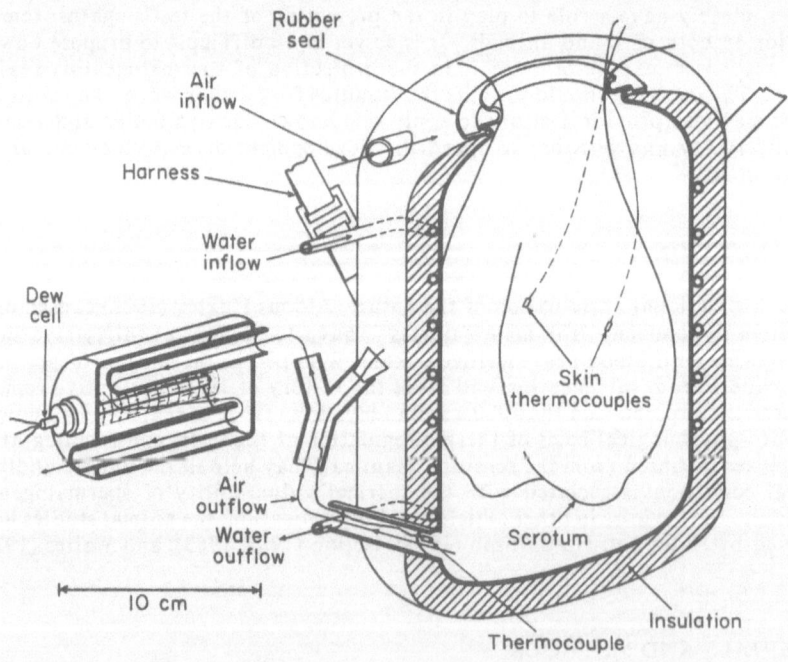

Fig. 1. Temperature-controlled chamber for detecting and measuring fluid production and skin temperature responses to warming the scrotum of the ram (From Waites and Voglmayr, 1963).

10

1970; Waites and Voglmayr, 1963). Raising the temperature of scrotal skin from its normal value of around 32°C to above a threshold of 35 to 36°C evoked a dramatic polypneic response from fully-fleeced rams with respiratory rates in excess of 200 per min. Heating equivalent areas of mid-side skin was without effect. If prolonged, the panting led to sharp falls in deep body temperature of up to 2°C. The polypneic response was lost when the body wool was removed and skin temperature on the general body surface was thereby cooled to below 35°C (Fig. 2).

Hales and Hutchinson (1971) established that the pattern of effector mechanisms recruited by scrotal warming in the ram was considerably modified by body skin temperature. When the body was warm the cooling achieved through panting was supported by peripheral vaso-dilation in the skin of the ears, mid-side and limbs, without any alteration in metabolic rate. Previously removing the fleece so that body skin temperature fell caused the rams to elevate their oxygen consumption. When the scrotum was then warmed, metabolic rate was reduced without any alteration in respiratory rate and with only variable and disorganized vasomotor responses. Thus stimulation of warm receptors in the scrotum of bare sheep caused a reduction of body temperature (Fig. 2) by regulating metabolic rate, the polypnea response presumably being inhibited by the thermal input from the cold receptors in general body skin.

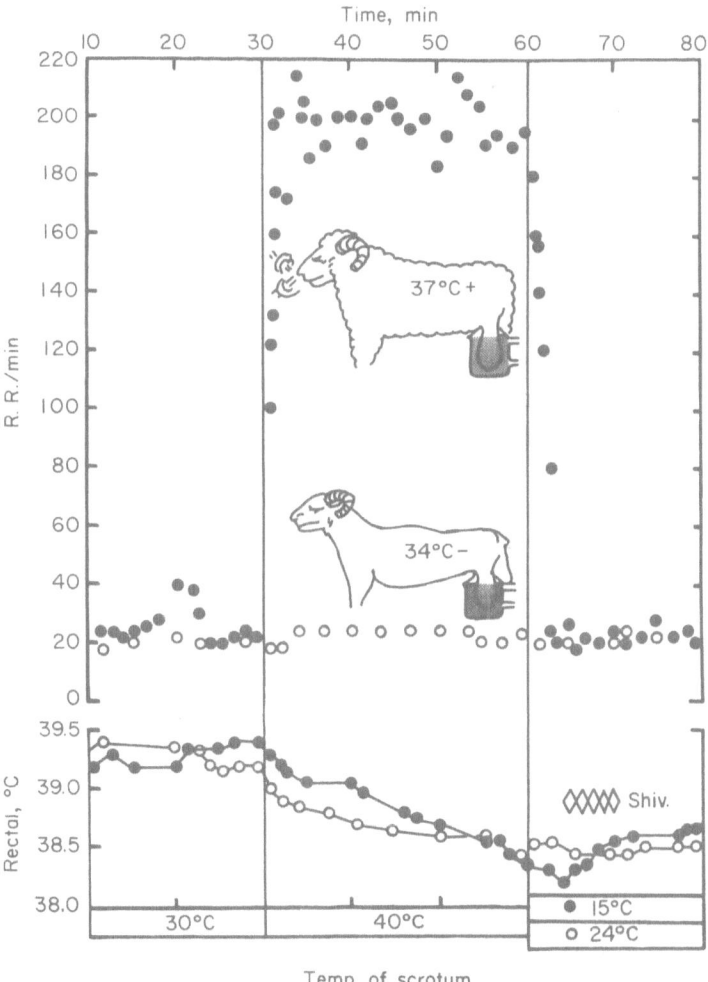

Fig. 2. Respiratory and rectal temperature changes induced by scrotal heating to 40°C in wool-covered (closed symbols) and bare (open symbols) rams (Waites 1961, 1962).

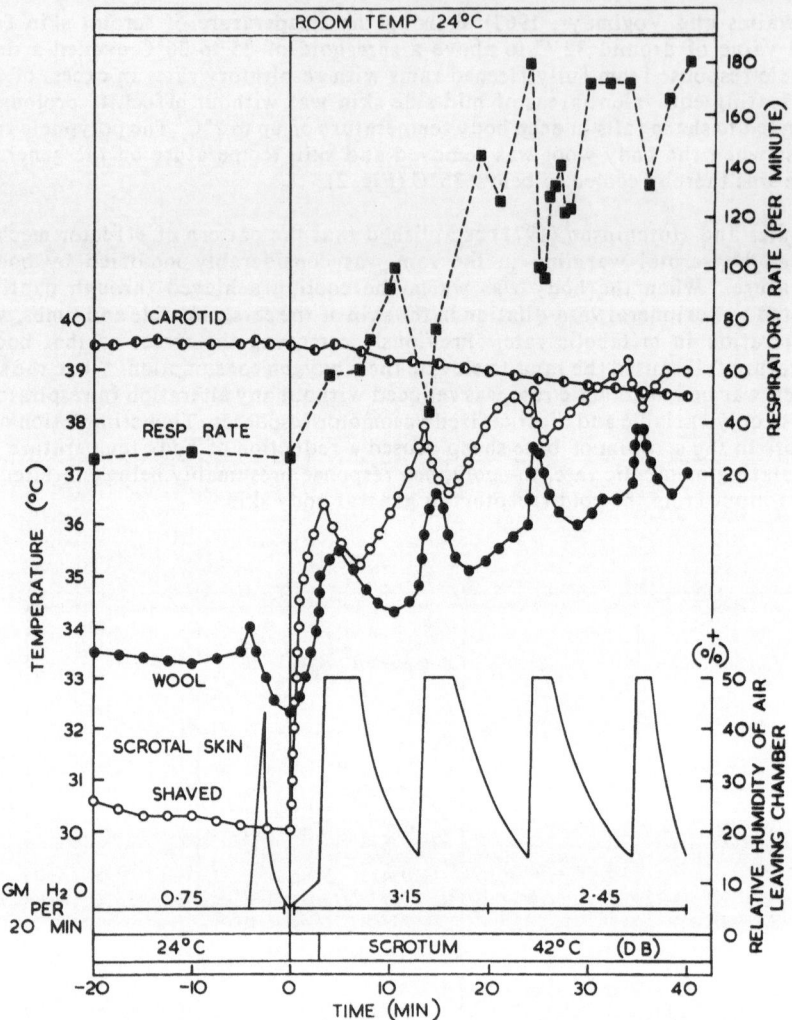

Fig. 3. Effects of heating the scrotum (at 0 min) on the temperature of bare (open circles) and of wool-covered scrotal skin (closed circles) of a ram. Note that the dew cell recorded 4 sweat discharges during the heating period and that the polypneic response was interrupted during each of the last 3 discharges. The wool on the 5cm² patch was 30 mm long (From Waites and Voglmayr, 1963).

The effector responses recruited when the scrotum of the naturally- naked, domestic pig was heated also depended on the temperature of its general body surface (Ingram and Legge, 1972). At ambient temperatures above 30°C, the pig selected panting but at lower air temperatures, oxygen consumption was depressed and blood flow through the tail, as an indicator of peripheral vasomotor control, was substantially increased. In this species, there were clear indications that thermally-sensitive central neurones also intervened in the responses. Cooling the hypothalamus or regions of the spinal cord of the pig by means of implanted thermodes reduced the responses to heating the scrotum, whereas warming these central regions potentiated these responses. Thus, not only did the thermal state of different regions of the body surface modify the responses to scrotal warming, but there was also an interaction with thermally-sensitive neurones in the central nervous system.

When scrotal skin temperature was raised to above 35°C, the large apocrine sweat glands of the scrotum discharged sweat synchronously, usually at intervals of 2-14 min. Evaporation of the sweat when wool on the scrotum was short caused sharp falls in skin

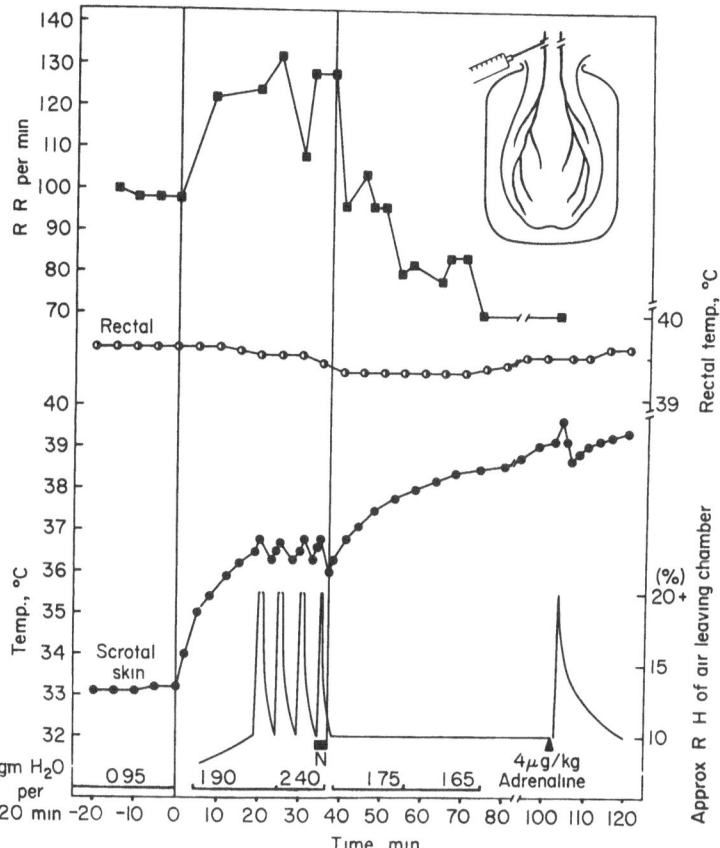

Fig. 4. Abolition of scrotal reflexes by the percutaneous anaesthesia of the superior perineal nerves (at N; from Waites and Voglmayr, 1963).

temperature of up to 2.6°C. When wool was present, there was an exothermic response between the sweat and the wool keratin and the skin temperature first rose and then fell (Fig. 3, Waites and Voglmayr, 1963).

All heat-induced sweat discharges and the reflex respiratory responses were abolished by locally anaesthetizing the superior perineal nerves which carry sensory fibres from the scrotal skin in the ram, after which skin temperatures rose smoothly to above 39°C. The integrity of the sweat glands remained intact during the anaesthesia and responded to intravenous injections of adrenergic agents (Fig. 4). It was concluded that the cutaneous receptors initiating the general reflex responses, e.g., panting, also initiated local and reflex sweating through a neural adrenergic mechanism.

### Central Processing of the Thermosensory Input from the Scrotum

By recording from single fibres in the scrotal nerve of the rat, Iggo (1969) demonstrated that the cutaneous receptors of the scrotum were more abundant than in other skin areas and comprised two populations: warm receptors excited by temperatures in the range 35 to 45°C, and cold receptors stimulated by scrotal temperatures from 33 to 13°C (Fig. 5; for review see Hensel, 1983).

Hellon and his colleagues traced the neural pathways for the scrotal reflexes in the rat and demonstrated their uniqueness compared to other sensory afferent pathways. In the region of the spinal cord where the scrotal nerve enters, about half of the dorsal horn neurones were responsive to changes of scrotal temperature, increasing their firing rate

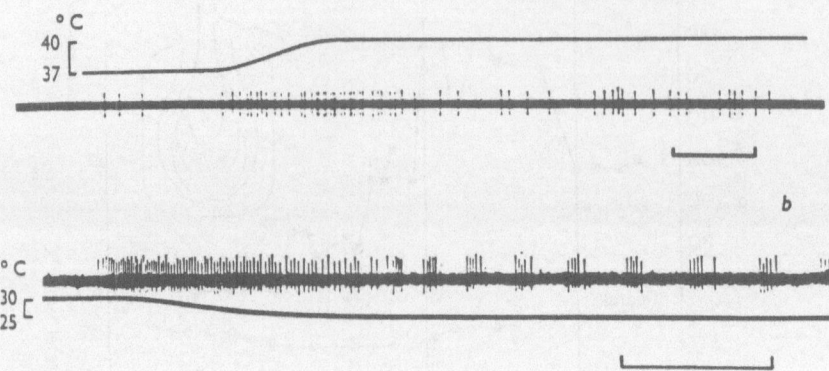

Fig. 5. Impulse discharges in single fibres of the rat scrotal nerve: (a) responding to warming, 37 to 40°C; (b) responding to cooling, 30 to 25°C. Time calibration, 1 sec. (From Hellon et. al., 1975).

during scrotal warming (Fig. 6, lower). This thermal input was traced to the ventrobasal complex of the thalamus where neurones were excited by warming the scrotum, mostly in the range 33 to 38°C (Fig. 6, centre). The responses were explosive, increases of scrotal temperatures of only 0.5 to 2°C evoking eight-fold changes in the firing rates of thalamic neurones. Indeed, Hellon and Misra (1973) commented that, in the cat, the thalamic representation of scrotal temperature appeared to be greater than that for the rest of the body surface (Martin and Manning, 1971). Finally, the thermo-sensory input from the scrotum being processed in the spinal cord and thalamus is also linked to cells in the somatosensory cortex. The firing rate of about 40% of the cortical cells examined by Hellon et al. (1973) abruptly changed when scrotal temperature was raised by 0.5 to 2°C within the range 33 to 41°C. It would seem that thalamic excitation gives rise to cortical suppression because the majority of the thermally responding cortical cells decreased their firing rate when the scrotum was heated (Fig. 6, upper).

More recent studies clearly demonstrated that the ascending sensory pathway signalling the thermal state of the rat scrotum passed through and relayed in the nucleus raphe magnus of the brainstem before ascending to the thalamus and hypothalamus (Taylor, 1982). In these regions, scrotal temperature increases of only 0.5°C or less induced abrupt responses called "switching responses" by Hellon and Taylor (1982). These responses depended on a feed-back loop with the cerebral cortex since cooling the surface of the cortex abolished the firing of thalamic neurones switched on by scrotal warming (Fig. 7).

CONCLUSION

There is now ample evidence that the skin of the scrotum of animals provides thermal monitoring through neural pathways with unique features compared to thermal inputs from other regions of the body. The central synaptic connections are unorthodox with processing occurring in several regions of the central nervous system leading to abrupt "switching" changes in impulse activity, thus transforming wide-range scrotal receptor responses into narrow-range central responses. By contrast, thermal information from the face and trunk reaches the thalamus relatively unchanged (reviewed by Hellon, 1983). Much less is known about the connections to the hypothalamus, which presumably exist to account for the reflex efferent responses observed in conscious animals.

At the local level, thermoregulatory responses of the scrotum would benefit the testis through the heat exchange in the pampiniform plexus which acts to thermally isolate the scrotum and testes from the body (see Waites, 1970). The central responses, depending on absolute scrotal temperature levels, may be ideal for signalling a thermal

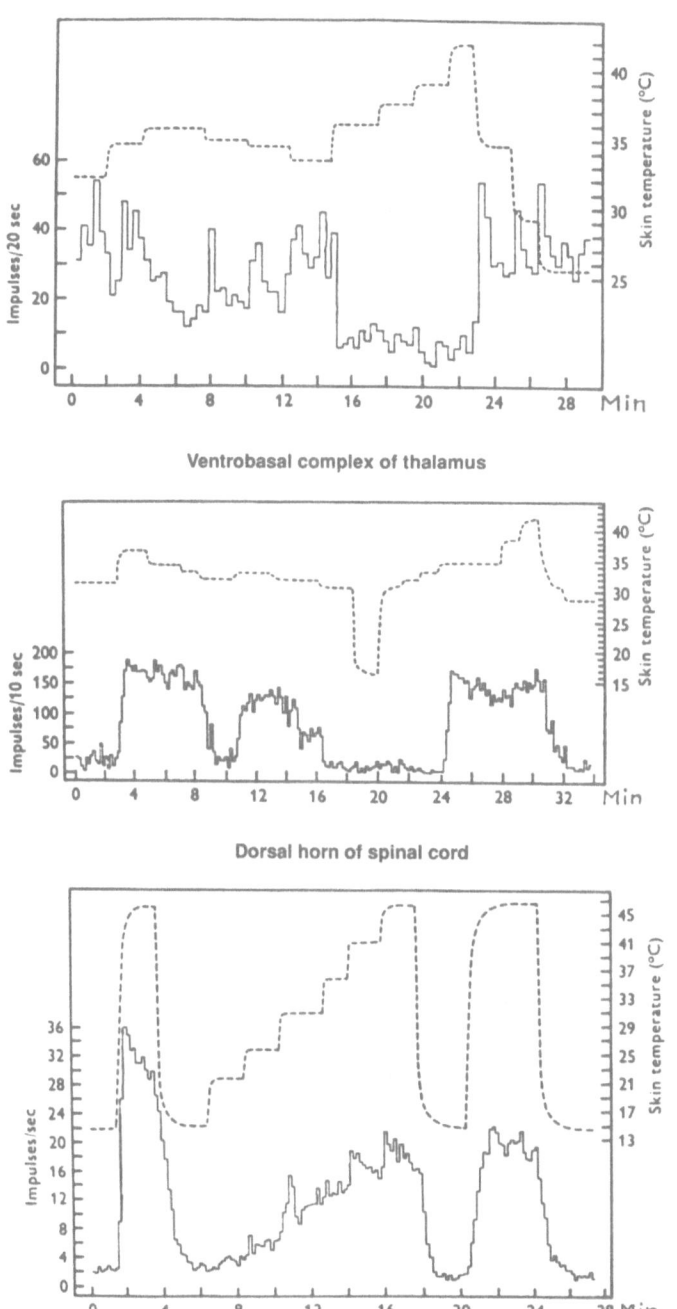

Fig. 6. Impulse frequency "switching" of central nervous system neurones during warming and cooling of the scrotum: a cell in the dorsal horn of the spinal cord (lower); abrupt and maintained response of a cell in the ventrobasal thalamus (centre); and suppression of the activity of a cell in the somatosensory cortex (upper). (From Hellon and Misra, 1973; Hellon et al., 1973).

15

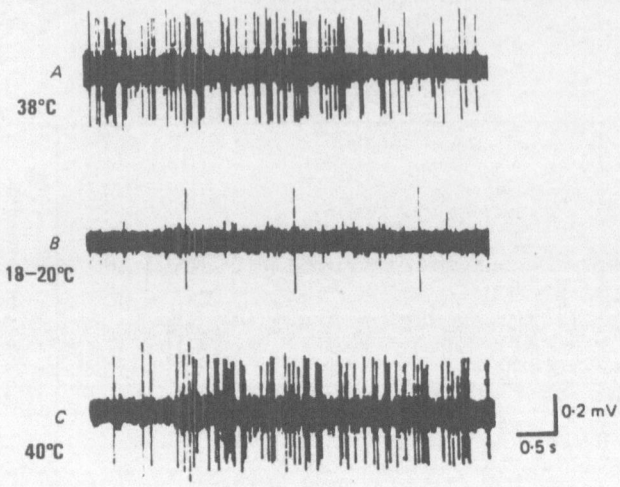

Fig. 7. Recordings from a neurone in the ventrobasal thalamus of the rat showing maximal firing rate induced at a scrotal skin temperature of 39°C, before (A) and after (C) being abolished by cooling the cerebral cortex (B; From Hellon and Taylor, 1982).

threat from the environment. Where the efferent responses lead to falls in body temperature, there would also be a benefit for general thermoregulation in exposed mammals. However, transposing this benefit to man, has proved to be difficult. The proceedings of this meeting should help us to link up the studies on animals with the clinical observations in man.

ACKNOWLEDGMENT

I wish to thank Ms. Maud Keizer for her secretarial assistance.

REFERENCES

Hales, J.R.S. and Hutchinson, J.C.D. 1971. Metabolic, respiratory and vasomotor responses to heating the scrotum of the ram. J. Physiol., 212: 353-375.
Hellon, R.F. 1982. Thermoreceptors, In Handbook of Physiology - The Cardiovascular System III, Chapter 18, 659-673, Williams and Wilkins Co, Baltimore.
Hellon, R.F. 1983. Central projections and processing of skin-temperature signals. J. therm Biol., 8: 7-8.
Hellon, R.F., Hensel, H. and Schäfer, K. 1975. Thermal receptors in the scrotum of the rat. J. Physiol., 248: 349-357.
Hellon, R.F. and Misra, N.K. 1973. Neurones in the ventrobasal complex of the rat thalamus responding to scrotal temperature changes. J. Physiol., 232: 389-399.
Hellon, R.F., Misra, N.K. and Provins, K.A. 1973. Neurones in the somatosensory cortex of the rat responding to changes in scrotal skin temperature. J. Physiol., 232: 401-411.
Hellon, R.F. and Taylor D.C.M. 1982. An analysis of a thermal afferent pathway in the rat. J. Physiol., 326: 319-328.
Hensel, H. 1983. Recent advances in thermoreceptor physiology. J. therm. Biol., 8: 3-6.
Huxley, A. 1932. Brave New World, Chapter 1.
Iggo, A. 1969. Cutaneous thermoreceptors in primates and sub-primates. J. Physiol., 200: 403-415.
Ingram, D.L. and Legge, K.F. 1972. The influence of deep body and skin temperatures on thermoregulatory responses to heating of the scrotum in pigs. J. Physiol., 224: 477-487.
Martin, H.F. and Manning, J.W. 1971. Thalamic "warming" and "cooling" units responding to cutaneous stimulation. Brain Res., 27: 377-388.

Taylor, D.C.M. 1982. The effects of nucleus raphe magnus lesions on an ascending thermal pathway in the rat. J. Physiol., 326: 309-318.

Waites, G.M.H. 1961. The polypnœa evoked by heating the scrotum of the ram. Nature, 190: 172-173.

Waites, G.M.H. 1962. The effect of heating the scrotum of the ram on respiration and body temperature. Q.J. Exp. Physiol., 47: 314- 323.

Waites, G.M.H. 1970. Temperature regulation and the testis. In: The Testis, A.D. Johnson, W.R. Gomes and N.L. Van Demark (eds), Academic Press, New York, p. 241.

Waites, G.M.H. 1973. Ambient temperature. J. Reprod. Fertil. Suppl. 19: 151-154.

Waites, G.M.H. 1976. Temperature regulation and fertility in male and female mammals. Isr. J. Med. Sci., 12: 982-993.

Waites, G.M.H. and Voglmayr, J.K. 1963. The functional activity and control of the apocrine sweat glands of the scrotum of the ram. Aust. J. Agric. Res., 14: 839-851.

# EFFECTS OF ELEVATED TEMPERATURE ON THE

# EPIDIDYMIS AND TESTIS: EXPERIMENTAL STUDIES

J. Michael Bedford

Cornell University Medical College
Departments of Obstetrics and Gynecology
  and Cell Biology and Anatomy
1300 York Avenue
New York, NY 10021

## INTRODUCTION

The testis has been the natural focus of attention in considering the effects of elevated temperatures on male reproduction. In all mammals studied, including man, the germinal epithelium of the scrotal testis is acutely sensitive to an increase of only a very few degrees, and much effort has been devoted to analysis of this response (Waites and Setchell, 1969; Vandemark and Free, 1970; Harrison, 1975; Kandeel and Swerdloff, 1988). Considerable attention has been given also to the broader biological question as to why the testes of a majority of mammals should need to function at temperatures somewhat below that of the body. However, a focus on the physiology of the testis per se has produced no satisfactory explanation of this phenomenon.

It has been known for many years that if the cauda epididymidis is elevated to the abdomen with its testis, there is also an immediate suppression of its ability to maintain spermatozoa there in a viable state (e.g., Heller, 1929; Glover, 1960). The basis for this effect was not clear, but there have been few efforts to clarify the bearing of temperature on the epididymal function. Our interest in the effects of temperature on the epididymis was stimulated by the concept that the sperm storage function of the cauda epididymidis may have been a major selective factor in the evolution of the scrotum (Bedford, 1977a; 1978a). That idea, initially based on comparative anatomical observation for the most part, has formed the basis for a variety of investigation of the bearing that temperature may have on several different functions of the male tract.

This chapter reviews briefly some testicular responses to temperature elevation reported in certain naturally cryptorchid experimental species. Then it describes in outline, studies in scrotal species of the effects of physiological temperature elevation on sperm transport and sperm maturation in the epididymis, on the function of the epithelium and storage capacity of the cauda epididymidis, and on the numbers and capacitation characteristics of spermatozoa ejaculated by males whose epididymides are maintained artificially at body temperature. Finally, it considers the possible bearing of such findings on reproductive function in men.

## ANIMAL STUDIES

### Effects of Limited Temperature Elevation on Testis Function

The lower temperature of the scrotum is not a mandatory condition for testis function in mammals. According to species, the normal location of the testis varies from

*Temperature and Environmental Effects on the Testis*
Edited by A. W. Zorgniotti, Plenum Press, New York, 1991

an abdominal position in animals such as the elephant, hyrax and some whales (testicondids), to a state of partial descent to the inguinal region (cryptorchids), or further externally to the scrotum proper. In testicondid mammals, the epididymis typically passes as a variably corrugated duct directly down from its testis connection to the pelvic outlet (Short et al., 1967; Glover and Sale, 1968). In a majority of both cryptorchid and scrotal species, by contrast, the epididymis follows the testis outline to the external pole and variably beyond it. There it forms the swelling of the cauda epididymidis and, continuing as vas deferens, returns at first along the same path and diverges into the pelvic cavity. Such a curious loop configuration serves to position the cauda, the sperm storage site, at a point that in cryptorchids is moderately cooler, and in some scrotal species is significantly cooler than the adjacent testis (Brooks, 1973; Bedford, 1978).

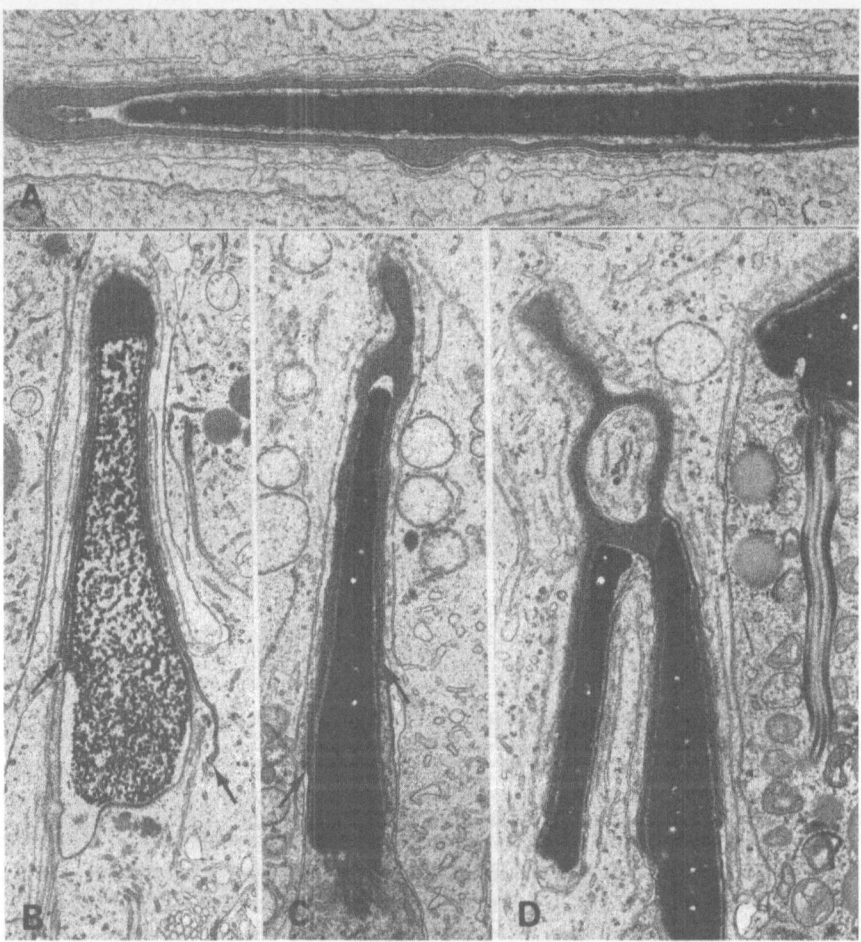

Figure 1. Electronmicrographs of late spermatids in the testis of the degu, Octodon degus. The spermatid in (A) is typical of the normal form seen in the control (sub-integumental) testes. Examples of abnormally formed spermatids were seen commonly and only in the contra-lateral testes which had been elevated to the abdominal cavity at an ambient temperature about 1.5°C higher than that of the normal testis. The spermatid in (B) displays a dorsoventral asymmetry of the posterior border of the acrosome (arrows) and an irregular outline of the condensing nucleus. That in (C) displays a similar acrosomal asymmetry and an unusual rostral extension of the acrosome. In (D) two spermatids share one cystic acrosome, and there is an abnormal undulation of tail dense fibers in the adjacent spermatid (from Bedford, Berrios and Dryden, 1982).

Animals which are natural cryptorchids, with inguinal or subintegumental testes, lend themselves to the study of a small temperature rise on testis function, since surgical repositioning from an inguinal to a truly abdominal location raises the testis temperature only slightly. Such procedures, performed unilaterally in the naturally cryptorchid musk shrew (Suncus murinus) and degu (Octodon degus), raised the ipsi-lateral testis temperature approximately 1.5°C above normal, as measured by copper:constantan thermocouples (Bedford et al., 1982). When evaluated some 12 to 18 weeks later, the weight of the abdominal testis was reduced by a mean of 27% in the shrew and 52% in the degu, yet spermatogenesis had continued in such testes at a somewhat higher temperature than normal. However, this occurred at a lower rate and with a significantly greater number of structurally abnormal spermatozoa formed (Fig. 1).

Such results obtained with naturally cryptorchid species demonstrate what may be an important principle in evaluating the state of the testis in man: that the response of the germinal epithelium to temperature elevation is not all-or-none. The germinal epithelium may be only partially suppressed by a rise in temperature, according to the extent of that elevation.

## Epididymal Spermatozoa Develop the Ability to Fertilize at Abdominal Temperatures

Possible effects of abdominal temperature on sperm maturation in the epididymis have been studied in rats, hamsters and rabbits by establishment of a cryptepididymal state. As shown in Fig. 3, this involves surgical reflection of the epididymis up through the inguinal canal to be loosely anchored within the abdomen, yet remaining connected to a normally functioning scrotal testis. Such a situation suppresses the storage function of the cauda epididymidis (Bedford, 1978b). Nonetheless, rats rabbits and hamsters whose epididymides are reflected to the abdomen leaving the testis in the scrotum, maintain a level of fertility sufficient to induce pregnancies for many months (Bedford, 1978b; Wong et al., 1982; Bedford and Yanagimachi, 1991). However, as discussed below, a suppressed state of the abdominal (rat) cauda is reflected in a poorer overall quality of the spermatozoa released from it in the sexually rested male.

## Effects of Abdominal Temperature on Epididymal Sperm Transport

Notwithstanding the fact that bilaterally cryptepididymal males remain fertile, the quality of the ejaculate collected with an artificial vagina from long-term cryptepididymal rabbits declined after about 11 months. This was manifested by more spermatozoa that had neck droplets and a reduced % motility (Bedford, 1978b). The possibility that this might reflect an inappropriately rapid rate of epididymal transport was tested by injection of $^3$H thymidine into 3 control males, and into 6 experimental males in which the epididymis had been reflected to the abdomen some 11 months earlier. In agreement with previous reports (Orgebin-Crist, 1965; Amann et al., 1965) labelled spermatozoa appeared first in the ejaculate of normal control males at 41 days and peaked at 45-46 days. However, they appeared as early as 33-34 days and peaked at 37 days in the cryptepididymal rabbits (Bedford, 1978b). Thus the time required for sperm passage from the testis to the ejaculate is reduced from about 10 days in the normal control to about 3-4 days in the rabbit whose epididymides are maintained at the temperature of the abdomen.

## Temperature Effects on the Epithelium and Environment of the Cauda Epididymidis

The epithelium of the epididymis possesses a functional regionality. Its activities support specific features of the maturation process in the upper segment, but create specific conditions for storage in the cauda region. Since maturation continues normally at deep abdominal temperature in the rat, hamster and rabbit, there is no reason to suspect that temperature effects of interest occur in the caput or corpus epithelium. However, the cauda's ability to maintain spermatozoa in a viable state is rapidly eliminated at body temperature (Heller, 1929; Glover, 1960; Bedford, 1978b). Our investigations variously using rats and rabbits indicate that the special environment in the cauda epididymidis is maintained by a temperature-sensitive epithelium, the dysfunction of which at deep body temperature is soon reflected by changes in its fluid milieu.

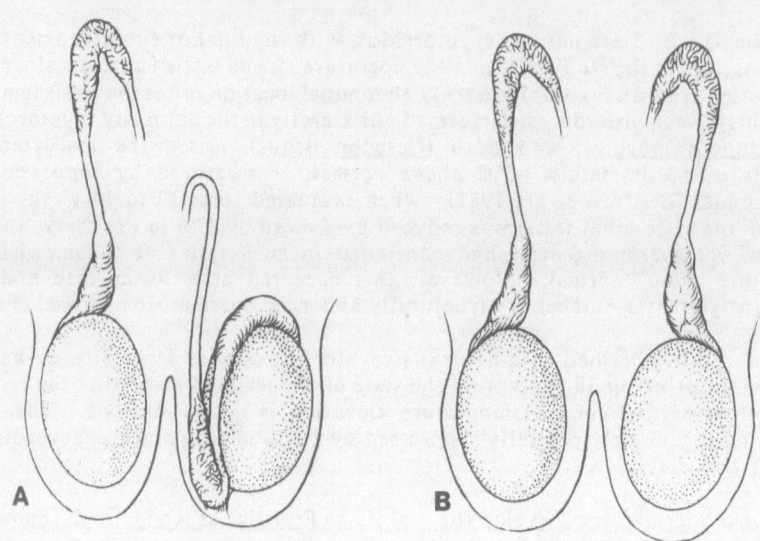

Figure 2.  In the cryptepididymal animal model, the epididymis is surgically reflected uni- or bilaterally up into the abdomen.  Normal numbers of spermatozoa pass into it from an unchanged functional scrotal testis.

First, ultrastructural and especially epididymal perfusion studies have demonstrated that the water and ion transport mechanisms of the cauda epithelium are disrupted by abdominal temperature (Wong et al., 1982; Rasweiler and Bedford, 1982). In the rat, $H_2O$, $Na^+$, and $Cl^-$ resorption, and $K^+$ secretion, are significantly depressed after only 2 days at deep body temperature, and more so after 7 days; and this changes the ionic composition of the cauda fluid (Wong et al., 1982). Temperature effects are reflected also in the protein content of that fluid.  PAGE analysis of fluid obtained without contamination from the rat cauda epididymidis reveals that several characteristic proteins disappear from that fluid between 6 and 15 days after the imposition of abdominal temperature (Esponda and Bedford, 1986). Similarly, epithelial explants from the rabbit abdominal cauda epididymidis pulsed in vitro with $^{35}S$-methionine, produced a pattern of protein synthesis different in several respects from that in explants from the scrotal cauda (Esponda et al., 1990).

Although the bilaterally cryptepididymal male remains fertile, the suppression/alteration of the special environment created in the cauda by deep body temperature, is reflected in a relatively poor overall motility in the sperm population released from it.  In the unilaterally cryptepididymal rat, there is no detectable difference in the motility of the populations released from the lower corpus region of the scrotal versus the abdominal epididymis. Consistently, however, that released from the cryptic cauda of the inactive male is characterized by tangled clumps of immotile cells; and overall only 10-30% display active motility as compared with 60-90% motility in the populations released from the scrotal cauda (J.M. Bedford and H. Kim, in preparation).

If it is populated by normal spermatozoa from the corpus region, why does such a mixed inferior population arise in the cryptic cauda?  It is known that sperm death occurs in less than 6 days in the androgen-deficient or heat-suppressed rat cauda (Bedford, 1978b), and individual spermatozoa are cleared from the cauda at different rates (Orgebin-Crist, 1965).  Since the cryptic/abdominal epididymis of the rat can maintain sperm viability there for less than 6 days, the dead segment of the population in sexually rested males probably represents those not cleared from the cryptic cauda before that brief limit of their intrinsic viability.  Indeed, mating at 3-day intervals markedly improves the motility of the population housed in the cryptic cauda.

In conclusion, the cauda epididymidis is a temperature-sensitive organ.  Deep body temperature brings rapid disruption of the normal absorptive/secretory function of the cauda epithelium with consequent change in the composition of cauda fluid. This presumably underlies the early degeneration of spermatozoa retained in the abdominal cauda of the sexually <u>inactive</u> male.

## Temperature Effects on the Epididymal Storage Capacity and on Ejaculate Sperm Numbers

In the cryptepididymal situation shown in Figure 2, the cauda epididymidis maintained in the abdomen becomes significantly smaller than its sham-operated scrotal counterpart within a few days (Bedford, 1978b). However, the size and weight of the ipsi-lateral testis remains normal and the sperm number in the caput of the cryptic epididymis equals that in the caput of the scrotal epididymis (Foldesy and Bedford, 1982). Thus, the smaller size of the cryptic cauda does not reflect lower sperm numbers passing into it, but is the result of an immediate reduction in the size of the tubule, particularly the large diameter segment in the distal cauda and vas deferens which provides the bulk of the ejaculate. Histological sections taken 2-3 weeks after imposition of abdominal temperature indicate that the length as well as the diameter of that duct segment is reduced in the cryptic cauda epididymidis (Fig. 3). That response is paralleled by a reduction in the number of spermatozoa housed in it. Already lower after only 2 days at abdominal temperature, by 16 days the number in both the proximal and distal segments of the rat cauda is reduced by 75-80%, despite normal testis production and normal numbers in the associated caput (Foldesy and Bedford, 1982).

Because elevated temperature diminishes the storage capacity of the cauda epididymidis, it also reduces sperm numbers in the ejaculate. Notwithstanding a constant testicular sperm production rate, imposition of deep body temperature on the cauda epididymidis reduces the total number of spermatozoa that can be ejaculated. It also changes the pattern of their distribution in successive ejaculates. Though the individual performance is variable, we have observed that the normal mature male rat generally has the capacity for 6-8 successive emissions, each produced every 15 minutes or so. Equivalent mean numbers were present in each of the first four ejaculates (Fig. 4), the

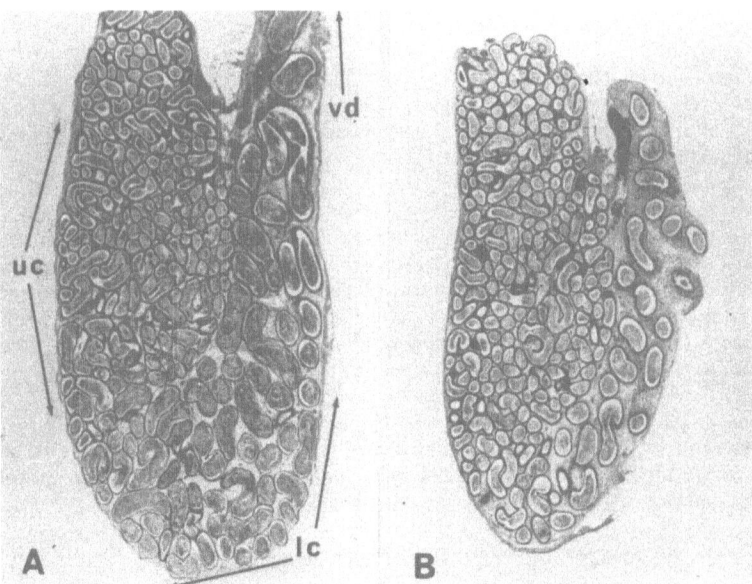

Figure 3. Longitudinal sections of (A) the scrotal cauda epididymidis, and (B) the cauda of the contra-lateral epididymis reflected to reside in the abdomen for some months while remaining functionally connected to a scrotal testis, as in Fig. 2A. Note the smaller diameter and fewer cross-sections of the duct in the cryptic cauda and vas deferens. UC: upper cauda; LC: lower cauda; VD: vas deferens.

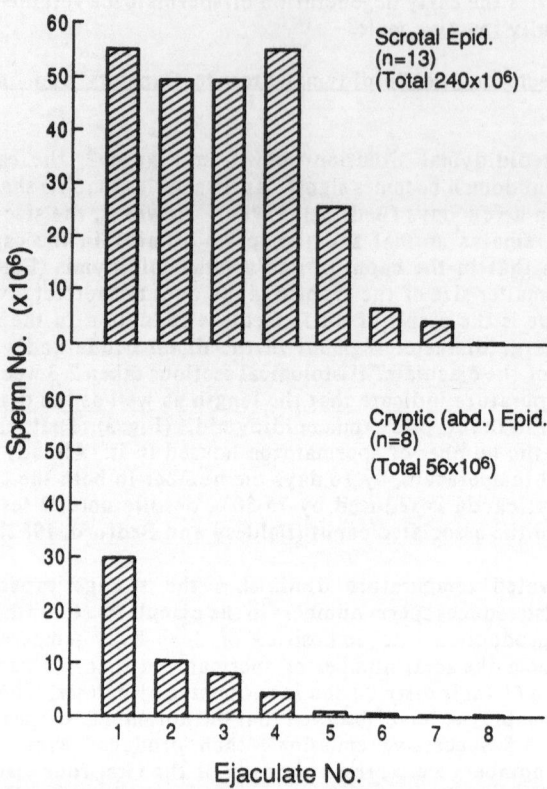

**RAT**

Scrotal Epid.
(n=13)
(Total 240x10$^6$)

Cryptic (abd.) Epid.
(n=8)
(Total 56x10$^6$)

Sperm No. (x10$^6$)

Ejaculate No.

Figure 4. Ejaculation sequence to exhaustion in normal (upper panel) and in bilaterally cryptepididymal rats mated to normal females. The normal sequence produces several ejaculates of approximately similar quality before the number diminishes. This stands in contrast to males with a suppressed cauda epididymidis though normal testis production, in which a maximal number appears in the first ejaculate and the overall sperm reserve available is approximately one-fifth of that in the normal state.

number falling significantly thereafter, with very few present in the 6th-8th samples; but together these contain a total of about 250 X 10$^6$ spermatozoa. By contrast, bilaterally cryptepididymal males can ejaculate a total of only some 50 X 10$^6$ spermatozoa, about 20% of the number ejaculated by normal males, despite a comparable testis production. Moreover, more than 40% of that total is produced in the first sample, with subsequent ejaculates containing progressively fewer spermatozoa (Fig. 4).

In summary, imposition of deep body temperature on the epididymis alone dramatically reduces the total sperm number available for ejaculation, notwithstanding a normal production by the testis. This also changes the pattern of the numbers in sequential ejaculates.

### Reversibility of the Temperature-Induced Suppression of the Cauda Epididymidis

We were interested to discover whether temperature-induced suppression of the cauda epididymidis is reversible. To examine this, a bilaterally cryptepididymal state was established in rats. Then, one testis and its associated epididymis were removed one month later for counting of the cauda-/vas sperm population, and at the same time the remaining abdominal epididymis was returned to the scrotum. After a further six weeks, that remaining testis was weighed and the number of spermatozoa in the vas deferens, distal and proximal regions of the replaced cauda was counted. Whereas the sperm number was typically low in one abdominal cauda at four weeks, it was essentially normal in the other cauda six weeks later after its return to the scrotum. Moreover, the population released from it exhibited excellent motility. Clearly, suppression of the (rat) cauda epididymidis by abdominal temperature is reversible and its capacity and function can be restored to a normal level after return for some weeks to the temperature of the scrotum.

## The Cauda Epididymidis and Sperm Capacitation

There is some reason now to believe that the phenomenon of sperm capacitation in the female tract is a secondary but physiologically important consequence of the specific environment created within the cauda epididymidis. As discussed elsewhere, specifically, capacitation may wholly or in part constitute a necessary escape from a stable state imposed by sperm binding macromolecules in the caudal fluid (Bedford, 1990; 1991).

Since body temperature can modify the cauda fluid protein profile (Esponda and Bedford, 1986; Regalado et al., 1990), we have examined the possibility that temperature-induced suppression of the cauda epididymidis may also affect the capacitation characteristics of the spermatozoa that pass through it. An initial study performed in the hamster demonstrates that spermatozoa released from the abdominal cauda cannot fertilize immediately, but require a shorter capacitation time. Thus, they undergo the acrosome reaction and hyperactivate sooner _in vitro_, and fertilize sooner _in vitro_ and _in vivo_ than do spermatozoa from the scrotal cauda (Bedford and Yanagimachi, 1991). The extent to which the findings in this hamster study will apply to other species, remains to be determined. In all, however, these first results suggest that interactions with factors in the caudal environment are one determinant of the capacitation characteristics of mammalian spermatozoa, and that some such elements are temperature sensitive.

## IMPLICATIONS OF ANIMAL STUDIES FOR MALE REPRODUCTIVE TRACT FUNCTION IN MAN

Judgment as to whether or not man's reproductive status is being influenced by elevated scrotal temperatures, is made difficult by an absence of baseline data about control populations living in an "evolutionary" state, i.e., naked or very lightly clothed. Moreover, even among the five genera that comprise the Hominoidea, there is wide variation in innate testis size and so sperm production rates (Short, 1979; Kenagy and Trombulak, 1986), the gorilla being especially unimpressive in that respect. Thus, without such controls, there are obvious dangers in extrapolation from other species to modern man.

As shown in Table 1, it appears, however, that the human male tract is indeed exposed to temperatures that may be high enough to modulate its function. We have measured the temperature of the skin of the scrotum at the low point over the cauda epididymidis in six men, using sensitive copper:constantan thermocouples attached to a recording device with an accuracy of ±0.1°C (J.M. Bedford and M. Berrios, unpublished

Table 1. Temperatures recorded with a copper:constantan thermocouple closely applied to the lower border of the scrotal surface overlying the cauda epididymidis in normal man

|  | Scrotal Surface Temperature |
| --- | --- |
| Without clothing | 29 - 30°C |
| Sitting clothed | 33 - 34°C |
| Sitting cross-legged, clothed (20 min) | 36 - 37°C |

N = 6

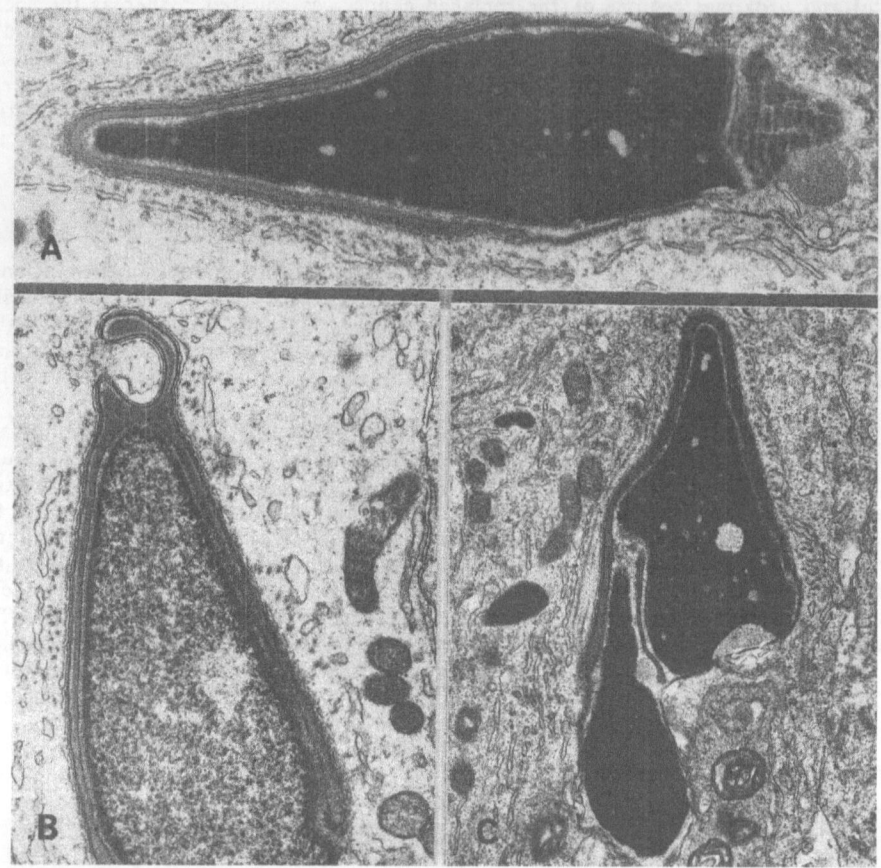

Figure 5. Late spermatids in testis biopsies from "normal" men (cf. Fig. 1). (A) shows a normal spermatid with regular proportions. By contrast that in (B) displays an abnormal rostral extension of the acrosome, and in (C) two spermatid nuclei share one acrosome. Such anomalies are found commonly in the human testis, adjacent to normal spermatids.

studies). In general agreement with Ehrenberg et al. (1957), a scrotal skin surface temperature of 29-30°C in the standing naked state at an ambient temperature of 24°C, increased by some degrees after dressing (jockey or boxer shorts, trousers, shirt and sweater), and somewhat higher if the subject sat with crossed legs for another 20-30 minutes (Table 1).

Such results indicate that the human testis and epididymis are subject almost uninterruptedly throughout life to a temperature several degrees higher than that at which the human male tract evolved. Clearly, there are no striking catastrophic effects of this on man's reproductive function. However, despite the lack of "control" subjects there is reason to suspect that some of the effects recorded in temperature-conditioned animals may be mirrored to a variable individual degree in modern men. The parameters that appear to reflect this include (a) the germinal epithelium and so finally both the number and quality (motility and morphology) of spermatozoa produced by the testis, then (b) the rate of their transport through the epididymis, (c) the number in sequence and quality of those ejaculated, and possibly (d) their capacitation characteristics.

It is interesting, first, in view of the outcome of experiments in small naturally cryptorchid animals (Bedford et al., 1982) that sperm production by the human testis occurs at a relatively low rate of 4-5 X $10^6$/gm testis parenchyma/day, or only 20-50% of that in animals (Amann, 1981); and that human testis biopsies display a spectrum of abnormal spermatids among others of normal morphology (Fig. 5) (Bedford et al., 1973; Holstein, 1975). Although pleomorphism has been noted in several mammals (Cummins, 1990), it seems significant that a comparable picture, that is, a lower production and the appearance of similar abnormalities in a significant proportion of spermatids and spermatozoa, occurs in both the musk shrew and degu following a moderate prolonged rise in the temperature of the testis (Figs. 1 and 5).

In regard to epididymal transport, there are two notable exceptions to a rather uniform timing of this among mammals, the cryptepididymal rabbit and man, in both of which transport appears exceedingly fast (Fig. 6). That general picture raises the possibility that the rapid rate in man (Johnson and Varner, 1988) is not entirely innate, and may reflect to some degree the operation of a temperature-conditioned system, as it does in rabbits.

As temperature has consistent negative effects on the size and storage capacity of the cauda epididymidis in animals (Figs. 3 and 4), so there are three possible indicators that this occurs in a subtle way in man. Notwithstanding a low testis production rate, the available sperm reserve/testis production ratio is lower in men than in normal animals, at sexual rest or with almost daily ejaculation (Amann, 1981). Second, the human cauda appears as a minimally developed "organ" compared to that in a majority of other scrotal mammals (Fig. 7). Third, the ejaculate sequence profile in man (Fig. 8) closely resembles that in the cryptepididymal rat (Fig. 3), where reduced cauda storage is reflected negatively in the sperm number and sequence profile, despite a normal testicular production. Of the mean total spermatozoa in six sequential ejaculates produced by young men within a 24 h period, more than 50% appeared in the first ejaculate, with a marked progressive decrement in each of the succeeding samples (Fig. 8).

Interpretation of the basis of this human ejaculation pattern should be tentative in view of the lack of human "controls". Moreover, observations of this type are limited to

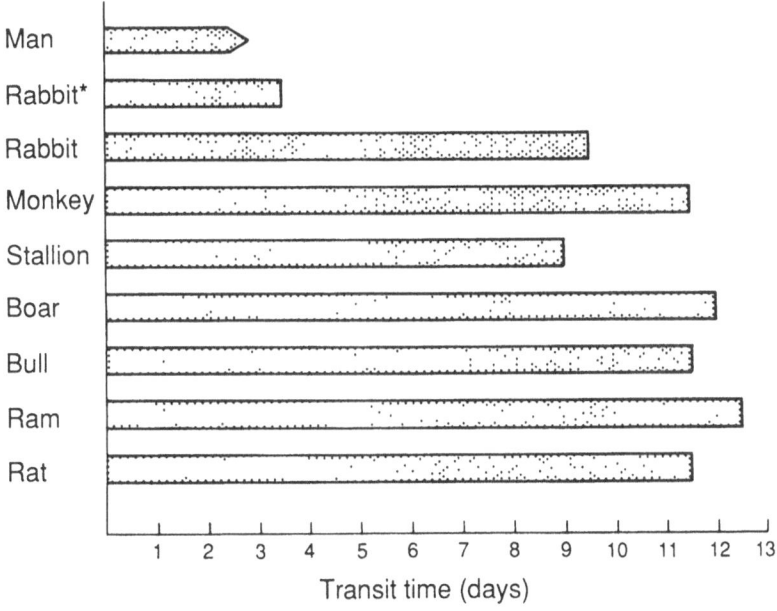

* Epididymis at abd temperature

Figure 6. Transit time for sperm passage through the epididymis. The time is similar in most animals, notwithstanding significant species differences in duct length. Notable exceptions include the cryptepididymal rabbit and man. Modified after Robaire and Hermo (1988).

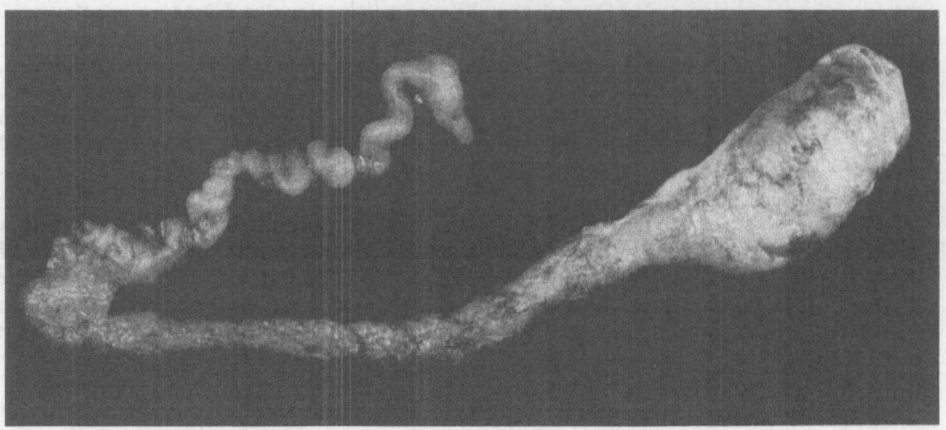

Figure 7. Epididymis from a middle-aged man. Dissected free of connective tissue, this displays a bulbous caput occupied in large part by efferent ducts. Typically, the cauda epididymis is poorly developed compared to that in a majority of mammals.

a very few species, and the ejaculation sequence profile undoubtedly will vary among different mammals. Nevertheless, there is some reason to suspect that the storage capacity of the human cauda epididymidis may often be subtly suppressed by the scrotal temperatures that present in modern men. The indication (see above) that many rat spermatozoa degenerate within a few days if not cleared from the cryptic cauda, also may have links to the situation that manifested in some men. The motility index of the human ejaculate often varies within a single individual and a picture in sub-fertile men of 10-20% highly motile spermatozoa in an otherwise sluggish or immotile population is not uncommon. Although preliminary, our observations on repeat semen samples obtained after an interval of only 2-3 days, suggest that a significant improvement in the percentage motility may be achieved in some men by this simple manoeuvre. Such timed

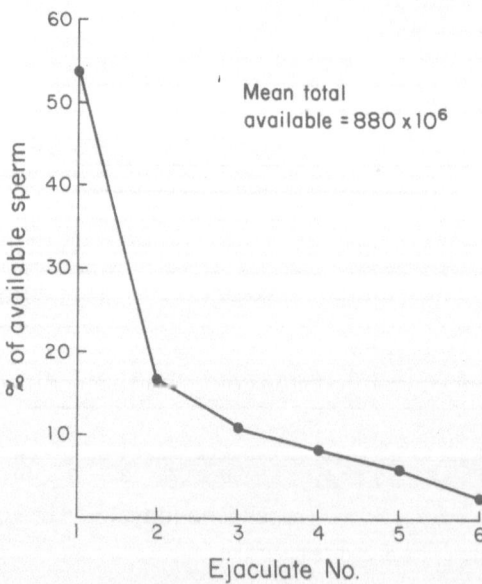

Figure 8. The pattern of sperm numbers in successive ejaculates produced over a 24 h period by young men. Each point is expressed as a mean percentage of the total available sperm reserve (N = 10).

ejaculations may prove useful also as a way of distinguishing cauda dysfunction from athenospermia/necrospermia of testis origin where ejaculation at 3-day intervals would not enhance % motility.

A final issue in the present context is that of capacitation. Though not able to fertilize immediately, some ejaculated human spermatozoa express characteristics that in other species seem to reflect some element of the capacitated state. They recognize only the hominoid zona pellucida, failing to adhere to that of subhominoid primates and other mammals (Overstreet and Hembree, 1976; Bedford, 1977), a specificity of attachment that in laboratory animals may become more marked as a correlate of capacitation. Washed ejaculated human spermatozoa also readily penetrate intact cumulus oophorus (White et al., 1990), again an ability that in the rabbit and hamster appears in capacitated spermatozoa (Overstreet and Bedford, 1974; Corselli and Talbot, 1987). Finally, while sperm capacitation times may vary from one man to another (Perreault and Rogers, 1982), some undergo functional capacitation within one hour in appropriate conditions (Overstreet et al., 1980). This brief timing is not unprecedented (e.g., cat: Goodrowe et al., 1989). However, it is distinctly faster for ejaculated spermatozoa than in most mammals. In light of the shorter capacitation time required by hamster spermatozoa from the cryptic cauda (Bedford and Yanagimachi, 1991), one must wonder whether what appears as an imminent or "primed" capacitation state of some human spermatozoa is innate, or if this is partly a consequence of chronic elevation of the temperature of the cauda epididymidis.

## SUMMARY

The effects of temperature on the male tract have been examined in the rat, rabbit and hamster, as well as other species that include a naturally cryptorchid rodent, the degu, and an insectivore, the musk shrew.

In principle, a small increase in the temperature of the testis does not destroy the germinal epithelium; however, it reduces testis weight and sperm production and brings a greater incidence of morphologically abnormal spermatids and spermatozoa. This demonstrates that the testis can be partially suppressed yet remain functional in the face of moderate elevation in its ambient temperature.

Selective imposition of abdominal temperature on the epididymis alone does not suppress sperm maturation there, and bilaterally cryptepididymal males remain fertile for long periods. However, deep body temperature changes at least the ionic and protein composition of cauda fluid by virtue of effects on the cauda epithelium, and it eliminates the special ability of the cauda to store and prolong the life of spermatozoa. Additionally, deep body temperature also immediately curtails the storage capacity of the cauda epididymidis, an effect that is reflected in a reduced diameter and apparently length of that segment of the cauda as well as vas deferens, which contracts during orgasm to provide the bulk of the ejaculate. One consequence of this, notwithstanding a normal sperm production by the testis, is a much smaller number of spermatozoa in the first ejaculate, and an atypically steep decline in the number in subsequent ejaculates produced by cryptepididymal males. Further effects of deep body temperature on the epididymis are seen in significantly faster (rabbit) sperm transport through it, and a reduction in the time required for capacitation of (hamster) spermatozoa, in vitro and in vivo.

Man's scrotal surface temperature is chronically elevated by several degrees in the clothed state. Although observations are lacking for human "control" populations, certain of the temperature-related phenomena described in animals are nevertheless suggested variably in different measurable functions of the human male reproductive tract. These include a relatively low number produced/gm of testis and a poorer quality of spermatozoa released from it, a rapid epididymal transport and minimally developed sperm storage system in the cauda epididymidis. Finally, the character of the ejaculated spermatozoa in several respects may imply an imminent state of capacitation. In all, this circumstantial evidence makes it seem possible that the human epididymis as well as the testis often exists in a state of temperature-induced partial suppression.

# REFERENCES

Amann, R.P. 1981. A critical review of methods for evaluation of spermatozoa from seminal characteristics. J. Androl., 2: 37.

Amann, R.P., Koefoed-Johnson, H.H. and Levi, H. 1965. Excretion pattern of labelled spermatozoa and the timing of spermatozoa formation and epididymal transit in rabbits injected with thymidine $^3$H. J. Reprod. Fertil., 10: 169.

Bedford, J.M. 1977a. Sperm/egg interaction: the specificity of human spermatozoa. Anat. Rec., 188: 477.

Bedford, J.M. 1977b. Evolution of the scrotum: the epididymis as the prime mover? In: "Evolution and Reproduction", J.H. Calaby and C.H. Tyndale-Biscoe (eds.), Aust. Acad. Sci., Canberra, p. 171.

Bedford, J.M. 1978a. Anatomical evidence for the epididymis as the prime mover in the evolution of the scrotum. Am. J. Anat., 152: 483.

Bedford, J.M. 1978b. Influence of abdominal temperature on epididymal function in the rat and rabbit, Am J. Anat., 152: 509.

Bedford, J.M. 1990. Fertilization mechanisms in animals and man: current concepts. In: "Establishment of a Human Pregnancy", R.G. Edwards (ed.), Raven Press, New York, p. 115.

Bedford, J.M. 1991. Co-evolution of mammalian gametes. In: "A Comparative Overview of Mammalian Fertilization", B.S. Dunbar and M.G. O'Rand (eds.), Plenum Press, New York, in press.

Bedford, J.M., Calvin, H. and Cooper, G.W. 1973. The maturation of spermatozoa in the human epididymis. J. Reprod. Fertil. Suppl. 18: 199.

Bedford, J.M., Berrios, M. and Dryden, G.L. 1982. Biology of the scrotum, IV. Testis location and temperature sensitivity. J. Exp. Zool., 224: 379.

Bedford, J.M. and Yanagimachi, R. 1991. Epididymal storage at abdominal temperature reduces the time required for capacitation of hamster spermatozoa. J. Reprod. Fertil., In Press.

Brooks, D.E. 1973. Epididymal and testicular temperature in the unrestrained conscious rat. J. Reprod. Fertil., 35: 157.

Corselli, J. and Talbot, P. 1987. In vitro penetration of hamster oocyte-cumulus complexes using physiological numbers of sperm. Dev. Biol., 122: 222.

Cummins, J.M. 1990. Evolution of sperm form: levels of control and competition. In: "Fertilization in Mammals", B.D. Bavister, J.M. Cummins and E.R.S. Roldan (eds.), Serono Symposia, U.S.A., p. 51.

Ehrenberg, L., Ehrenstein, G. and Hedgran, A. 1957. Gonad temperature and spontaneous mutation rate in man. Nature, London, 180: 1433.

Esponda, P. and Bedford, J.M. 1986. The influence of body temperature and castration on the protein composition of fluid in the rat cauda epididymidis. J. Reprod. Fertil., 78: 505.

Esponda, P., Regalado, F. and Nieto, A. 1990. Effects of temperature and castration on the synthesis of secretory proteins in the rabbit cauda epididymidis. In: "Fertilization in Mammals", B.D. Bavister, J.M. Cummins and E.R.S. Roldan (eds.), Serono Symposia, U.S.A., p. 416.

Foldesy, R.G. and Bedford, J.M. 1982. Biology of the scrotum, 1. Temperature and androgen as determinants of the sperm storage capacity of the rat cauda epididymidis. Biol. Reprod., 26: 673.

Glover, T.D. 1960. Spermatozoa from the isolated cauda epididymidis of rabbits and some effects of artificial cryptorchidism. J. Reprod. Fertil., 1: 121.

Goodrowe, K.L., Howard, J.G., Schmidt, P.M. and Wildt, D.E. 1989. Reproductive biology of the domestic cat with special reference to endocrinology, sperm function and in vitro fertilization. J. Reprod. Fertil. Suppl. 39: 73.

Harrison, R.G. 1975. Effect of temperature on the mammalian testis. In: "Handbook of Physiology, Section 7: Endocrinology, Volume V, Male Reproductive System", D.W. Hamilton and R.W. Greep (eds.), American Physiological Society, Washington, D.C., p. 219.

Holstein, A.F. 1975. Morphologische studien an abnormen spermatiden und spermatozoen des menschen. Virchows Arch (Pathol. Anat.), 367: 93.

Johnson, L. and Varner, D.D. 1988. Effect of daily spermatozoon production but not age on transit time of spermatozoa through the human epididymis. Biol. Reprod., 39: 812.

Kandeel, F.R. and Swerdloff, R.S. 1988. The role of temperature in regulation of spermatogenesis and the use of heating as a method for contraception. Fertil. Steril., 49: 1.

Kenagy, G.J. and Trombulak, S.C. 1986. Size and function of mammalian testes in relation to body size. J. Mammal., 67: 1.

Orgebin-Crist, M.-C. 1965. Passage of spermatozoa labelled with thymidine $^3$H through the ductus epididymidis of the rabbit. J. Reprod. Fertil., 10: 241.

Overstreet, J.W. and Hembree, W.C. 1976. Penetration of the zona pellucida of non-living human oocytes by spermatozoa in vitro. Fertil. Steril., 27: 815.

Overstreet, J.W., Gould, J.E., Katz, D.F. and Hanson, F.W. 1980. In vitro capacitation of human spermatozoa after passage through a column of cervical mucus. Fertil. Steril., 34: 604.

Perreault, S.D. and Rogers, B.J. 1982. Capacitation patterns of human spermatozoa. Fertil. Steril., 38: 258.

Rasweiler, J.J. and Bedford, J.M. 1982. Biology of the scrotum. III. Effects of abdominal temperature upon the epithelial cells of the rat cauda epididymidis. Biol. Reprod., 26: 691.

Robaire, B. and Hermo, L. 1988. Efferent ducts, epididymis, and vas deferens: structure, functions and their regulation. In: "The Physiology of Reproduction", Vol. I, E. Knobil and J.D. O'Neill (eds.), Raven Press, New York, p. 999.

Short, R.V. 1979. Sexual selection and its component parts, somatic and genital selection, as illustrated by man and the great apes. Adv. Stud. Behav., 9: 131.

Short, R.V., Mann, T. and Hay, M.F. 1967. Male reproductive organs of the African elephant, Loxodonta africana. J. Reprod. Fertil., 13: 517.

Vandemark, N.L. and Free, M.J. 1970. Temperature effects. In: "The Testis", Vol. III, A.D. Johnson, W.R. Gomes and N.L. Vandemark (eds.), Academic Press, New York, p. 233.

Waites, G.M.H. and Setchell, B.P. 1969. Some physiological aspects of the function of the testis. In: "The Gonads", K.W. McKerns (ed.), Appleton, New York, p. 649.

White, D.R., Phillips, D.M. and Bedford, J.M.  1990.  Factors affecting the acrosome reaction in human spermatozoa.  J. Reprod. Fertil., 90: 71.

Wong, P.Y.D., Au, C.L. and Bedford, J.M.  1982. Biology of the scrotum, II.  Suppression by abdominal temperature of transepithelial ion and water transport in the cauda epididymidis.  Biol. Reprod., 26: 683.

# EFFECTS OF TEMPERATURE ON THE BIOCHEMISTRY OF THE TESTIS

Anna Steinberger

Department of Obstetrics, Gynecology and
Reproductive Sciences, The University of Texas
Medical School at Houston, Houston, Texas 77030

## INTRODUCTION

In most mammalian species, including human, the male gonads are located outside the body cavity in a highly specialized skin pouch -- the scrotum. Generally, the scrotal and testicular temperatures are below the core body temperature, although the differences between abdominal and testicular temperatures vary among species 8.3 - 8.5°C in the rat and mouse to approximately 2°C in the rhesus monkey (Harrison and Weiner, 1948). The lower temperature is of great importance for the maintenance of normal testicular functions, particularly the process of spermatogenesis. Clearly, the existing mechanisms for controlling testicular temperature must have developed through evolutionary processes to assure the survival of the species; however, the reason as to why nature chose this particular form of adaptation is not clear.

Although it has been recognized for a long time that above-scrotal temperatures (i.e., cryptorchidism, febrile illness, varicocele, etc.) interfere with normal sperm production, often resulting in lack of fertility, the biochemical mechanisms underlying the heat-induced damage are still not known (Van Demark and Free, 1970; Steinberger and Steinberger, 1972; Keel and Abney, 1980, 1981; Kandeel and Swerdloff, 1988; Abney and Keel, 1989). Most likely, multiple indirect mechanisms are involved in addition to a possible direct effect of heat on the germ cells. In some cases the effects of heat may be additive to other pathological conditions, rather than being the primary cause of the testicular abnormalities. It is relevant to point out, however, that in greater than 83% of subfertile men, the intrascrotal temperature was found to be greater than 34.1°C regardless of clinical diagnosis (Zorgniotti and Sealfon, 1988).

The complexity of the spermatogenic process and the existence of different cell types composing the tubules (Sertoli cells, germ cells, and peritubular cells) and those located in the interstitium (i.e., Leydig cells) make the biochemical investigation of individual cell types and the interaction between specific cell types in the testis inherently difficult. Also, the creation within the tubules of basal and adluminal compartments (containing fluids with distinct biochemical differences), by specialized tight junctions between adjacent Sertoli cells, adds additional complexity to the biochemical investigations of testicular functions. Advances in methods for physical separation of seminiferous tubules from the interstitial tissue, tubule segments in the same stage of the spermatogenic cycle, and of individual cell populations, made it possible to explore in greater detail their biochemistry, and the effects of various adverse conditions such as elevated temperatures. Another approach has been to use immature animals with various degrees of incomplete spermatogenesis, or to alter the germ cell

*Temperature and Environmental Effects on the Testis*
Edited by A. W. Zorgniotti, Plenum Press, New York, 1991

population in experimental animals by using treatments which result in preferential loss of specific germ cell types, i.e., spermatogonia (irradiation) or more advanced germ cells (hypophysectomy, cryptorchidism, local heating).

A major drawback in using these treatments is that, besides eliminating certain germ cell populations, the conditions employed also often affect some biochemical parameters in the remaining cells making interpretation of the results more difficult. Recent studies have clearly demonstrated extensive cell-to-cell interactions in the testis and paracrine influences of one cell type on another. Thus, even subtle biochemical changes in one cell type may dramatically affect other testicular cells. Moreover, the various animal models, used to investigate the effects of elevated temperature on testicular functions, may not truly represent the conditions existing in the human male. However, despite these shortcomings, much has been learned about testicular biochemistry and heat-induced damage using experimentally altered animal testes as well as human testicular tissues. Extensive literature on these subjects can be found in publications by Johnson and Gomes (1970, 1977), Steinberger and Steinberger (1972), Fritz (1973), Blackshaw (1977), Kandeel and Swerdloff (1988), Abney and Keel (1989), as well as in other chapters of these proceedings.

The basic mechanism by which higher temperatures induce damage to the spermatogenic process is not known. Earlier studies in animals, mainly rats, (Chowdhury and Steinberger, 1964, 1970; Steinberger and Dixon, 1969; Blackshaw and Hamilton, 1970; Van Demark and Free, 1970; Blackshaw et al., 1973) showed that acute local heating (40°C - 43°C) of the testis results in spermatogenic damage with the late pachytene spermatocytes and round spermatids being most sensitive to elevated temperatures. These cells are also damaged in the cryptorchid testes (Blackshaw, 1973; Parvinen, 1973), whereas the initiation of spermatogenesis is not inhibited by the cryptorchid milieu (Chowdhury and Steinberger, 1972). The reasons for the greater temperature sensitivity of pachytene spermatocytes and round spermatids remain to be clarified but could be due to greater membrane instability (discussed later). Moreover, although testicular somatic cells were for a long time considered to be relatively resistant to heat, more recent studies indicate significant temperature-related changes in both the Sertoli and Leydig cells.

The main intent of this chapter was to focus on the more recent findings relating to the effects of temperature on testicular biochemistry rather than to provide an exhaustive bibliographic review of this subject. Thus, references to some relevant publications may not have been included.

EFFECTS OF TEMPERATURE ON TESTICULAR BIOCHEMISTRY

Biosynthesis of Protein, DNA and RNA

Attempts have been made to correlate morphologic changes occurring in the cryptorchid testis with metabolic events observed in incubates of testicular tissue or its components; however, the results obtained are not in good agreement. Incorporation of lysine into total protein was lower at 37°C than at 32°C in some species (i.e., rats, mice and hamsters) but higher in other species (rabbits, guinea pigs and dogs) (Byer and Davis, 1966). In seminiferous tubules of adult C57BL inbred mouse testis incubated at different temperatures, the incorporation of $^3$H-uridine into total RNA was greatly increased at 38.5°C compared to those incubated at 32°C, whereas incorporation of $^{14}$C-serine into total protein was not significantly affected. On the other hand, incorporation of $^3$H-thymidine into DNA at 38.5°C was only 50% of that observed at 32°C. Since in autoradiographs of testicular tissue $^3$H-thymidine incorporation was observed in the spermatogonia and resting primary spermatocytes, but not in more advanced germ cells, the data suggested that DNA synthesis in preleptotene germ cells is most sensitive to elevated temperature (Nishimune and Komatsu, 1972, 1977). Using Sertoli cells isolated from immature rats, Hall et al. (1985) found increased protein synthesis, and the production of cyclic AMP and lactate at 38°C when compared to 34°C, whereas Heindel et al. (1982) found no difference in cyclic AMP responses to FSH by immature Sertoli cells cultured at 32°C or 37°C.

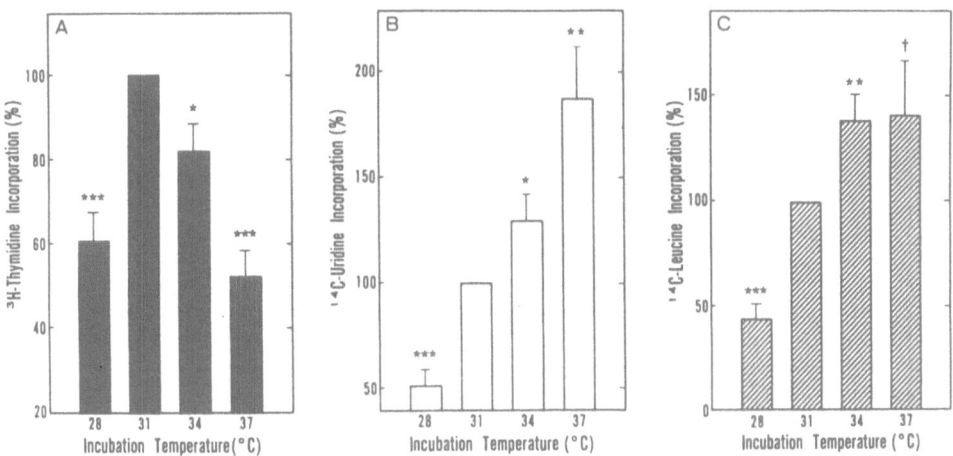

Figure 1.  Effect of incubation temperature on precursor incorporation into macromolecules of cultured human testis.  The incorporations of $^3$H-thymidine (A), $^{14}$C-uridine (B) and $^{14}$C-leucine (C) at 31°C were 23,994 ± 5953 dpm/A$_{260}$ for the DNA fraction, 3436 ± 287 dpm/A$_{260}$ for the RNA fraction, and 4875 ± 503 dpm/mg protein, respectively (means of 6 subjects ± SE).  The results are expressed as percentage of those at 31°C.  The bars show standard error of the mean.  + not significant; * p<0.05; ** p<0.02; *** p<0.01, compared with the value at 31°C by the t-test.  The difference in $^3$H-thymidine incorporation between 34°C and 37°C is statistically significant by the t-test (p<0.05) but that of $^{14}$C-uridine incorporation is not.  (From Nakamura et al., Arch. Andrology, 20: 41, 1988; with permission.)

Nakamura et al. (1988) investigated *in vitro* the effect of temperatures on the synthesis of DNA, RNA and protein in human testicular tissue (Fig. 1).  Testicular fragments (approximately 1 mm in size) from the non-affected contralateral testis of patients with varicocele were incubated at 28°C, 31°C, 34°C and 37°C, and the incorporation of $^3$H-thymidine, $^{14}$C-uridine and $^{14}$C-leucine into DNA, RNA and protein fractions, respectively, were determined after acid precipitation.  The incorporation of $^3$H-thymidine was maximal at 31°C and was significantly (P<0.05) inhibited by small temperature changes.  In contrast, the incorporation of $^{14}$C-uridine and $^{14}$C-leucine was highest at 34°C or 37°C.  These results indicate that DNA synthesis is most sensitive to relatively small temperature changes and may play a critical role in the spermatogenic damage occurring at above-scrotal temperatures.  Autoradiography of tissue sections after exposure to $^3$H-thymidine (Nakamura et al., 1987) showed label localization in the spermatogonia and resting pachytene spermatocytes but not in the more advanced germ cells.  These data, however, are difficult to interpret, since pachytene spermatocytes and round spermatids, cells that readily degenerate at higher temperatures, are not actively involved in DNA synthesis.  Thus, impairment of DNA synthesis in preleptotene germ cells may not be the underlying direct cause of temperature-induced spermatogenic damage but may interfere with the normal mitotic process.  It appears that different metabolic parameters could be affected by elevated temperatures at various stages of germ cell development.

### Testicular Enzymes

One of the first studies concerning the effect of cryptorchidism on a specific testicular enzyme was that by Steinberger and Nelson (1955) which showed a significant decrease of hyaluronidase activity in adult rat testes following surgical bilateral cryptorchidism.  Changes in enzymes involved in DNA synthesis were investigated in testes of cryptorchid rats by Fujisawa et al. (1988).  The activities of DNA polymerase alpha and topoisomerase I did not change up to 7 days after surgically induced cryptorchidism, and showed no significant difference from those in control testes (sham-operated).  In contrast, the activity of DNA polymerase beta decreased by 43% at 5 days

(P<0.01) and by 47% at 7 days (P<0.001). Also, the activity of DNA polymerase gamma decreased by 46% at 3 days (P<0.02) and by 78% at 7 days (P<0.01) after surgery. The amount of mRNA for DNA polymerase beta decreased in parallel with the enzyme activity. Since the sensitivity to heat inactivation of testicular DNA polymerase beta was the same as that in the liver, the decrease in DNA polymerase beta activity in the cryptorchid testis could have been, at least in part, due to reduced biosynthesis of the enzyme. The accompanying morphological changes observed in cryptorchid testes suggested that the decrease in DNA polymerase beta and gamma activities might be related to the deleterious effects of elevated temperature on spermatogenesis. The activity of neutral cholesteryl ester hydrolase activity, a soluble cytoplasmic enzyme present in Sertoli cell cytoplasm, also decreased rapidly (24h) in the abdominal testis of unilaterally cryptorchid adult rats, but increased in prepubertal rats made unilaterally cryptorchid at birth (Hoffmann et al., 1989).

One of the factors that may render certain classes of germ cells heat-sensitive could be a difference in their lysosomal stability. In studies conducted by Lee and Fritz (1972), lysosome-rich particles isolated from normal adult rat testes released acid phosphatases, ß-N-acethylhexosaminadase and arylsulfatase at 37°C at a much higher rate than at 25°C compared to liver lysosomes. In contrast, lysosome-rich preparation from spermatozoa isolated from the vas deferens, regressed cryptorchid testes, or testis from 14-day-old testes containing mainly Sertoli cells and spermatogonia, showed no such increase in the release rate of these lysosomal enzymes. The authors concluded that spermatocytes and spermatids contain lysosomes which at 37°C release enzymes at an accelerated rate compared to other germ cells or non-germinal testicular cells. The implication of these observations is that degeneration of germ cells in the cryptorchid testes may be associated with an accelerated release of lysosomal enzymes, since a 2- to 5-fold increase in the activity of ß-N-acethylhexosaminidase and arylsulfatase was observed in testicular cytosol fractions within 40 hours of experimental cryptorchidism as compared to 120-140 hours needed to detect a decrease of testicular weight and protein content. It should be pointed out, however, that it is still not certain from these studies whether the lysosomal changes were the cause or consequence of germ cell degeneration, or if the lysosomal instability was only one example of a more generalized alteration of membrane permeability in these cells. Blackshaw and Hamilton (1971) interpreted the histological and histochemical changes observed in the rat testis after 30-60 min exposure to 42°C as indicative of lysosomal fragility in result of injury to cellular membranes at certain stages of germ cell maturation. Laufer and Davies (1969) proposed that injury to cell membranes leads to ionic imbalance, mitochondrial damage, and then activation of lysosomes. Thus, selective heat damage of late pachytene spermatocytes and round spermatids may be due to greater instability of their plasma membranes.

Several lines of evidence suggest that lysosomal enzymes change during spermatogenic development. This possibility is supported by findings that ß-glucuronidase activity is present in spermatogonia but not in primary spermatocytes or spermatids (Males and Turkington, 1971). In contrast, hyaluronidase activity is high in spermatids and is absent in spermatogonia (Males and Turkington, 1970). Thus, different characteristics of lysosomes in spermatogonia and those in the more mature germ cells may, at least in part, account for the increased temperature sensitivity of the spermatocytes and round spermatids. Moreover, testosterone was found to increase labilization of lysosomal membranes *in vitro* (Szego et al., 1971) suggesting that changing hormonal status during cryptorchidism may also contribute to the instability of lysosomes in certain cell populations.

Early changes in the permeability of cellular and lysosomal membranes in rat testes after local heating of the scrotum (41°C for 30 min) were investigated by Vunder and Murashev (1984). During subsequent incubation *in vitro* at 33°C for 1 hour, fragments of the heated testes released a large amount of substances that increased the optical density of the incubation liquid at 270 nm. This phenomenon could be recorded shortly after scrotum heating with further increases of optical density observed 12 hours and 24 hours after heating. Fragments of the heated testes also secreted a large amount of lysosomal enzymes, acid phosphatase and cathepsin D. It appears that post-thermal degeneration of the spermatogenic epithelium is preceded by a change in the permeability of cellular and lysosomal membranes.

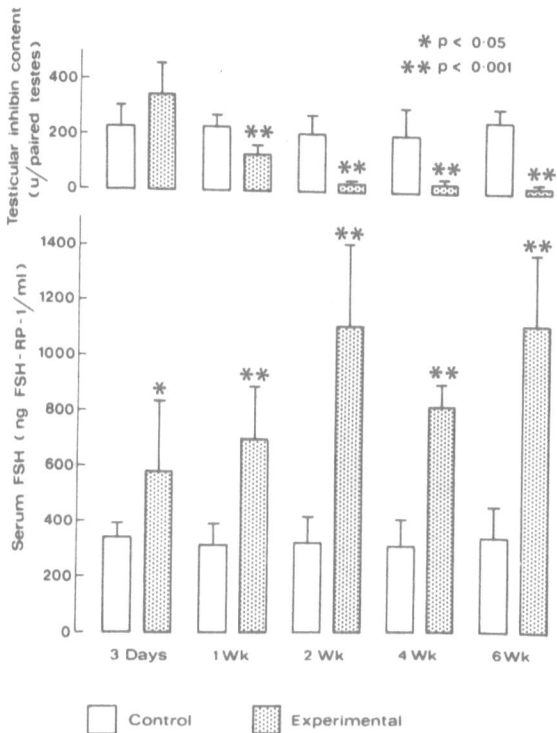

Figure 2. Changes in testicular inhibin content and serum FSH in sham-operated and bilaterally cryptorchid adult rats at different periods after surgery (mean + SD). Inhibin content was measured after 16 hours of duct ligation. (From Au, C.L., Robertson, D.M. and deKretser, D.M. <u>Endocrinology</u>, 112: 239, 1983; with permission.)

## Sertoli Cell Secretory Activity

Au et al. (1983) have demonstrated that experimentally induced cryptorchidism in adult rats caused damage to the germinal epithelium and a decrease in both testicular content and production rate of inhibin, as measured after efferent duct ligation (Fig. 2). Such changes were inversely correlated with an increase of circulating FSH level, presumed to reflect diminished feedback inhibition on the pituitary secretion of FSH by testicular inhibin. Reduced testicular content of inhibin in bilaterally cryptorchid rats was also reported by Demura et al. (1987). In rats made unilaterally cryptorchid for ten months, levels of immunoreactive inhibin in the interstitial fluid of the abdominal testis were reduced by 60% as compared with the contralateral scrotal testis (Sharpe et al., 1988).

Seethalakshmi and Steinberger (1983) provided evidence for altered secretion of inhibin by Sertoli cells isolated at various time intervals (21-35 days) after experimental bilateral cryptorchidism and after 14 and 42 days of scrotal recovery in adult rats (Fig. 3). The cultures were maintained for 6 days at 32°C with fresh medium added after 3 days. The amount of inhibin accumulated in the medium during the last 3 days of incubation was measured by *in vitro* pituitary cell bioassay. Sertoli cells isolated after testicular exposure to abdominal temperature showed a dramatic decline in their ability to secrete inhibin *in vitro*. A decline of bioassayable inhibin was previously observed in cultures of Sertoli cells from rats that were bilaterally cryptorchid for 14 or 25 days (LeGac-Jegou and deKretser, 1980; Steinberger, 1980).

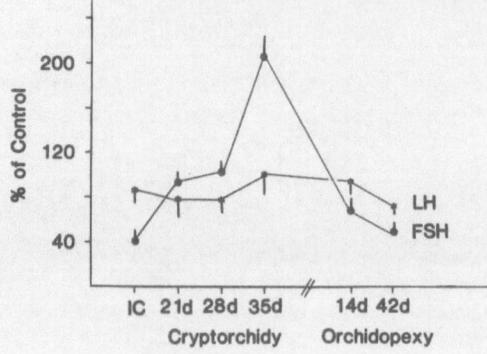

Figure 3. Effect of bilateral cryptorchidism (21d, 28d, 35d) followed by orchidopexy (14d, 42d) in adult rats on the secretion of inhibin *in vitro*. Sertoli cells were isolated at the indicated times and cultured for 6 days at 32°C. Inhibin activity accumulated in the medium during the last three days of incubation was bioassayed in pituitary cell cultures. The results are mean + SE data from three separate cultures expressed as percent of control gonadotropin values (no inhibin). IC: cultures of Sertoli cells from intact control testes. (Based on data from L. Seethalakshmi and A. Steinberger, J. Andrology, 4: 131, 1983.)

The damaging effect of cryptorchidism was reversed after returning the testis to the scrotum (Seethalakshmi and Steinberger, 1983). Inhibin secretion was restored partially after 14 days, and was only slightly below normal values after 42 days of orchidopexy. The decline of inhibin secretion was accompanied by a rise in circulating FSH and LH levels and a progressive loss of spermatids and spermatocytes. Conversely, gonadotropin levels in the blood at 42 days after orchidopexy declined to near-normal values, and many seminiferous tubules showed increased number of spermatocytes, cap-phase spermatids and spermatozoa. Au et al. (1987) demonstrated that brief heating (43°C for 15 min) of the rat scrotum also caused a significant decrease in testicular inhibin content, and inhibin production rate, as measured after efferent duct ligation. The effect of heat became apparent after about one week, and coincided with elevated circulating levels of both FSH and LH, decreased testis weight and spermatogenic damage. Maximal effects were observed after two weeks, and thereafter (17 weeks) testicular inhibin content and production rate, testicular weight, and seminiferous tubule fluid production returned to control values.

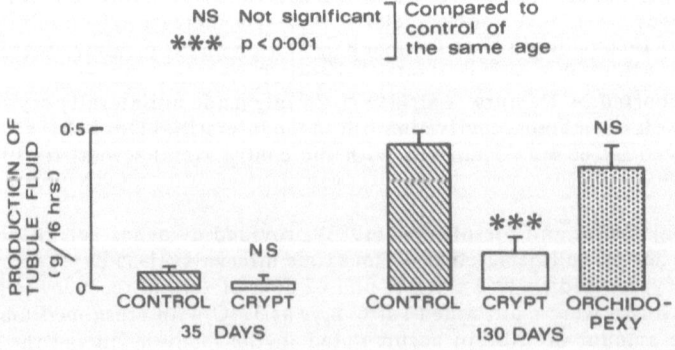

Figure 4. The effect of bilateral cryptorchidism in immature rats induced at 14d of age and subsequent orchidopexy (35d) on the production of tubule fluid at 35d and 130d of age (mean + SE, n = 5). (From Jegou, B., Peake, R.A., Irby, D.C. and deKretser, D.M. Biol. Reprod., 30: 179, 1984a; with permission.)

Inhibin is not unique in its decline at above-scrotal temperatures. Other secretory products of the Sertoli cells, i.e., seminiferous tubule fluid (Fig. 4) and androgen binding protein (ABP) (Fig. 5) were also reported to be decreased in the cryptorchid animals (Hagenas and Ritzen, 1976; Hagenas et al., 1978; Karpe et al., 1984; Bergh et al., 1984b; Jegou et al., 1983, 1984a). The production of ABP by the rat testis was significantly reduced at 14 and 26 days after heating at 43°C for 15 min, but then recovered (Jegou et al., 1984b). However, other investigators (Bartlett and Sharpe, 1987) found a significant rise of ABP in the interstitial fluid of rat testes exposed to 43°C for 30 min. The latter findings, however, may have reflected a change in the directionality of ABP secretion following germ cell destruction, as was observed after hypophysectomy (Gunsalus et al., 1980). All secretory products of Sertoli cells investigated in two-compartment culture chambers were found to be secreted bidirectionally (Janecki et al., 1986, 1987, 1988). Studies in the lamb (Monet et al., 1987) also showed decline of ABP secretion after bilateral cryptorchidism and recovery following orchidopexy. Seminiferous tubules from abdominal testis of unilaterally cryptorchid adult rats secreted less lactate, but similar amounts of tissue plasminogen activator, as did tubules from scrotal testis (Bergh et al., 1987). On the other hand, Sertoli cells isolated from bilaterally cryptorchid immature and adult rats had increased and similar cyclic AMP responses to FSH, respectively, as compared to Sertoli cells from age-matched control animals (Heindel et al., 1982).

It remains unclear if the changes of Sertoli cell functions *in vivo* were caused by direct heat-induced damage in the Sertoli cells, or were an indirect consequence of germ cell depletion, Leydig cell damage, etc. Studies with isolated rat Sertoli cells *in vitro* clearly demonstrated that inhibin secretion can be directly suppressed by elevated temperatures. When cultures of rat Sertoli cells from normal scrotal testes were maintained at 32°C and 38°C (Fig. 6), inhibin secretion declined to near non-detectable

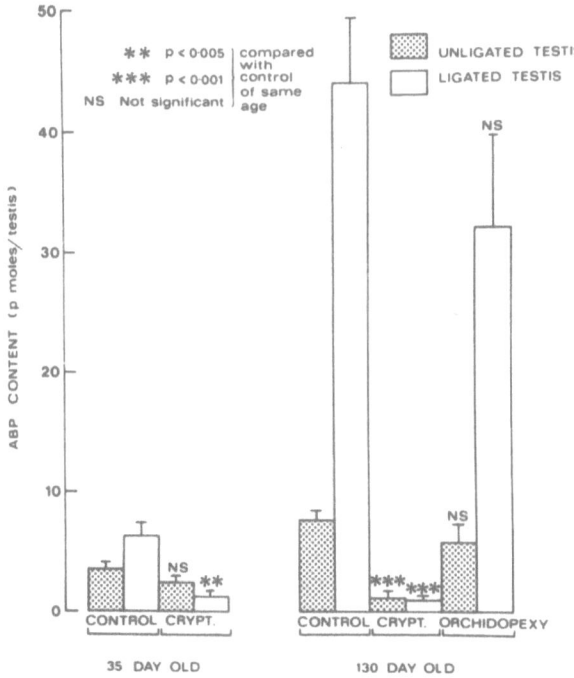

Figure 5. The effect of bilateral cryptorchidism in immature rats (induced at 14d of age) and subsequent orchidopexy (35d) on the ABP content and accumulation after 16h of unilateral efferent duct ligation. Volumes represent mean + SE (n = 5). (From Jegou, B., Peake, R.A., Irby, D.C. and deKretser, D.M. Biol. Reprod., 30: 179, 1984a; with permission.)

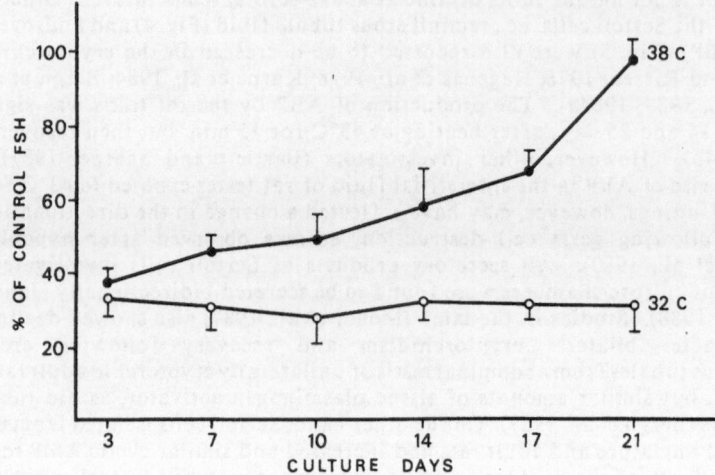

Figure 6. Effect of incubation temperature on the secretion of inhibin by Sertoli cells
isolated from intact adult rats. The cultures were incubated for 21 days at 32°C or
38°C. The media collected at 3- to 4-day intervals were bioassayed for inhibin
activity. The results are mean + SE of data from three separate cultures expressed
as percent of FSH released in control pituitary cell cultures (no inhibin). (From A.
Steinberger, 1980, see References.)

level after 21 days of incubation at 38°C but remained unaltered at 32°C (Steinberger,
1980). These findings indicate a possible direct damaging effect of elevated temperature
on Sertoli cell secretions *in vivo*.

On the other hand, studies by Parvinen (1982) showed that many Sertoli cell
functions vary with the stage of spermatogenic cycle, possibly due to the presence of
different germ cell populations. Thus, a modulatory influence of germ cells (or other
testicular cell types) on Sertoli cell secretions cannot be ruled out. Addition of pachytene
spermatocytes to Sertoli cell cultures was shown to modulate the secretion of ABP and
Trf *in vitro* (Galdieri et al., 1984; Janecki et al., 1988; Castellon et al., 1989). Based on
results from *in vivo* and *in vitro* experiments, Gonzales et al. (1989) suggested a dual effect
of elevated temperature on inhibin secretion, an initial rise due to direct effect of heat
on the Sertoli cells followed by a decline due to disruption of spermatogenesis. Some of
the heat-induced changes in Sertoli cell functions could also be due to altered
permeability of the blood testis-barrier (Turner et al., 1982).

Leydig Cells and Steroidogenesis

Bergh et al. (1984a) compared steroid biosynthesis and Leydig cell morphology in the
scrotal and abdominal testes of adult rats made unilaterally cryptorchid. The area
occupied by Leydig cells in the abdominal testis was reduced; the cytoplasm contained
an increased number of lipid droplets, and the endoplasmic reticulum was more sparse
than in scrotal Leydig cells. Measurements of intratesticular concentrations of
progesterone, 20 α-dihydroprogesterone, 17 α-hydroxyprogesterone and testosterone, both
basal and after LH stimulation, revealed a decrease of testosterone biosynthesis in the
abdominal testes and a block at the 17 α-hydroxylase level. Observations after LH
stimulation indicated that earlier steps in testosterone biosynthetic pathway were also
affected. Steroid biosynthesis in the abdominal testes revealed several alterations
compared to that in scrotal testis. The conversion mediated by 17 α-hydroxylase and 17
ß-ketosteroid reductase was lower, and the conversion mediated by 20 α-dehydrogenase
was higher in the abdominal testis. These results support previously reported
observations on the impaired conversion of progesterone by testicular tissue from
bilaterally cryptorchid adult rats (Ficher and Steinberger, 1982). The underlying cause
of the disturbed steroid biosynthesis in the abdominal testes is unknown, but was thought

to be related to increased concentration of estradiol in the abdominal testes and/or altered paracrine influences by the damaged seminiferous tubules, or changes in the microcirculation of the testis (Bergh et al., 1984a).

In a related study by Bergh and Damber (1984), unilateral cryptorchidism was induced in adult rats for 24 hours, and its effect on testicular morphology and intratesticular testosterone concentration after hCG-stimulation was investigated. In seminiferous tubules from abdominal testes, an increased number of degenerating germ cells was noted in stages XIV-III of the spermatogenic cycle, and the Sertoli cells in stages XIV-VIII contained increased amount of lipid droplets. However, both germ cells and Sertoli cells at other stages of the cycle appeared unaffected. In scrotal testes, the size of peritubular Leydig cells varied with the stages of spermatogenic cycle. The largest cells were found adjacent to stages VII-VIII and the smallest adjacent to stages XI-XII, whereas in the abdominal testes no stage-dependent variation in the size of peritubular Leydig cells was seen. Interestingly, perivascular Leydig cells were of equal size in the abdominal and scrotal testes. Testicular testosterone concentration following stimulation with a low dose of hCG was significantly lower in the abdominal testes. The authors suggested that the seminiferous tubules locally modulate Leydig cell function, and that the paracrine stimulatory influence from stages VII-VIII is rapidly lost during experimental cryptorchidism. In unilaterally cryptorchid rats, the eutopic testis often shows increased testosterone content relative to intact control testes, suggesting a compensatory mechanism for the reduced testosterone content of the abdominal testis (Fig. 7) (Keel and Abney, 1981). The content of estradiol receptors in the abdominal and scrotal testes was similar (Keel and Abney, 1980).

Cryptorchidism also affects steroidogenesis in immature rat testes. In one set of experiments by Bergh et al. (1984b), testicular descent was prevented unilaterally by cutting the gubernaculum testis in newborn rats. At 20 days of age, the unilaterally cryptorchid rats were injected intraperitoneally with 2 μg bFSH per gram body weight and testicular testosterone and estradiol concentrations were determined 6 hours later. The increase in estradiol was subnormal in the abdominal testes. When the efferent ducts were ligated bilaterally in 18-day-old unilaterally cryptorchid rats and the animals killed 48 hours later, the weight increase due to accumulation of seminiferous tubule fluid was significantly greater in the abdominal testes, whereas the ABP content of the abdominal epididymis was reduced. In another set of experiments, unilateral orchidectomy was performed four days after 16-day-old rats were made unilaterally cryptorchid, and intratesticular testosterone and estradiol, and plasma FSH and LH concentrations were compared to those in 20-day-old control unilaterally cryptorchid rats. Removal of the

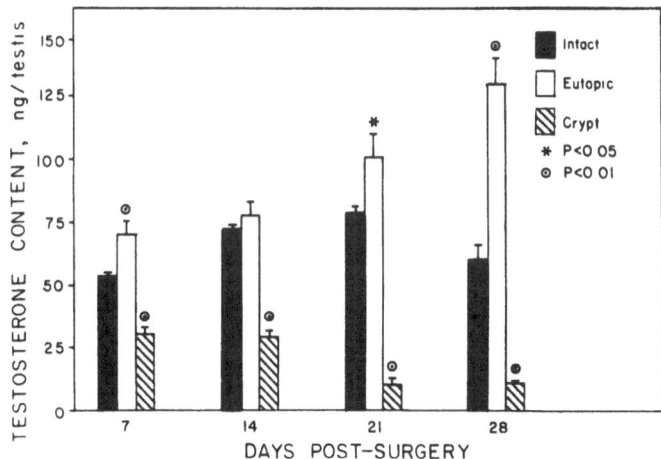

Figure 7. Effects of unilateral cryptorchidism on testicular testosterone content in adult rats (mean + SE). Statistical analysis refers to a comparison between the experimental and intact control groups. (From Keel, B.A. and Abney, T.O., Proc. Soc. Exp. Biol. Med., 166: 489, 1981; with permission.)

abdominal testis resulted in increased plasma FSH and intratesticular estradiol concentrations, whereas plasma levels of LH and intratesticular levels of testosterone remained unaffected. On the other hand, removal of the scrotal testis resulted in increased plasma levels of both FSH and LH coupled with increased intratesticular testosterone and estradiol. It was concluded that Sertoli and Leydig cell functions are influenced by cryptorchidism at the time when the morphological differences between the abdominal and scrotal testes are minimal.

To investigate the possibility that increased temperature can lead to a pattern of testicular steroidogenesis that results in increased estradiol synthesis, Munabi et al. (1984) examined the effects of temperature changes on the activities of four key steroidogenic enzymes: 3 ß-hydroxysteroid dehydrogenase, 17-hydroxylase, 17,20-desmolase and aromatase. The enzyme activities were measured in microsomal fractions of rat, pig and horse testes incubated at 34°C, 36°C, and 38°C. The activities of all four enzymes increased with each 2°C temperature elevation in roughly proportional amounts, indicating that even minor elevations in temperature can increase the activity of testicular steroidogenic enzymes, including aromatase activity which converts testosterone to estradiol.

Jansz and Pomerantz (1986) investigated Leydig cell responses to disrupted gametogenesis *in vitro* using tissue slices and collagenase dispersed Leydig cells from testes of unilaterally or bilaterally cryptorchid rats, or following efferent duct ligation. Four weeks after surgery, androgen secretion per mg of tissue, or per Leydig cell, in response to maximal LH stimulation was significantly greater in testes from the treated than sham-operated animals. It was concluded that disruption of spermatogenesis resulted in Leydig cells that were more responsive to LH stimulation *in vitro*. Unilateral lesions produced different responsiveness of Leydig cells from the ipsilateral and contralateral testes, supporting the hypothesis that Leydig cell function is modulated locally by the seminiferous tubules and that this modulation is altered when the seminiferous tubules are impaired. Stimulated androgen production by Leydig cells from the contralateral, nonligated testis did not differ from that of sham-operated controls, whereas Leydig cells from the scrotal testis of unilaterally cryptorchid animals were hyper-responsive compared to those from sham-operated control. This suggests different mechanisms of Leydig cell hyper-responsiveness to LH in the cryptorchid and ligated testes, and possible influence of aspermatogenesis-induced cryptorchidism on the contralateral testis.

Hormone Receptor Levels

Differences were observed in hCG regulation of steroidogenesis and hormone receptor level between the scrotal and abdominal testes of unilaterally cryptorchid rats (Huhtaniemi et al., 1984). In rats made unilaterally cryptorchid by cutting the gubernaculum testes at birth and injected with 600 IU hCG/kg at 100 days of age, a biphasic testosterone response was seen in the scrotal testis with peaks at 1 hour and 3 days after the injection. The initial peak of testosterone was of similar magnitude in the abdominal testis, but the secondary peak was absent. The response of testicular progesterone concentration to hCG stimulation showed a maximum at 1 day in both gonads, but it was 10- to 20-fold higher (P<0.01) in the abdominal testis. The testicular content of LH, FSH, and PRL receptors was decreased in the abdominal testis. After hCG injection, the loss of available LH receptors was faster in the abdominal testis. Likewise, the recovery of LH binding was faster in the abdominal testis; at day 10 of the experiment, it was 102 ± 5% of the starting levels compared to 52 ± 4% in the scrotal gonad (P<0.01). The administration of hCG did not affect FSH binding in the scrotal testis, but induced a transient drop of 25-35% on day 1 in the abdominal testis (P<0.05). On days 3-10, FSH binding of the abdominal testes was 20-40% higher compared to the contralateral gonad (P<0.05-0.01). In PRL binding, heterologous down-regulation of 50-80% was found in both testes between 12-24 h. Thereafter, PRL receptors in the abdominal testis showed a transient elevation of 25-70% (P<0.05-0.01) on day 3, which was not seen in the scrotal testis. Thus, the abdominal testis displays a dramatically enhanced blockade of C21 steroid side-chain cleavage upon hCG stimulation. The kinetic changes in testicular LH receptors after hCG stimulation are also faster at the abdominal site. Only the abdominal testes displayed hCG-induced changes in FSH binding and a transient up-regulation of PRL receptors. The altered tropic regulation of Leydig and

Sertoli cells in the abdominal testis are indicative of direct functional changes in these cells at elevated temperature and/or of changes in paracrine testicular regulation.

A decline of FSH and androgen receptors was found in bilaterally cryptorchid lamb testes followed by recovery to normal levels after orchidopexy (Monet-Kuntz et al., 1987). Decreased levels of FSH and LH receptors were also observed in human cryptorchid testes (Hovatta et al., 1986). On the other hand, Namiki et al. (1987) found no difference in FSH and hCG receptors between the left and right testis of varicocele patients and testis of normal men.

Some of the changes in circulating and intratesticular testosterone concentrations in cryptorchidism could be due, at least in part, to altered blood flow at elevated temperature (Damber et al., 1978; Gomes et al., 1976). Studies by Galil and Setchell (1988) showed that exposure of one or both rat testes to 43°C for 30 min resulted in a significant reduction in testicular blood flow, as measured by using microspheres. The effects on the testes of unilateral and bilateral heating were similar, although the changes in peripheral blood FSH levels were in general less marked after unilateral exposure to heat. Testicular blood flow fell along with testicular weight, and was lowest at 14-21 days after heating. Both parameters began to recover 35 days post-heating and blood flow per testis was normal by 46 days, although testicular weights were still slightly reduced. Heating one or both testes to 42°C produced similar, but smaller responses 21 days later, whereas temperatures of 41°C, or lower, did not affect the parameters measured, except for some rises in serum LH and FSH. With slight reductions in blood flow, there were corresponding increases in testicular venous testosterone concentration so that testosterone secretion was unaffected; however, further reductions in blood flow at 14 and 21 days after heating were not fully compensated by elevated testosterone concentration in testicular venous blood, resulting in decreased testosterone secretion.

CONCLUDING REMARKS

There is sufficient evidence now that most cell types in the testis are either directly or indirectly affected by above-scrotal temperatures. Whereas the changes noted after acute heat exposure (40-43°C) are more dramatic and have provided valuable clues as to the relative heat sensitivity of different testicular cell types, the more meaningful experiments biologically are those using subtle temperature changes that are closer to normal body temperature. Despite extensive studies and accumulation of considerable information concerning the effects of heat on various biochemical parameters, the underlying biochemical lesion(s) caused by elevated temperature remains to be determined. It is hoped that additional biochemical studies using enriched populations of individual cell types, and examining key enzymes and metabolic pathways in testicular tissues from different animal models as well as from human testes, will ultimately clarify the picture. It is also hoped that modern methods of molecular biology will be utilized in the study of heat-induced testicular lesions. Due to complex cell-to-cell interactions and the dependence of spermatogenesis on normal function of all testicular components, a subtle change in one cell type may dramatically affect sperm production and, thus, male fertility.

REFERENCES

Abney, T.O., and Keel, B.A., eds., 1989, "The Cryptorchid Testis", CRC Press, Boca Raton, Florida.

Au, C.L., Robertson, D.M., and de Kretser, D.M., 1987, Changes in testicular inhibin after a single episode of heating of rat testes, Endocrinology, 120:973.

Au, C.L., Robertson, D.M., and de Kretser, D.M., 1983, In vitro bioassay of inhibin in testes of normal and cryptorchid rats, Endocrinology, 112:239.

Bartlett, J.M., and Sharpe, R.M., 1987, Effect of local heating of the rat testis on the levels in interstitial fluid of a putative paracrine regulator of the Leydig cells and its relationship to changes in Sertoli cell secretory function, J. Reprod. Fertil., 80:279.

Bergh, A., Ason-Berg, A., Damber, J.E., Hammar, M., and Selstam, G., 1984a, Steroid biosynthesis and Leydig cell morphology in adult unilaterally cryptorchid rats, Acta Endocrinol. (Copenh)., 107:556.

Bergh, A., and Damber, J.E., 1984, Local regulation of Leydig cells by the seminiferous tubules. Effect of short-term cryptorchidism, Int. J. Androl., 7:409.

Bergh, A., Damber, J.E., Jacobsson, H., and Nilsson, T.K., 1987, Production of lactate and tissue plasminogen activator in vitro by seminiferous tubules obtained from adult unilaterally cryptorchid rats, Arch. Androl., 19:177.

Bergh, A., Damber, J.E., and Ritzen, M., 1984b, Early signs of Sertoli and Leydig cell dysfunction in the abdominal testes of immature unilaterally cryptorchid rats, Int. J. Androl., 7:398.

Blackshaw, A.W., 1977, Temperature and seasonal influences, in "The Testis", A.D. Johnson and W.R. Gomez, eds., Vol. IV, pp. 517-545, Academic Press, New York.

Blackshaw, A.W., 1973, Testicular enzymes and spermatogenesis, J. Reprod. Fertil., Suppl., 18:55.

Blackshaw, A.W., and Hamilton, D.J., 1971, Early histological and histochemical changes in the heated and cryptorchid rat testis, J. Reprod. Fertil., 24:151.

Blackshaw, A.W., Hamilton, D. and Massey, P.F., 1973, Effect of scrotal heating on testicular enzymes and spermatogenesis in the rat, Austr. J. Biol. Sci., 26:1395.

Byer, R., and Davies, J.R., 1966, Species variation in the effect of temperature on the incorporation of L-lysine-U-C$^{14}$ into protein of testis slices, Comp. Biochem. Physiol., 17:151.

Castellon, E., Janecki, A., and Steinberger, A., 1989, Influence of germ cells on Sertoli cell secretory activity in direct and indirect co-culture with Sertoli cells from rats of different ages, Mol. Cell. Endocrinol., 64:169.

Chowdhury, A.K., and Steinberger, E., 1970, Early changes in the germinal epithelium of rat testes following exposure to heat, J. Reprod. Fertil., 22:205.

Chowdhury, A.K., and Steinberger, E., 1972, The influence of cryptorchid milieu on the initiation of spermatogenesis in the rat, J. Reprod. Fertil., 29:173.

Chowdhury, A.K., and Steinberger, E., 1964, A quantitative study of the effect of heat on germinal epithelium of rat testes, Am. J. Anat., 115:509.

Damber, J.E., Bergh, A., and Janson, P.O., 1978, Testicular blood flow and testosterone concentrations in the spermatic venous blood in rats with experimental cryptorchidism, Acta Endocrinol. (Copenhagen), 88:611.

Demura, R., Suzuki, T., Nakamura, S., Komatsu, H., Jibiki, K., Odagiri, E., Demura, H., and Shizuma, K., 1987, Effect of uni- and bilateral cryptorchidism on testicular inhibin and testosterone secretion in rats, Endocrinol. Japon., 34:911.

Ficher, M., and Steinberger, E., 1982, Conversion of progesterone by testicular tissue of cryptorchid rats, Acta Endocrinol., 101:301.

Fritz, I.B., 1973, Selected topics on the biochemistry of spermatogenesis, Current Topics in Cellular Regulation, 7:129

Fujisawa, M., Matsumoto, O., Kamidono, S., Hirose, F., Kojima, K., and Yoshida, S., 1988, Changes of enzymes involved in DNA synthesis in the testes of cryptorchid rats, J. Reprod. Fertil., 84:123.

Galdieri, M., Monaco, L., and Stefanini, M., 1984, Secretion of androgen binding protein by Sertoli cells is influenced by contact with germ cells, J. Andrology, 5:409.

Galil, K.A., and Setchell, B.P., 1988, Effects of local heating of the testis on testicular blood flow and testosterone secretion in the rat, Int. J. Androl., 11:73.

Gomes, D., Kein, N.D., and Hamlin, R.L., 1976, Testicular blood flow and testosterone secretion rates in normal and cryptorchid rats, Physiologist, 19:208.

Gonzales, G.F., Risbridger, G.P., and de Krester, D.M., 1989, In vivo and in vitro production of inhibin by cryptorchid testes from adult rats, Endocrinology, 124:1661.

Gunsalus, G.L., Musto, N.A., and Bardin, C.W., 1980, Bidirectional release of a Sertoli cell product, androgen binding protein, into the blood and seminiferous tubule, in: "Testicular Development, Structure and Function", A. Steinberger and E. Steinberger, eds., p. 291, Raven Press, New York.

Hagenas, L., and Ritzen, E.M., 1976, Impaired Sertoli cell function in experimental cryptorchidism in the rat, Mol. Cell Endocrinol., 4:25.

Hagenas, L., Ritzen, E.M., Svenson, J., Hansson, V., and Purvis, K., 1978, Temperature dependence of Sertoli cell function, Int. J. Androl., Suppl. 2, 449.

Hall, P.F., Kew, D., and Mita, M., 1985, The influence of temperature on the functions of cultured Sertoli cells, Endocrinology, 116:1926.

Harrison, R.G., and Weiner, J.S., 1948, Abdomino-testicular temperature gradients, J. Physiol. 107:48P.

Heindel, J.J., Berkowitz, A., Steinberger, A., and Strada, S.J., 1982, Modification of Sertoli cell responsiveness to FSH by cryptorchidism and hypophysectomy in immature and adult rats, J. Androl., 3:337.

Hoffmann, A.M., Bergh, A., and Olivecrona, T., 1989, Changes of testicular cholesteryl ester hydrolase activity in experimentally cryptorchid rats, J. Reprod. Fertil., 86:11.

Hovatta, O., Huhtaniemi, I., and Wahlstrom, T., 1986, Testicular gonadotrophins and their receptors in human cryptorchidism as revealed by immunohistochemistry and radioreceptor assay, Acta Endocrinol. (Copenh.), 111:128.

Huhtaniemi, I., Bergh, A., Nikula, H., and Damber, J.E. 1984. Differences in the regulation of steroidogenesis and tropic hormone receptors between the scrotal and abdominal testes of unilaterally cryptorchid adult rats, Endocrinology, 115: 550.

Janecki, A., Jakubowiak, A., and Steinberger, A., 1988, Effect of germ cells on vectorial secretion of androgen binding protein and transferrin by immature rat Sertoli cells in vitro., J. Androl., 9:126.

Jansz, G.F., and Pomerantz, D.K., 1986, A comparison of Leydig cell function after unilateral and bilateral cryptorchidism and efferent- duct-ligation, Biol. Reprod., 34:316.

Jegou, B., Laws, A.O., and de Kretser, D.M., 1984b, Changes in testicular function induced by short-term exposure of the rat testis to heat: further evidence for interaction of germ cells, Sertoli cells and Leydig cells, Int. J. Androl., 7:244.

Jegou, B., Peake, R.A., Irby, D.C., and de Kretser, D.M., 1984a, Effects of the induction of experimental cryptorchidism and subsequent orchidopexy on testicular function in immature rats, Biol. Reprod., 30:179.

Jegou, B., Risbridger, G.P., and de Kretser, D.M., 1983, Effects of experimental cryptorchidism on testicular function in adult rats, <u>J. Androl.</u>, 4:88.

Johnson, A.D., and Gomes, R.R., eds., 1977, "The Testis," Vol. II, 1970, and Vol. IV, 1977, Academic Press, New York.

Kandeel, F.R., and Swerdloff, R.S., 1988, Role of temperature in regulation of spermatogenesis and the use of heating as a method for contraception, <u>Fertil. Steril.</u>, 49:1.

Karpe, B., Ploen, L., and Ritzen, E.M., 1984, Maturation of the juvenile rat testis after surgical treatment of cryptorchidism, <u>Int. J. Androl.</u>, 7:154.

Keel, B.A., and Abney, T.O., 1980, Influence of bilateral cryptorchidism in the mature rat: alterations in testicular function and serum hormone levels, <u>Endocrinology</u>, 107:1226.

Keel, B.A., and Abney, T.O., 1981, Alterations of testicular function in the unilaterally cryptorchid rat, <u>Proc. Soc. Exp. Biol. Med.</u>, 166:489.

Laufer, A., and Davies, A.M., 1969, Experimental granulomatous myocarditis: genesis and immunological aspects, <u>Ann. N.Y. Acad. Sci.</u>, 156:91.

LeGac-Jegou, F., and deKretser, D.M., 1980, Studies on isolated Sertoli cells from normal and cryptorchid testes [Abstr. #684], <u>Program of 6th International Congress of Endocrinology</u>, p. 551, Melbourne, Australia.

Lee, L.P.K., and Fritz, I.B., 1972, Studies in spermatogenesis in rats. V. Increased thermal lability of lysosomes from testicular germinal cells and its possible relationship to impairments in spermatogenesis in cryptorchidism, <u>J. Biol. Chem.</u>, 247:7956.

Males, L.P.K., and Turkington, R.W., 1970, Hormonal regulation of hyaluronidase during spermatogenesis in the rat, <u>J. Biol. Chem.</u> 245:6329.

Males, L.P.K., and Turkington, R.W., 1971, Hormonal control of lysosomal enzymes during spermatogenesis in the rat, <u>Endocrinology</u>, 88:579.

Monet-Kuntz, C., Barenton, V., Locatelli, A., Fontaine, I., Perreau, C., and Hochereau de Reviers, M.T., 1987, Effects of experimental cryptorchidism and subsequent orchidopexy on seminiferous tubule functions in the lamb, <u>J. Androl.</u>, 8: 148.

Munabi, A.K., Cassorla, F.G., D'Agata, R., Albertson, B.D., Loriaux, D.L., and Lipsett, M.B., 1984, The effects of temperature on the activity of testicular steroidogenic enzymes, <u>Steroids</u>, 43:325.

Nakamura, M., Namiki, M., Okuyama, A., Koh, E., Kondoh, N., Takeyama, M., Fujioka, H., Nishimune, Y., Matsumoto, K., and Matsuda, M., 1988, Optimal temperature for synthesis of DNA, RNA and protein by human testis <u>in vitro</u>, <u>Archives of Andrology</u>, 20:41.

Nakamura, M., Namiki, M., Okuyama, A., Matsui, T., Doi, Y., Takeyama, M., Fujioka, H., Nishimune, Y., Matsumoto, K., Sonoda, T., 1987, Temperature sensitivity of human spermatogonia and spermatocytes <u>in vitro</u>, <u>Arch. Androl.</u>, 19:127.

Namiki, M., Namiki, M., Okuyama, A. et al., 1987, Influence of temperature on the function of Sertoli and Leydig cells of human testes, <u>Fertil. Steril</u>. 47: 475.

Nishimune, Y., and Komatsu, T., 1977, Temperature-sensitivity of mouse testicular DNA synthesis <u>in vitro</u>, <u>Exp. Cell Res.</u>, 75:514.

Parvinen, M., 1973, Observations on freshly isolated and accurately identified spermatogenic cells of the rat. Early effects of heat and short-time experimental cryptorchidism., Virchows. Arc. B. 13:38.

Parvinen, M., 1982, Regulation of the seminiferous epithelium, Endocr. Rev., 3:404.

Seethalakshmi, L., and Steinberger, A., 1983, Effect of cryptorchidism and orchidopexy on inhibin secretion by rat Sertoli cells, J. Androl., 4:131.

Sharpe, R.M., Swanston, I.A., Cooper, I., Tsonis, C.G., and McNeilly, A.S., 1988, Factors affecting the secretion of immunoactive inhibin into testicular interstitial fluid in rats, J. Endocrinol., 119:315.

Steinberger, A., 1980, Factors affecting in vitro secretion of inhibin by isolated Sertoli cells, in: "Endocrinology," I.A. Cumming, J.W. Funder, and F.A.O. Mendelsohn, eds., pp. 259-262, The Australian Academy of Science, Canberra, Australia.

Steinberger, E., and Dixon, W.J., 1969, Some observations of the effect of heat on the testicular germinal epithelium, Fertil. Steril., 10:578.

Steinberger, E., and Nelson, W.O., 1955, Effect of hypophysectomy, cryptorchidism, estrogen and androgen upon the level of hyaluronidase in the rat testis, Endocrinology, 56:429.

Steinberger, E., and Steinberger, A., 1972, Basic and clinical aspects, in: "Reproductive Biology", H. Balin and S. Glasser, eds., p. 144, Excerpta Med. Found., Amsterdam.

Szego, C.M., Seeler, B.J., Steadman, R.A., Kimura, A.K., and Roberts, J.A., 1971, Biochem. J., 123:523.

Turner, T.T., D'Addario, D.A., Forrest, J.B., and Howards, S.S., 1982, The effects of experimental cryptorchidism on the entry of [$^3$H]-inulin and [$^3$H]-horseradish peroxidase in the lumen of the rat seminiferous tubules, J. Androl., 3:178.

VanDemark, N.L., and Free, M.J., 1970, Temperature effects, in: "The Testis", A.D. Johnson, W.R. Gomes, and N.L. VanDemark, eds., Vol. 3, p. 233, Academic Press, New York.

Vunder, P.A., and Murashev, A.N., 1984, Early changes in the permeability of cellular and lysosomal membrane in rat testes after local heating of the scrotum, Bull. Exptl. Biol. Med., 98:608.

Zorgniotti, A.W., and Sealfon, A.I., 1988, Measurement of intrascrotal temperature in normal and subfertile men. J. Reprod. Fertil., 82:563.

# EPIDEMIOLOGIC ASPECTS OF THE RELATIONSHIP BETWEEN

# TEMPERATURE AND MALE REPRODUCTION

Alfred Spira

INSERM U 292 Hôpital de Bicêtre
78, rue du Général Leclerc
94275 Le Kremlin - Bicêtre Cédex (France)

The present knowledge about the relationship between temperature and male reproductive functions comes mainly from animal studies, from in vitro experiments and from clinical observations. In animals such as the rat, rabbit, dog, sheep, ram and bull, procedures such as experimental cryptorchidism, scrotal insulation, acute febrile illness, increased ambient temperature and experimental varicocele all alter the spermatogenic functions and inhibit spermatogenesis[1].

The great sensitivity of human spermatogenesis to temperature increase has been shown by in vitro studies. For instance, Nakamura et al.[2] have studied both the number and DNA synthesis of germ cells cultured at 31°C and 37°C. It appeared that at normal body temperature (37°C), differentiated germ cells such as spermatids and spermatozoa are fragile and that DNA synthesis of spermatogonia and resting primary spermatocytes is delayed. On this basis, it seems interesting to undertake a review of literature on the present knowledge of the effects of temperature on human spermatogenesis in the human, and then to explore the possible epidemiologic studies of temperature-related risk factors for male reproduction.

## PRESENT KNOWLEDGE

In this area about all of the knowledge we have today comes from clinical studies, and not from epidemiological studies (i.e., including controls) which are very few. A complete review including this topic has been recently published by Donald R. Kandeel and Ronald S. Swerdloff[1], and we will follow very closely their overview. Many clinical studies showing an harmful effect of elevated temperature on spermatogenesis have been published.

## FEVER

The first evidence comes from the study of spermatogenesis in men after febrile diseases. As shown on Fig. 1, the gradient of body to intrascrotal temperature is decreased in febrile patients, and sperm production is also altered in febrile patients. This effect has also been studied in the WHO survey[3] of infertile couples (Table 1), showing that pyrexia over 38°C may result in temporary impairment of semen quality and reduction of fertility. The patients reporting a history of high fever in the last six months before consultation were found more frequently to have abnormal semen, in particular reduced sperm concentration, than men without such a history. The prevalence of azoospermia was not different in the two groups.

*Temperature and Environmental Effects on the Testis*
Edited by A. W. Zorgniotti, Plenum Press, New York, 1991

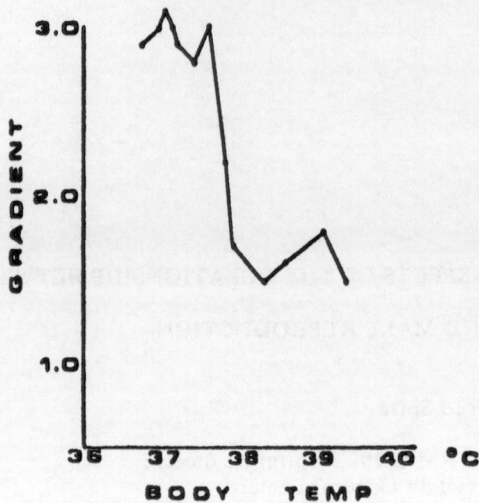

Figure 1. Body/intrascrotal temperature gradients in febrile patients. Mean gradients were plotted against body temperatures of six febrile patients. A sharp drop in the gradient was observed when body temperature rose to between 37.6 and 37.9°C. (From Lazarus and Zorgniotti, 1975; reproduced with permission of the publisher).

CRYPTORCHIDISM

A second set of evidences comes from the alteration of spermatogenesis from such diseases as cryptorchidism or retractile testes. Among other effects, they tend to increase the temperature of the testes. Indeed, a French study has shown that cryptorchidism provokes an impairment of both sperm concentration and morphology (Table 2). Of course, the subsequent fertility of these men is altered by these abnormalities.

ENVIRONMENT

Temperature variations of the testes can come from environment modifications. In a recently published paper, Laven et al.[4] have studied the influence of occupation and living habits on semen quality in men. The 56 patients of this study were categorized into two groups according to their occupational and sleeping habits. Those with normal heat dissipation during the day or during the night were classified as "cool workers" and "cool sleepers". The second group categorized as "warm workers" and "warm sleepers", showed evidence for day and night scrotal insulation. This study (Fig. 2) clearly indicates that patients in the first group consistently exhibited a greater sperm motility

Table 1. Prevalence of azoospermia, abnormal semen classification and odds ratio according to a recent history of high fever (from WHO, Investigation of the infertile couple)

| High Fever | Number of cases | Azoospermia (%) | Sperm Present N | Abnormal semen (%) | Odds Ratio | Concentration (mill/ml) |
|---|---|---|---|---|---|---|
| No | 6426 | 7.3 | 5955 | 46.4 | | 53.8 |
| Yes | 77 | 5.2 | 73 | 64.4 | 2.09 (1.29-3.39) | 35.1 |

Table 2. Sperm characteristics in cryptorchid patient as compared to controls (Jouannet and Spira, 1988)

| | Number | Volume (ml) | Concen-tration (mil/ml) | Motility (%) | Morpho-logy (%) | Azoo-spermia (%) |
|---|---|---|---|---|---|---|
| Cryptorchidism | 31 | 4.1 | 32.4 | 45.4 | 34.8 | 6.5 |
| Unexplained infertility | 182 | 3.5 | 90.2 | 51.9 | 46.0 | 4.4 |
| Fertile | 139 | 3.8 | 116.5 | 56.2 | 51.0 | 0.7 |

than the second group (p<0.001), either when considering the concentration of spermatozoa, or their total number in the ejaculate.

OCCUPATION

Very few studies have been devoted to the effect of elevated temperature inside a working environment on reproductive functions and men's subsequent fertility. Truck driving and associated occupations are suspected to influence scrotal temperature. The main reason is that truck drivers usually seat right above the engine for many hours. The second reason is that, being seated most of the day, the contact surface between scrotum and both thighs is extended. However, to our knowledge, no specific study of these workers has been undertaken to date. The only well described phenomenon is the fall of scrotal temperature of about 1°C when normal subjects stand up after previously being supine[5].

Table 3 shows a very short list of various jobs which could be related to increased scrotal temperature and possible consequences on spermatogenesis. However, to our knowledge, no study of these particular conditions has been published to-date.

ARTIFICIAL HEAT

The effect of dry heat and hot saunas has been more extensively studied. Over forty years ago, "fever treatment" consisted in elevating body temperature to 40°C to 41°C. As early as 1941, MacLeod and Hotchkiss published the effects of hyperpyrexia

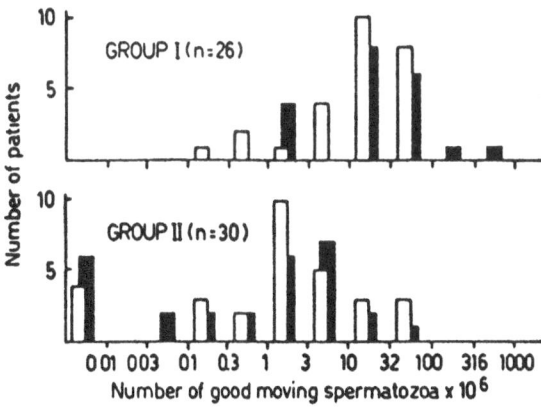

Figure 2. Graph showing the number of good moving spermatozoa per ml in each group (open bars = first sample; closed bars = second sample). (From Laven et al., 1988).

Table 3. Various occupations which could be related to increased scrotal temperature

Drivers, Truck Drivers
Bakers
Smelters
Miners

upon spermatozoa counts in men[6]. They showed that a single "fever" treatment from 40.5°C to 41°C reduced sperm counts some 40 days after the exposure to heat in most subjects and that the fall in sperm count lasted for an average of 32 days. Later on, it was shown by Robinson and Rock[7] that the immersion of the body below the neck in baths at temperatures of 38 to 43°C induced a reversal in the scrotal-rectal temperature differential. At temperatures above 39°C, scrotal temperature began to exceed rectal temperature. This phenomenon can indicate a rapid thermal exchange across the scrotal wall that could overwhelm the regulatory mechanism responsible for the maintenance of the normal negative scrotal-rectal differential.

Another interesting condition is that of sauna heating. Procope[8] could observe a transient but reversible decrease in sperm count in 12 normal men exposed to hot saunas during an average total time of 2 hours and 24 minutes over a 2-week period. Between 30 and 39 days after cessation of treatments, a fall of 50% from initial values occurred.

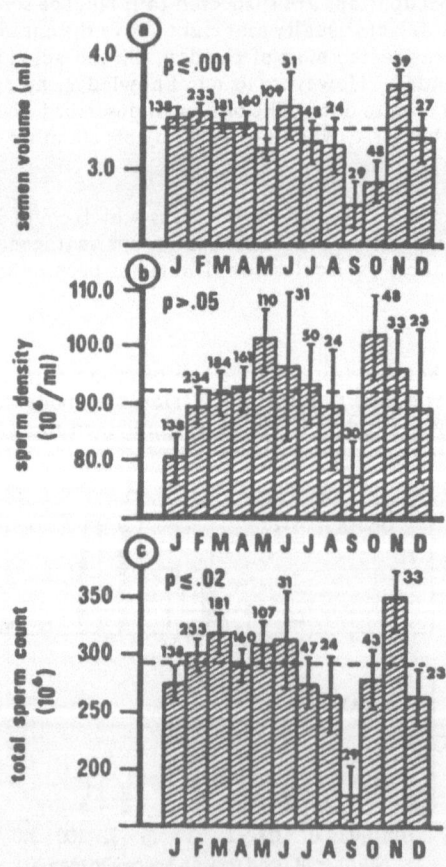

Figure 3. Monthly means (±SEM) of (a) semen volume, (b) sperm density and (c) total sperm count of all ejaculates.

However, it is reasonable to think that this effect is at least transient, since populations using saunas as bathrooms, like Finns or Turks, would have encountered serious reproductive problems, which is not actually the case.

AMBIENT TEMPERATURE

Ambient temperature modifications can result from seasonal variations or climatic conditions. To our knowledge no relationship between the average temperature in a given country and any of the reproductive outcomes, from spermatogenesis to birth rate, has ever been established. On the contrary, it is quite striking to observe that for all the studies performed under the same laboratory conditions, sperm count is a very stable parameter. For example, the mean sperm count of fertile men, when they are selected in the same way, always lies around 100 million spermatozoa per ml. This remarkable stability is a strong argument in suggesting that there exists a very efficient regulatory process of sperm production, or at least that its sensitivity to environmental modifications is quite small.

Many studies have been published concerning the seasonal variations of testicular production, and R. Levine will review them for us during this conference. I have had the opportunity to analyze a set of unique data collected by J. MacLeod[9]. These data came from 52 medical students who each provided 1 to 120 sperm samples over a 1 to 3 year period. The results of Figs. 3 and 4 clearly show that sperm count undergoes a tendency towards a semiannual rhythm, with two minima (January and September) and two maxima (May and October). Motility and morphology show a peak in late summer (August-September) and a valley in late winter early spring (February-March-April). Of course, there is no indication about the origin of these modifications. These could be temperature related but could also be consequences of other seasonal factors, such as the length of daytime for example.

LOCAL HEAT MODIFICATIONS

Many studies have favored the local application of heat: wearing an athletic supporter with insulating quality, for example. Robinson and Rock[7] noted a depression

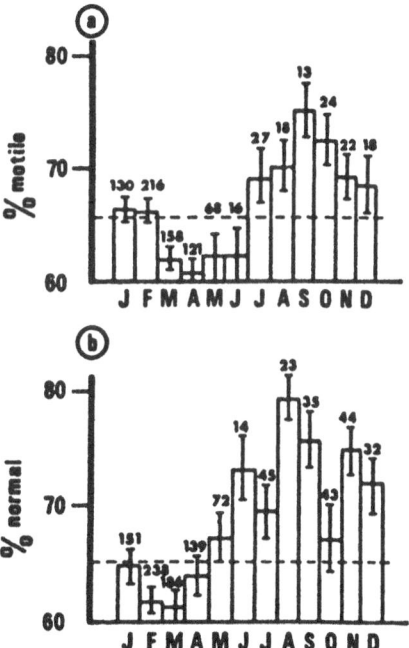

Figure 4. Monthly means (±SEM) of (a) progressively motile and (b) morphologically normal spermatozoa in all ejaculates.

in mean total sperm count on the third week, which reached its lowest level between the fifth and ninth weeks. A sperm count recovery was observed at previous levels after a period of 3 to 8 weeks. Mieusset et al.[11] have shown the inhibiting effect of artificial cryptorchidism on spermatogenesis obtained by raising the testicles into the inguinal canal during the day in adult volunteers. A similar technique better maintaining the testicles in the inguinal canal resulted in a greater reduction of spermatogenesis. The mean inhibitory effect of hyperthermia was at least 97% after 2 months using this technique.

A last way of heat exposure has been the scrotal heating for 30 minutes, at 43°C to 47°C. This technique used by Tokuyama[12] daily for 12 consecutive days on 5 subjects, induced a fall in sperm counts reaching near azoospermia. Concomitant testicular biopsies showed an arrest of spermatogenesis at the stage of primary spermatocytes.

## PROPOSED EPIDEMIOLOGIC STUDIES

### Measures

Epidemiological studies are designed to statistically assess and quantify the possible relationship between risk factors and outcome variables. The independent variables and the possible outcome variables for the study of the effect of temperature modifications on male reproduction are listed on Table 4. The independent variables can be the direct measurement of scrotal temperature performed under standardized conditions, the measurement of temperature in the testicular environment or in the subject's environment, the knowledge of the subject's exposure to heat modifications, such as occupation, fever, etc. The outcome variables can be the measurement of testicular production in terms of numbers and quality, sperm count, motility, morphology, ability to penetrate cervical mucus, zona-free hamster eggs, human eggs, etc. Other possible outcomes are the measurement of testicular endocrine secretion (testosterone, inhibin, etc.) but also the ability of the man to impregnate and fertilize a woman. This outcome can be measured through couple fertility, incidence of spontaneous abortions, rate of congenital malformation, birth rates, etc.

### Designs

Among all the available epidemiologic study designs, very few of them can be proposed for the subject under study here. The main difficulty comes from the fact that exposure to elevated heat is a very rare phenomenon in the general population. For this reason, cohort studies are not adequate at all. The most useful design for the study of the effect of heat on spermatogenesis seems to be the "exposed-non-exposed" one. This kind of design allows the comparison of a group of subjects with a given characteristic, e.g., occupation in a hot atmosphere, to a control group matched for general characteristics, but not exposed to increased heat. These two comparable groups can be compared in a retrospective, transversal or prospective manner, according to one of the chosen outcomes, for example, the incidence of spontaneous abortions in both groups.

Table 4. Possible exposure and outcome variables

| Risk Factors | Outcomes |
| --- | --- |
| Scrotal temperature | Sperm count, motility and morphology |
| Ambient temperature | Sperm function |
| Exposure to elevated heat: fever, occupation | Hormonal status<br>Fertility<br>Spontaneous abortions<br>Congenital malformation |

54

After taking into account, by using a multivariate analysis, the possible confounders such as age, smoking and drinking habits or sexually transmitted diseases experience, such protocols would permit to calculate, for each given risk factor, the attached relative risk or its estimation (odds ratio). The most important difficulty for this kind of non-experimental studies will be to take into account all the possible confounders; the usual way for doing this is to select the more appropriate controls.

Another attempt to study the effects of increased heat on sperm production is the use of case-control studies. For example, if one suspects that increased scrotal heat may be the source of infertility of male origin, a protocol can compare the scrotal temperature of two sets of fertile and infertile men. Such a study has been undertaken by Mieusset et al.[11], one of the main difficulties of this design being to find adequate controls and to perform all the comparative measurements "blindly". Moreover, if the aim of this kind of study is to evaluate the effect of exposure to heat, for example occupation in a hot atmosphere, the prevalence of exposure to the risk factor may be so small in the controls and in the cases that an unreachable number of subjects in each group may be necessary. Such case-control studies could be undertaken for all the different outcomes listed on Table 4.

It finally appears that, even if it is suspected to be very important, the effect of elevated temperature on spermatogenesis is difficult to assess appropriately. Clinical studies can only suggest new hypothesis; epidemiologic studies are difficult. It is possible that experimentation, using cooling or heating devices will bring new clues in this research area; however, it is possible that ethical considerations will also limit these investigations.

REFERENCES

1. Kandeel, R. and Swerdloff, R.S. 1988. Role of temperature in regulation of spermatogenesis and the use of heating as a method for contraception. Fertil. Steril., 49: 1.

2. Nakamura, N. et al. 1989. Impaired DNA synthesis by early pubertal cryptorchid testis in vitro. Arch. Androl., 22: 137.

3. WHO: Investigation of the infertile couple.

4. Laven, J.S.E., Haverkorn, M.J. and Bots, R. 1988. Influence of occupation and living habits on semen quality in men (scrotal insulation and semen quality). Europ. J. Obstet. Gynecol. Reprod. Biol., 29: 137-141.

5. Zorgniotti, A.W. and MacLeod, J. 1973. Studies in temperature, human quality and varicocele. Fertil. Steril., 24: 854.

6. MacLeod, J. and Hotchkiss, R.S. 1941. The effect of hyperpyrexia upon spermatozoa counts in men. Endokrinologie, 28: 780.

7. Robinson, D., Rock, J. and Menkin, M.E. 1968. Control of human spermatogenesis by induced changes of intrascrotal temperature. Jama, 204: 290.

8. Procope, B.J. 1965. Effect of repeated increase of body temperature on human sperm cells. Int. J. Fertil., 10: 333.

9. Spira, A. 1984. Seasonal variations of sperm characteristics. Arch. Androl., 12: 23.

10. Rock, J. and Robinson, D. 1967. Effect of induced intrascrotal hypothermia on testicular function in man. Am. J. Obstet. Gynecol., 29: 217.

11. Mieusset, R., Bujan, L. and Mansat, A. 1987. Hyperthermia and human spermatogenesis: enhancement of the inhibitory effect obtained by "artificial cryptorchidism". Int. J. Androl., 10: 57.

12. Tokuyama, I. 1963. Quoted by Leblon, C.P., Steinberger, E. and RoosenRunge, E.C. in: Spermatogenesis: Mechanisms concerned with conception. Edited by C.G. Hartman, Oxford, Pergamon, p. 1.

# SECTION 2

# ENVIRONMENTAL FACTORS

# SEASONAL PATTERNS OF BIRTHS AND CONCEPTION

# THROUGHOUT THE WORLD

Stan Becker

Department of Population Dynamics
Johns Hopkins University
Baltimore, MD

Seasonal patterns of births in human populations were first described in the 19th century for the populations of Europe (Villerme, 1831; Quetelet, 1869), and India (Hill, 1888). Earlier in this century, Gini (1912), Huntington (1938) and later Cowgill (1966a, 1966b) documented seasonal patterns in much of the world; since then a large number of studies have been done and a seasonal pattern has been observed virtually everywhere researchers have looked.

This paper has six sections. The first section contains a brief description of data sources and methods used in analysis of seasonal patterns. A review of seasonal patterns of births throughout the world is presented in the second section. Socio-economic and environmental causes proposed for the observed patterns are considered in the third and fourth sections respectively. Seasonal patterns of intermediate fertility variables are explored in the fifth section and conclusions are given in the final section.

## DATA AND METHODS

Vital registration systems provide the best source of data for studies of seasonality of births; these systems are usually organized at the national level. In some instances survey data can be used; for example, in areas where vital registration is nonexistent or of poor quality. However, frequent errors and heaping in retrospective reporting of dates make use of these data problematic (Som, 1973; Becker, 1984a). Counts of births in hospitals are another possible data source; however, these time series may reflect seasonal differences in access to hospitals in addition to true seasonal patterns of births, so their value is debatable. Thus, numbers of registered live births tabulated by month of occurrence usually provide the raw data for seasonality studies. Since at the national level these are typically very large numbers, sampling variability is not a concern.

Since calendar months have differing numbers of days, an adjustment of the data must be made; frequently this is done by multiplying the observed number of births in a month by $(365/12)/z$ where $z$ is the number of days in the month in question. The simplest approach at the next step is to calculate a monthly average for an interval of a multiple of 12 months and then find percentage deviations from this average for each month. This method is inadequate if there is a long or medium term trend in births, though in such cases monthly deviations can be calculated separately for each year. The magnitude of seasonal fluctuations in births can be estimated from the difference between the maximum and minimum of the percent deviations.

To better adjust for trends, time series analyses of various sorts are employed (Box and Jenkins, 1976). A simple approach is to assume a model of the form:

$$B_t = P_t \, S_t \, X_t$$

where $B_t$ is the adjusted number of births in month t, $P_t$ represents the trend component, $S_t$ the seasonal component and $X_t$ the residual (Hannan, 1963, Lam and Miron, 1988). Taking logs gives:

$$\ln(B_t) = b_t = p_t + s_t + x_t$$

In this form $p_t$ is usually expressed as a polynomial equation in t, and $s_t$ is written as

$$s_t = \sum_{j=1}^{12} \alpha_j \, d_t^j$$

where $d_t^j$ takes the value of 0 or 1 to indicate the given month of observation (t), and $\alpha_j$ gives the seasonal effect for the month. The latter are then the parameters of interest and the magnitude or amplitude of the seasonal pattern can be estimated as the difference between the maximum and minimum values of $\alpha_j$, and the magnitude can be interpreted as a percentage because of the relationship between $\alpha_j$ and the derivative of $\ln[B_t]$.

## PATTERNS

There are several reviews of seasonal patterns of births across the globe (MacFarlane, 1970; Crook and Dyson, 1984; Calot and Blayo, 1982). Table 1 summarizes seasonal patterns documented for a number of populations. The patterns vary throughout the world and sometimes different patterns are seen for population subgroups in the same geographic area (e.g., Malaysia in second panel of Table 1). In some areas the seasonal variation is so large that counts of births in certain months are double those in other months, e.g., Matlab, Bangladesh (Becker, 1981). Higher magnitudes are usually found in developing nations. In developed nations which typically have high levels of contraceptive use, such patterns persist but usually with lower magnitude; the decline in seasonal magnitude in several nations has been documented (Seiver, 1985; Harris and Mathers, 1983). In addition seasonal patterns have shifted in some nations (third panel of Table 1).

The explanations of birth seasonality fall into two groups of causes: socio-economic and climatological. These are discussed in the next two sections.

## SOCIO-ECONOMIC CAUSES

### Marriages

It is sometimes argued that the seasonality of marriages found in many nations leads to seasonality of births. However, this explanation is of limited utility since seasonal patterns are found for all birth orders and not just first births. (For an example, see Becker, 1981.)

### Agricultural and economic cycles

Several authors have studied seasonal patterns of births in relation to agricultural cycles. In 19th century peasant villages in Germany, Mosher (1979) found high conceptions during times with high workload. But studies in Mexico, Uganda, and Nigeria found high conceptions in times of low work (Thompson and Robbins, 1973, Ogum, 1979). Also agricultural cycles are highly correlated with temperature cycles, so the two factors are confounded. Given these problems and the fact that seasonal patterns persist in urban settings, this explanation is also of limited use.

### Holidays

The September peak of births in the U.S.A., Canada, and most Latin American nations (Table 1) is often attributed to greater frequency of sexual intercourse during the Christmas and New Year's holidays (Ashley, 1988; Cowgill, 1966a). Similarly, the spring peak of births in Europe is often related to assumed increases in sexual intercourse

Table 1. Summary of seasonal patterns reported in the literature since 1960[a]

| Continent and Nation | Period of data (years) | Month of Conception | | | Reference |
| | | Peak | Trough | Magnitude (%) | |
| --- | --- | --- | --- | --- | --- |
| NATIONAL LEVEL DATA | | | | | |
| _America_ | | | | | |
| Barbados | 57-59 | 12 | 9 | 26 | Dyson & Crook |
| Dominica | 60-62 | 1 | 9 | 15 | " |
| Jamaica | 59-61 | 12 | 10 | 11 | " |
| Trinidad & Tobago | 72-74 | 12 | 10 | 21 | " |
| Costa Rica | 73 | 1 | 10 | 16 | " |
| Guatemala | 58-59 | 12 | 9 | 16 | " |
| Panama | 71-73 | 2 | 10 | 17 | " |
| Ecuador | 64-66 | 12 | 2-3 | 16 | " |
| Venezuela | 71-72 | 2 | 4,7 | 15 | " |
| Bolivia | 44-45 | 3 | 6 | 17 | " |
| Mexico | 73-75 | 12 | 8-9 | 17 | " |
| Argentina | 58-60 | 12 | 8 | 16 | " |
| Chile | 67-69 | 12 | 3 | 13 | " |
| Uruguay | 58-60 | 12 | 8 | 21 | " |
| Canada | 60-84 | 6-7 | 3-4 | ≈10 | Ashley |
| U.S.A. | 47-76 | 12 | 7 | ≈10 | Seiver |
| _Asia_ | | | | | |
| Sri Lanka | 64-66 | 9 | 11 | 16 | Dyson & Crook |
| Malaysia | 73-74 | 4 | 3 | 22 | " |
| Philippines | 72-74 | 1 | 9 | 23 | " |
| Singapore | 74-76 | 1 | 5-6 | 12 | " |
| Thailand | 70-72 | 5 | 10 | 17 | " |
| India | 69-71 | 2 | 8 | 18 | " |
| Iraq | 58 | 3 | 7 | 58 | " |
| Jordan | 72-74 | 5 | 1 | 29 | " |
| Japan | 71-73 | 5,10-11 | 2 | 8 | " |
| Cyprus | 74-76 | 5 | 7 | 25 | " |
| Israel | 74-76 | 12 | 8 | 17 | " |
| Lebanon | 63,69-70 | 4 | 3 | 34 | " |
| Hong Kong | 52-61 | 12-1 | 7-8 | >15 | Chang |
| Pakistan | 59-65,71-4 | 12 | 8 | ≈35 | Krotki |
| _Africa_ | | | | | |
| Ghana | 76 | 9 | 4 | 23 | Dyson & Crook |
| Mauritius | 74-75 | 7 | 3 | 21 | " |
| Mozambique | 71-73 | 10 | 4 | 64 | " |
| Egypt | 54-56 | 4 | 8 | 18 | " |
| Libya | 72-74 | 4 | 8 | 29 | " |
| Algeria | 69 | 4 | 8 | 41 | " |
| Tunisia | 58-60 | 5 | 10 | 39 | " |
| Kenya | 79-81 | 12 | 3 | 11-20 | Ferguson |
| _Europe_ | | | | | |
| Iceland | 75-79 | 8 | 3 | 22 | Calot & Blayo |
| Norway | 75-79 | 7 | 3 | 26 | " |
| Sweden | 75-79 | 6 | 3 | 31 | " |
| Finland | 75-79 | 6 | 2 | 19 | " |
| Ireland | 74-78 | 8 | 3 | 13 | " |
| Denmark | 75-79 | 7 | 3 | 22 | " |
| England & Wales | 75-79 | 12 | 3 | 12 | " |

_(continued next page)_

| Europe (cont.) | | | | | |
|---|---|---|---|---|---|
| Netherlands | 75-79 | 9 | 3 | 16 | Calot & Blayo |
| Belgium | 74-78 | 8 | 2 | 15 | " |
| Germ. Dem. Rep. | 75-79 | 6 | 1 | 19 | " |
| Poland | 75-79 | 9 | 3 | 16 | " |
| Fed. Rep. Germany | 75-79 | 12 | 2 | 10 | " |
| Czechoslovakia | 75-79 | 7 | 1 | 18 | " |
| Austria | 75-79 | 5 | 3 | 9 | " |
| Hungary | 75-79 | 8 | 2 | 16 | " |
| Switzerland | 75-79 | 6 | 2 | 17 | " |
| France | 75-79 | 8 | 2 | 19 | " |
| Rumania | 75-79 | 12 | 3 | 16 | " |
| Yugoslavia | 75-79 | 4 | 3 | 20 | " |
| Italy | 75-79 | 8 | 3 | 17 | " |
| Spain | 74-78 | 8 | 3 | 10 | " |
| Portugal | 75-79 | 8 | 2 | 10 | " |
| Greece | 75-79 | 10 | 3 | 22 | " |

## SUBGROUPS

| | | | | | |
|---|---|---|---|---|---|
| Malaysia | | | | | |
| (Chinese) | 64-69 | 1 | 5 | 14 | Johnson et al. |
| (Malays) | 64-69 | 4 | 3 | 75 | " |
| Canada | 1778-1940 | 6 | 9-10 | 80 | Ehrenkranz |
| (Labrador Eskimos) | | | | | |
| Taiwan | 26-76 | 5 | 11 | 23 | Mosher |
| (fishing village) | | | | | |
| Nigeria (Oyo State) | 65-75 | 8 | 12 | 17 | Ayeni |
| Egypt (rural) | 63-70 | 3 | 9,12 | n/a | Levy |
| Finland (northern) | 66 | 8-9 | 5 | 40 | Rantakallio |

## SHIFTING PATTERNS OVER TIME

| | | | | | |
|---|---|---|---|---|---|
| Puerto Rico | 32-50 | 7-9 | --[b] | >10 | Vazquez, Cowgill |
| | 50-81 | 12-1 | 6-8 | >15 | " ('64) |
| Japan (Osaka) | 1755-1867 | 2 | -- | >20 | Shimura |
| | 1871-1955 | 4-5 | 8-9 | >20 | " |
| | 1960+ | 8-9 | -- | <10 | " |
| South Africa | | | | | |
| (Whites) | 57-59 | 12 | -- | ≈10 | Crook & Dyson |
| | 65-67 | 9 | -- | ≈10 | " |
| | 73-75 | 5 | 10 | 17 | " |
| (Colored) | 57-59 | 1 | 6-8 | >20 | " |
| | 65-67 | 1 | 7-8 | ≈20 | " |
| | 73-75 | 12 | 5 | 18 | " |
| Australia | 62-69 | 12 | 3 | ≈10 | Harris & Mathers |
| | 70-79 | 6 | 3 | ≈10 | " |

[a]Only studies using vital registration data with more than 200 births per month are included.

[b]Dashes indicate that a unique minimum was not present.

during the one month summer vacations common in those nations. This explanation also has counter-examples, e.g., summer vacations in the U.S.A. correspond with the <u>low</u> in conceptions.

## ENVIRONMENTAL FACTORS

Two environmental factors, heat and photoperiod, are discussed in this section. Humidity and air pressure, the latter associated with altitude, have also been posited as determinants of observed patterns, but space limitations prohibit adequate discussion of these latter issues.

## Temperature

Studies in other mammals have shown a deleterious effect of high temperatures on spermatogenesis (Blackshaw and Hamilton, 1970; Shiino and Rennels, 1971; Johnston and Branton, 1953; Newsome, 1973; Kandeel and Swerdloff, 1988). Also, urologists have long known the deleterious effects of increased scrotal temperature on sperm counts in men (e.g., Zorgniotti et. al., 1982). Recently researchers have documented a decline in sperm counts and motility in the summer months in the southern U.S.A. (Tjoa et al., 1982; Reinberg et al., 1988; Levine, 1988). At an ecological level, Seiver (1985) correlated the decline in the amplitude of the seasonal pattern in states in the southern U.S.A. during the 20th century with the rise of air conditioning. With data from multiple years, Seiver (1989), Miron (this volume), and Voranger (1953) have all shown that abnormally high summer temperatures in a given year lead to unusually low counts of births the following spring.

This paper will further examine the effect of high temperature in the U.S.A. and other selected nations. Figure 1, from Lam and Miron (1988) shows seasonal patterns of births in 18 states arranged from north to south in rows and west to east within a row. The September spike is evident in virtually every state. A trough in April-May is very pronounced in the southern states but is virtually absent in the northern states; states at mid latitudes have intermediate patterns.

As a result of the April-May trough, the patterns in the southern states have a greater amplitude. Figure 2a shows the amplitude plotted against the latitude (at geographic center) of each state (United States, 1970) with a quadratic curve and 95% confidence interval fit to the data (SAS, 1985). As noted from the previous figure, the amplitude decreases quite steadily as one moves northward.

Temperature varies quite directly with latitude. Since high temperature has the deleterious effect, maximum temperature is one meteorological variable of interest. Figure 2b shows the same seasonal birth amplitude for each state plotted against the mean maximum temperature in July, from reported data for each state (World Almanac, 1989). The quadratic fit and confidence interval exclude two very hot outliers: Nevada and Arizona. The states with low mean maximum temperatures have low amplitude.

To focus more specifically on the summer trough of conceptions, states were classified by whether the summer trough in conceptions was greater or less than 5% below the mean, and these categories were examined in relation to the July mean maximum temperature. For 25 out of 28 states with a trough of less than 5%, the mean maximum temperature was below 88°F and for 19 out of 22 states with troughs above 5%, the mean maximum temperature was above 88°F (Table 2).[1] (The cutoff of 88°F was selected after examining the data.) This provides a specific temperature hypothesis: that locations with mean maximum monthly temperatures above 88°F will show a trough in births approximately nine months later.

A next step is to test the hypothesis with data from elsewhere in the world. For this purpose two easily accessible sources of temperature data were used (Conway, 1963; Takahashi and Arakawa, 1981) and a compendium of seasonal patterns of births (Dyson and Crook, 1981). The nations with temperature data were grouped into three categories: those where no months had maximum temperatures above 88°F, those in which all months had maximum temperatures above 88°F, and those which had some months with a maximum temperature above 88°F. The hypothesis can only be tested with the latter group. Eleven nations fell into this group and also had seasonal birth data available; four of these were excluded as mean maximum temperature varied only slightly (by < 8°F). For plotting, the percent deviation in temperature is measured relative to 88°F (i.e., 88°F = 100%), the percent deviations of births from the mean level of 100% are also plotted.

---

[1]Indeed, examining the six states in which the summer trough and temperature classification appear anomalous, one discovers that the temperature reports in two of these states come from stations atypical of the state as a whole, (Denver in Colorado and Boise in Idaho) and that taking the average of all reporting meteorological stations in these states (National Climate Center, 1983), they would be classified as having temperatures below 88°F.

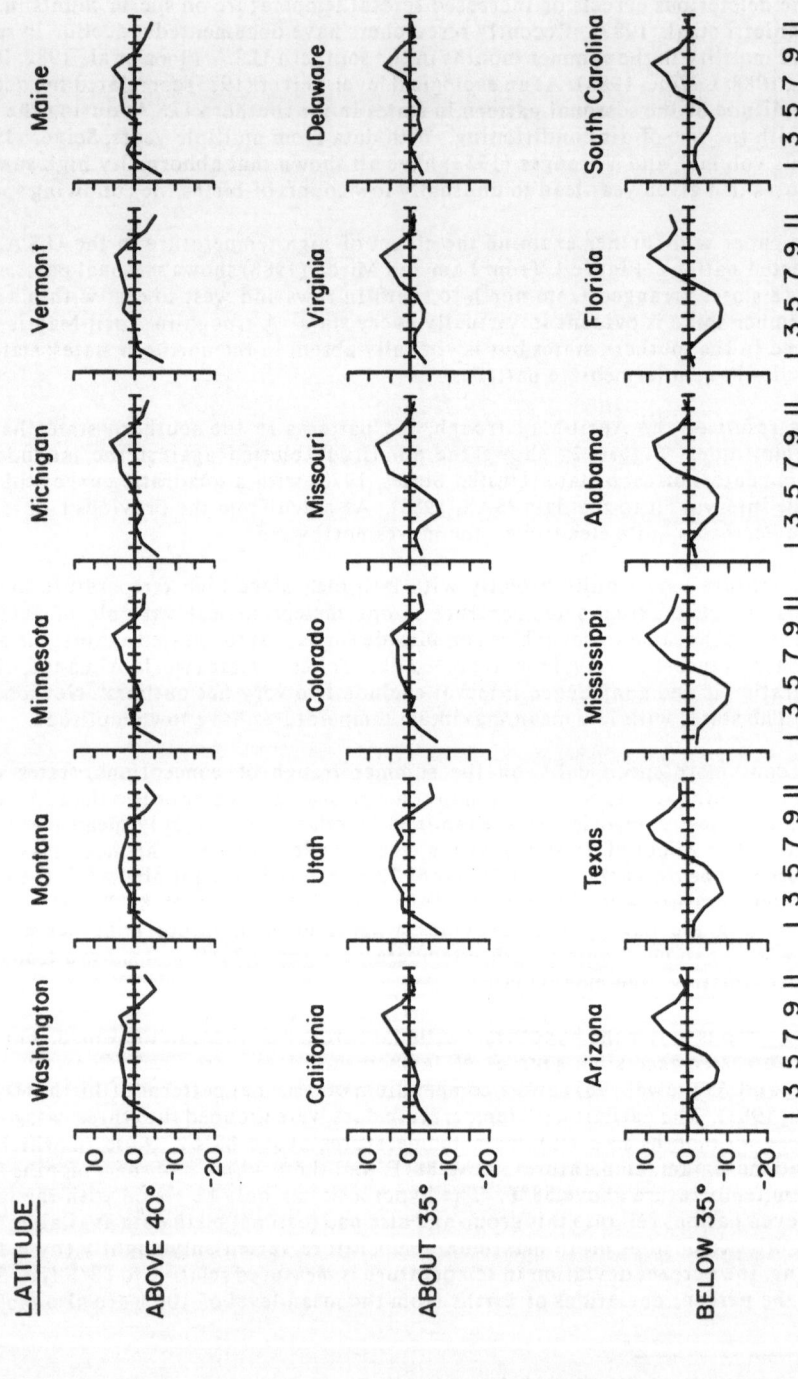

Figure 1. Seasonal patterns of birth for selected states. (Source: Lam and Miron, 1988).

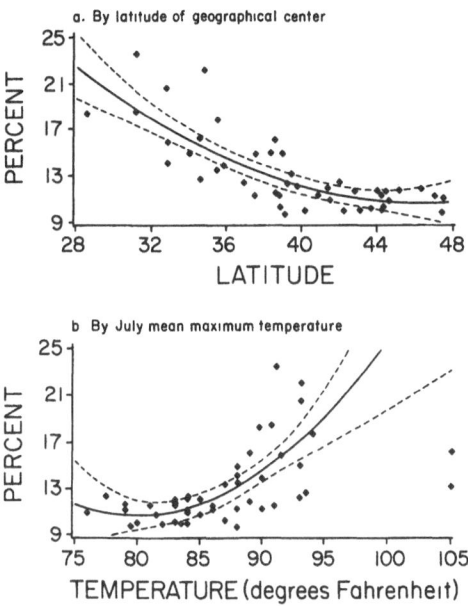

Figure 2. Seasonal amplitude (percent) of births for U.S. states. (Sources: Lam and Miron, 1988; United States, 1970; and World Almanac, 1989).

Table 2. Distribution of states in U.S.A. by magnitude of spring birth trough (1960-80) and level of summer maximum temperature

| Trough of births in April-May | July mean maximum temperature (°F) | |
| --- | --- | --- |
| | <88 | >88 |
| < 5% | AK,CA,FL,GA,IL,ME,MD MA,MI,MN,MT,NH,NJ,NY ND,OH,OR,PA,RI,SD,UT VT,WA,WI,WY | CO,ID,NV |
| > 5% | IN,IA,WV | AL,AZ,AR,CT,DE,HI KS,KY,LA,MS,MO,NE NM,NC,OK,SC,TN,TX,VA |

NOTE: Possible misclassification (on off-diagonal):

| Site | Location | °F |
| --- | --- | --- |
| CO | Denver average of 41 stations | 88 85.6 |
| ID | Boise average of 39 stations | 91 87.2 |

In IN, IA, WV, NV the averages for all stations are on the same side of 88 degrees as the single station value. (Sources: Lam and Miron (1988); World Almanac (1989); and National Climate Center (1983)).

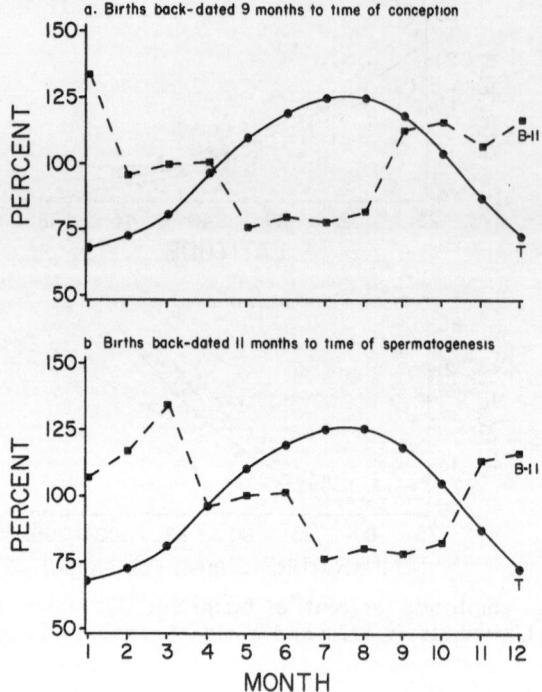

a. Births back-dated 9 months to time of conception

b Births back-dated II months to time of spermatogenesis

MONTH

Figure 3. Percentage levels of mean maximum temperature (T) and lagged births (B-11) in Iraq. (Sources: Dyson and Crook, 1981; Conway, 1963).

If the effect of heat on reproduction acts early in spermatogenesis, then we would expect the peak temperature to correspond with a trough in births 10 to 11 months later, rather than 9 months after conception. Figure 3 contrasts lags of 9 and 11 months for one nation, Iraq. The peak of temperature best corresponds with the trough of births shifted by 11 months (i.e., time of spermatogenesis). In the seven nations, the lagged correlations between temperature and births were higher for 11 months than for 9 or 10 months in most cases (data not shown).

Figure 4 shows the variation in temperature and births back-dated by 11 months for the other 6 nations. In all cases except Lebanon, the maximum temperature clearly occurs at the low point of births lagged 11 months. In Lebanon, lagged births are also declining at the point of maximum temperature, but there is a more marked trough corresponding to the cool season.

Photoperiod

Photoperiod has been shown to be a key environmental cue for annual reproductive cycles in various mammalian species including hamsters, mice, ferrets, deer and sheep (Bronson, 1988). Short day lengths lead to initiation of estrous cycles in ewes (via neuroendocrine pathways that have been documented (Goodman, 1988)) but lead to testicular regression in mice. A close correlation between latitude and season of conception in nonhuman primates has also been found, consistent with the hypothesis that photoperiod is an important cue (Ewing, 1982).

In human populations at latitudes far from the equator, several authors have documented low conception rates in the winter period. Such patterns have been found in Canada (Ehrenkranz, 1983), Northern Europe (Rantakallio, 1971), South Africa (Crook and Dyson, 1980), Chile (Hajek et al., 1981) and Southern Australia (Mathers and Harris, 1983). With peak conceptions in the summer (peak births in the spring) this pattern is thus

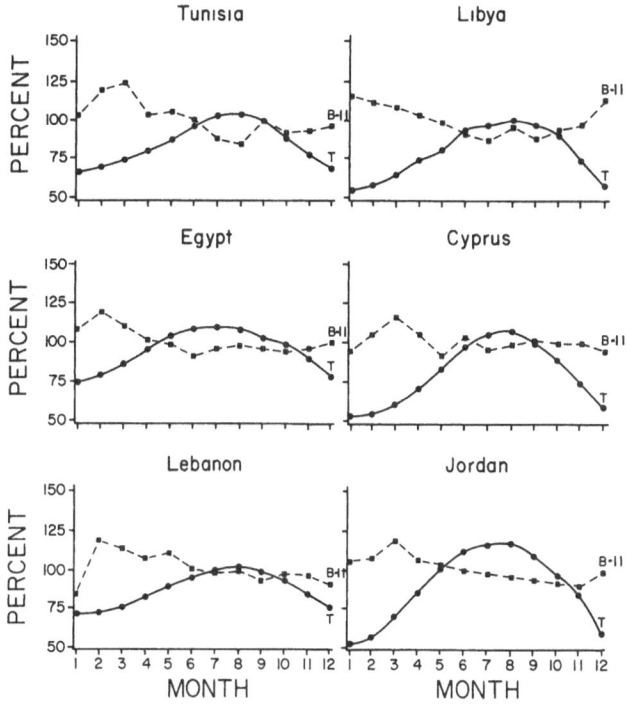

Figure 4. Percentage levels of mean maximum temperature (T) and births backdated 11 months to the approximate time of spematogenesis (B-11) in 6 nations. (Source: Dyson and Crook, 1981; Conway, 1963; Takahashi and Arakawa, 1981).

the reverse of that produced by high summer temperatures. This has engendered a photoperiod hypothesis that conceptions are diminished during seasons with less daylight.

In Australia and North America one can test both the temperature and photoperiod hypotheses since these continents span the temperate zone with seasonally high temperatures and also have zones at lower latitudes. Figure 5 (from MacFarlane et al., 1959) shows seasonality of conception rates in Australia by latitude. Indeed, at the lower latitudes a summer trough and winter peak of conception are seen while the pattern is reversed at the higher latitudes of Victoria and Tasmania. Similarly, in Canada, if we disregard the abrupt September peak associated with conceptions during the Christmas-New Years holidays, the pattern (Fig. 6) with high conception rates in the summer is reversed from that of the U.S.A. to the south. However, births in Alaska peak in September (data not shown), so there are exceptions.

INTERMEDIATE FERTILITY VARIABLES

Seasonal patterns of births are due to seasonal variations in one or more of the following factors:

Sexual intercourse frequency
Proportion of women ovulating
Number of sperm and/or their ability to fertilize
Failure to implant and other early subclinical losses of conceptions
Clinically observed fetal losses

Unfortunately, there are no studies which consider all these factors simultaneously in a population. Studies of seasonal patterns of the factors separately are reviewed here, but these are also rare.

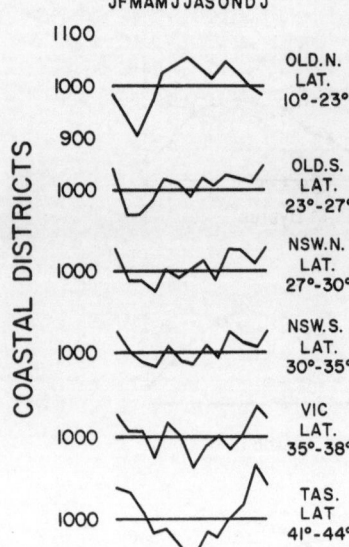

Figure 5. Seasonal patterns of conception in Australia by latitude. (Source: MacFarlane et al., 1959).

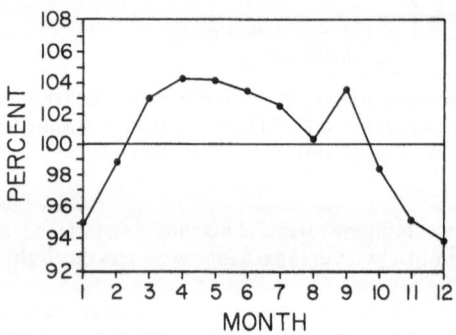

Figure 6. Seasonal pattern of births in Canada, 1960-1984 (mean monthly value = 100). (Source: Adapted from Ashley, 1988).

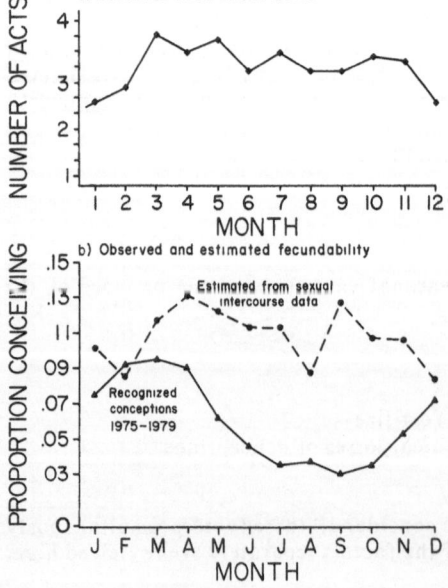

Figure 7. Sexual intercourse frequency and two estimates of fecundability in Matlab, Bangladesh, by month. (Source: Becker et al., 1986 for b; unpublished tables for a).

With respect to sexual intercourse frequency, one economic factor of importance is the seasonal separation of spouses typically due to seasonal migration of men for work. The effects of husband's absences on the seasonal pattern of births have been studied in Bangladesh (Becker, 1981); they accounted for a small part of the variation. In the same population, a seasonal pattern was found in the risk of return of menses postpartum, a good proxy for resumption of ovulation, but this also could only partially explain the observed seasonal pattern in fertility (Huffman et al., 1978; Chen et al., 1974).

Regarding the number of sperm and their ability to fertilize, studies of temperature effects on male fertility were reviewed above. Regarding early fetal loss, the rate of loss in humans may be up to 50% (Bongaarts and Potter, 1983). The rate is known to increase significantly with high temperatures in mammals (Alliston et al., 1965; MacFarlane et al., 1957; Sod-Moriah, 1971; Thwaites, 1969). However, non-invasive techniques for detecting losses in the first weeks after fertilization in humans have just recently become available (Wilcox et al., 1988). With respect to late fetal loss, some studies have found seasonal patterns and some have not (Slatis, 1969; McDonald, 1971; Warren et al., 1980), but the patterns that have been found are of insufficient magnitude to explain seasonal patterns of births.

Many researchers have assumed that seasonal patterns of birth reflect seasonal patterns of sexual intercourse frequency; longitudinal data on sexual intercourse frequency are rare. Udry and Morris (1967) examined daily records from 50 women in Indiana and did find a pattern but it did not correspond with the pattern of conceptions.

Fortunately, it was possible to measure sexual intercourse frequency in a prospective study in rural Bangladesh (Becker, 1984b). During the last year of a four-year study of birth interval dynamics, after rapport had been established between interviewer and interviewee, a question was added: "How many days has it been since you last had sexual intercourse?". Data from responses to this question can be converted into a probability that sexual intercourse occurs during the fertile period (Becker, 1985); these can be used in a mathematical model (Becker et al., 1986) to estimate fecundability, the probability of a woman conceiving in a month. These estimates can be compared with observed conception rates for menstruating women in the same study. Figure 7a shows the seasonal pattern in sexual intercourse: the estimated mean number of acts of coitus per month, and Figure 7b gives fecundability estimated from these data and from observed conceptions. The marked seasonal pattern in observed fecundability is apparent. The fecundability estimates agree closely at the beginning and end of the year, but during the period March-November there is a divergence of the curves with a maximum difference for the months between June and September. It is noteworthy that the mean maximum temperature in this area is highest in April and May (89°F) but June to September have only slightly lower mean maxima and these months have the highest values of mean minimum temperature (77°F). Obviously, one or more of the other intermediate factors also contributes to the seasonal patterns of births in Bangladesh and it could be the effect of high temperature on spermatogenesis.

CONCLUSION

Clearly there are multiple causes of the different seasonal patterns of births observed throughout the world. Indeed, within the same population there are multiple causes which interact and change over time so the patterns are not necessarily constant. For example, increased coitus during Christmas holidays probably explains the September peak in births in the U.S.A., while high temperatures can explain the pronounced April-May trough evident in the southern United States. The same conclusion of multiple interacting causes holds for other mammal species as well (Bronson, 1988; Ewing, 1982).

The temperature hypothesis was considered in some detail. The epidemiological criteria for establishing a causal link (Susser, 1973) between high environmental temperature and reduced births 9 to 11 months later seem to be met. First, there is biological plausibility from animal studies of effects of high temperatures on spermatogenesis and fertility. Second, the temporal sequence is established from the historical time series studies. Third, a dose-response relationship is seen since populations in temperate regions with the hottest summer temperatures generally have the most pronounced troughs of summer conceptions. Fourth the association is quite strong (e.g., Table 2). Fifth, the association is specific since the deterioration in fertility parameters of human sperm has been documented at high summer temperatures, parallel to the direct experimental evidence from animal studies. Sixth, the association is quite consistently found (e.g., Figure 4). One caution is needed, however: it is possible that high temperature is also affecting female reproductive performance, specifically in increasing unobserved early embryonic or fetal losses or increasing anovulatory cycles. The quantitative relationships between temperature and male and female fertility parameters have yet to be determined.

Though numerous studies have documented seasonal patterns of births and correlations of these with environmental and social factors at the ecological level, far too few have looked for seasonal patterns in intermediate fertility variables at the individual level that could explain these patterns. It has recently been demonstrated that reliable data can be collected on sexual intercourse frequency and such data are being collected (unfortunately only at one point in time) in more than 30 nations that are having Demographic and Health Surveys. Longitudinal studies of this factor need to be done in populations with marked seasonal patterns. More generally, studies of all the intermediate fertility variables simultaneously will be complex, but without these it will be difficult to scientifically investigate the causes of seasonal births.

## REFERENCES

Alliston, C.W., Howarth, Jr., B. and Ulberg, L. C. 1965. Embryonic mortality following culture in vitro of one- and two-cell rabbit eggs at elevated temperatures. J. Reprod. Fert. 9:337-341.

Ashley, M.J. 1988. Season of birth: Stability of the pattern in Canada. Canadian Journal of Public Health 79:101-103.

Ayeni, O. 1986. Seasonal variation of births in rural southwestern Nigeria. International Journal of Epidemiology 15:91-94.

Becker S. 1981. Seasonal patterns of fertility in Matlab, Bangladesh. Journal of Biosocial Science 13:97-105.

Becker, S. 1984a. A response bias in the reporting of month of birth in pregnancy history surveys. Working paper No. 5, Interuniversity Programme in Demography, Brussels, Belgium.

Becker, S. 1984b. Les variations saisonnieres des naissances au Bangladesh rural. Population 2:265-280.

Becker S. 1985. Estimation of fecundability from sexual intercourse data. Paper presented at IUSSP General Conference, Florence, Italy, 1985.

Becker, S., Chowdhury A.K.M.A. and Leridon, H. 1986. Seasonal patterns of reproduction in Matlab, Bangladesh. Population Studies 40:457-472.

Blackshaw, A.W. and Hamilton, D. 1970. The effect of heat on hydrolytic enzymes and spermatogenesis in the rat testis. J. Reprod. Fert. 22:569-571.

Bongaarts, J. and Potter, R.G. 1983. Fertility, Biology and Behavior: An Analysis of Proximate Determinants, Academic Press, New York.

Box, G.E.P. and Jenkins, G.M. 1976. Time Series Analysis: Forecasting and Control Holden Day, San Francisco.

Bronson, F.H. 1988. Seasonal regulation of reproduction in mammals. pp. 1831-1872, in: The Physiology of Reproduction, E. Knobil and J.D. Neill (eds), Raven Press.

Calot, G. and Blayo, C. 1982. Recent course of fertility in Western Europe. Population Studies 36:349-372.

Chang, K.S.F., Chan, S.T., Law, W.D. and Ng, C.K. 1963. Climate and conception rates in Hong Kong. Human Biology 35:366-375.

Chen, L.C., Gesche, M.C. and Mosley, W.H. 1974. A prospective study of birth interval dynamics in rural Bangladesh. Population Studies 28:277-297.

Conway, H.M. 1963. The Weather Handbook Conway Publications Inc., Atlanta, U.S.A.

Cowgill, U.M. 1964. Recent variations in the season of birth in Puerto Rico. Proceedings of National Academy of Science 1149-1151.

Cowgill, U.M. 1966a. Season of birth in man. Contemporary situation with special reference to Europe and the Southern Hemisphere. Ecology 47:614-623.

Cowgill, U.M. 1966b. The season of birth in man. Man 1:232-240.

Crook, N. and Dyson, T. 1980. Variations saisonnieres des evenements en Afrique du Sud: Contraste entre les races. Population 3:691-697.

Dyson, T. and Crook, N. 1981. Data on seasonality of births and deaths, pp. 141-148 and 241-247, in: Seasonal Dimensions to Rural Poverty, Chambers, R. et al. (eds), Frances Pinter, London.

Ehrenkranz, J.R.L. 1983. Seasonal breeding in humans: birth records of the Labrador Eskimo. Fertility and Sterility 40(4):485-489.

Ewing, L.E. 1982. Seasonal variation in primate fertility with an emphasis on the male. Am. J. Primatology Suppl. 1:145-160.

Ferguson, A.G. 1987. Some aspects of birth seasonality in Kenya. Soc. Sci. Med. 25:793-801.

Gini, C. 1912. Contributi statistici ai problemi dell'eugenica. Estr. Riv. ital. Sociol. 16:317.

Goodman, R.L. 1988. Neuroendocrine control of the ovine estrous cycle, pp. 1929-1970, in: The Physiology of Reproduction, E. Knobil and J.D. Neill (eds), Raven Press.

Hajek, E.R., Gutierrez, J.R. and Espinosa, G. 1981. Seasonality of conception in human populations in Chile. Int. J. Biometeor. 25:281-291.

Hannan, E.J. 1963. The estimation of seasonal variation in economic time series. Journal of American Statistical Assoc. 58:31-44.

Harris, R.S. and Mathers, C.D. 1983. Seasonal distribution of births in Australia. International Journal of Epidemiology 12:326-331.

Hill, S.A. 1888. The life statistics of an Indian province. Nature 38:245-250.

Huffman, S.A., Chowdhury, AKMA, Chakraborty, J. and Mosley, W.H. 1978. Nutrition and post partum amenorrhea in rural Bangladesh. Population Studies 32:251-260.

Huntington, E. 1938. The season of birth. John Wiley, New York.

Johnson, B., Malay, J.T., Ann, T.B. and Palan, V.T. 1975. Seasonality of births for West Malaysia's two main racial groups. Human Biology 47:295-307.

Johnston, James E. and Branton, Cecil 1953. Effects of seasonal climatic changes on certain physiological reactions, semen production and fertility of dairy bulls. J. Dairy Sci. 36:934-942.

Kandeel, F.R. and Swerfloff, R.S. 1988. Role of temperature in regulation of spermatogenesis and the use of heating as a method of contraception. Fertility and Sterility 49(1):1-23.

Krotki, K.J. 1978. Variations saisonnieres des naissances et des deces dans deux pays aseatiques et leurs significations pour la sante de la population. Cahier Quebecois de Demographie 3:55-78.

Lam, D.A. and Miron, J.A. 1988. The seasonality of births in human populations, Working Paper No. 87-114. University of Michigan.

Levine, R.J. et al. 1988. Deterioration of semen quality during summer in New Orleans. Fertility and Sterility 49:900-907.

Levy, V. 1986. Seasonal fertility cycles in rural Egypt: behavioral and biological linkages. Demography 23:13.

MacFarlane, V. 1970. Seasonality of conception in human populations. Biometerology 4:167-182.

MacFarlane, W.V., Pennycuik, P.R. and Thrift, E. 1957. Resorption and loss of foetuses in rats living at 35°C. J. Physiol. 135:451-459.

MacFarlane, W.V., Rennycuik, P.R., Yeates, N.T.M. and Thrift, E. 1959. Reproduction in hot environments, pp. 81-96, in: Encocrinology of Reproduction, C.W. Lloyd (ed), Academic Press.

Mathers, C.D. and Harris, R.S. 1983. Seasonal distribution of births in Australia. Int. J. Epidemiol. 12(3):326-331.

McDonald, Alison D. 1971. Seasonal Distribution of Abortions. Brit. J. Prev. Soc. Med. 25:222-224.

Mosher, S.W. 1979. Birth seasonality among peasant cultivators: The interrelationship of workload, diet, and fertility. Human Ecology 7:151-181.

National Climatic Center. 1983. Climate Normals for the U.S. (Base: 1951-80). Gale Research Company, Detroit, Michigan.

Newsome, A.E. 1973. Cellular degeneration in the testis of red kangaroos during hot weather and drought in central Australia. J. Reprod. Fert. 19:191-201.

Ogum, G.E.O. and Okorafor, A.E. 1979. Seasonality of births in south-eastern Nigeria. J. Biosoc. Sci. 11:209-217.

Quetelet, A. 1869. Physique Social 1:104.

Rantakallio, P. 1971. The effect of a northern climate on seasonality of births and the outcome of pregnancies. Kiripaino Osakeyhtio Kaleva, Finland.

Reinberg, A. et al. 1988. Annual variation in semen characteristics and plasma hormone levels in men undergoing vasectomy. Fertility and Sterility 49:309-315.

SAS. 1985. SAS/Graph User's Guide Version 5. SAS Institute, Cary, N.C.

Seiver, Daniel A. 1989. Seasonality of Fertility: New Evidence. Population and Environment 10:245-257.

Seiver, Daniel A. 1985. Trend and variation in the seasonality of U.S. fertility, 1947-1976. Demography 22:89-99.

Shiino, M. and Rennels, E.G. 1971. Influence of high ambient temperature on the reproductive function of the male rat. Texas Rep. Biol. Med. 29(3):313-330.

Shimura, M., Richter, J. and Miura, T. 1981. Geographical and secular changes in the seasonal distribution of births. Soc. Sci. Med. 15(D):103.

Slatis, H.M. and DeCloux, R.J. 1967. Seasonal variation in stillbirth frequencies. Human Biology 39:184-294.

Sod-Moriah, U. A. 1971. Reproduction in the heat-acclimatized female rat as affected by high ambient temperature. J. Reprod. Fert. 26:209-218.

Som, R.K. 1973. Recall lapse in demographic enquiries. Asia Publishing House, Bombay.

Susser, M. 1973. Causal Thinking in the Health Sciences: Concepts and Strategies of Epidemiology. Oxford University Press, London.

Takahashi, K. and Arakawa, H. 1981. Climates of Southern and Western Asia, Volume 9 of World Survey of Climatology. Elsevier Scientific Publishing Co., Amsterdam.

Thompson, R.W. and Robbins, M.C. 1973. Seasonal variation in conception in rural Uganda and Mexico. Am. Anthro. 75:676-686.

Thwaites, C. J. 1969. Embryo Mortality in the heat stressed ewe. II. Application of hot-room results to field conditions. J. Reprod. Fert. 19:255-262.

Tjoa, W.S. et al. 1982. Circanual rhythm in human sperm count revealed by serially independent sampling. Fertility and Sterility 38:454-459.

Udry, J.R. and Morris, N.M. 1967. Seasonality of coitus and seasonality of birth. Demography 4:673-679.

United States. 1970. National Atlas of the United States of America. U.S. Department of Interior, Geological Survey, Washington D.C.

Vazquez-Calzada, J.L. 1982. The seasonal pattern of live births in Puerto Rico. Puerto Rico Health Sciences Journal 1(4):167-171.

Villerme. 1831. De la distribution par mois des conceptions: Ann d'Hygiene 5.

Voranger. 1953. Influence de la meteorologie et de la mortalite sur les naissances. Population 1:93-102.

Warren, C.W., Gold, J., Tyler, C.W., Jr., Smith, J.C. and Paris, A.L. 1980. Seasonal variation in spontaneous abortions. Am. J. of Public Health 70:1297-1299.

Wilcox, A.J., Weinberg, C.R., O'Connor, J.F., Baird, D.D., Schlatterer, J.P., Canfield, R.E., Armstrong, E.G. and Nisula, B.C. 1988. Incidence of early loss of pregnancy. New England Journal of Medicine 319(4):189-194.

World Almanac and Book of Facts. 1989. Phoros Books, New York.

Zorgniotti, A.W., Sealfon, A.I. and Toth, A. 1982. Further clinical experience with testis hypothermia for infertility due to poor semen. Urology XIX:636-640.

# TEMPERATURE AND THE SEASONALITY OF BIRTHS

David A. Lam[1] and Jeffrey A. Miron[2]

[1]Department of Economics
University of Michigan
Ann Arbor, Michigan 48109
[2]Department of Economics
Boston University
Boston, Massachusetts 02215

## INTRODUCTION

The relation between human fertility and temperature has been the subject of extensive examination, with surprisingly inconclusive results. Despite the *a priori* plausibility that temperature affects fertility through behavioral or biological channels, existing attempts to identify these effects have had only limited success.

This paper considers the relation between temperature and the seasonal fluctuations in births. There is significant seasonality in births in every population for which data are available, and the presence of a seasonal pattern in any activity strongly suggests the influence of temperature. The differences in seasonal patterns across countries, however, provide at best ambiguous support for any temperature based explanation of birth seasonality. Some populations, such as the southern United States, exhibit a spring trough in births, consistent with the hypothesis that summer heat depresses conceptions. Other populations, however, such as those in Northern Europe, display a spring peak in births, a fact not easily reconciled with the simple summer heat hypothesis. While extreme summer heat may explain the trough in conceptions in warm climates, it does not explain the peak in conceptions in summer in other populations, and it does not explain the (local) September peak in births in almost all populations. Thus, while temperature is probably one important factor in determining birth seasonality, it is clear that other factors are also at work.

The paper has two purposes. The first is to survey existing evidence on the relation between temperature and the seasonality of births[1]. The second is to provide new evidence on possible effects of temperature on fertility by using actual monthly temperature data to estimate temperature's direct contribution in explaining the seasonal variation in monthly births. There are surprisingly few direct estimates of the relationship between temperature and births. Some historical studies, such as Lee (1981), and Richards (1983), have used annual or seasonally adjusted monthly data to look at the effect of temperature on fertility. Short-run physiological or behavioral effects of temperature will be captured very imperfectly in such analyses. Seiver (1989) conducts a direct test of the effect of monthly temperature on births, but he applies his test only

---

[1]A complementary survey, as well as additional evidence on the effects of temperature on fertility, is provided in the paper by Becker (1990) included in this volume.

to data for the United States. Our objective here is to determine whether cross country differences in temperature can explain the cross country differences in seasonal patterns.

The results that we present below show that temperature does have a quantitatively important influence on the seasonal variation in births in most of the populations we consider. In particular, there is a consistent tendency for temperature to depress summer time conceptions. Even after controlling for the effects of temperature, however, there is still significant seasonality in births in all populations, specifically a global spring peak, a local September peak, and a global winter trough. Thus, controlling for the effects of temperature resolves some of the cross country differences in birth seasonality, but it does not fully explain this seasonality. There is clearly some other variable or combination of variables influencing seasonal fluctuations in births.

The paper is organized as follows. Section 2 reviews the existing evidence on temperature and the seasonality of births. Section 3 presents estimates of the seasonal patterns in births in a variety of populations. In Section 4 we explain the estimation framework that we employ, while Section 5 presents results. Section 6 concludes.

## REVIEW OF THE LITERATURE

There are a number of hypotheses to explain birth seasonality in addition to possible temperature effects, including agricultural cycles, marriage cycles, holidays, and other economic variables. We have discussed and evaluated the possible role of these factors in previous work (Lam and Miron, 1987;1989)[2]. For the sake of brevity, we do not repeat that discussion here. It is important to keep in mind, however, that there is no reason to believe in a single factor explanation of birth seasonality, so the results we present should be interpreted in the light of other possible effects.

There are a number of possible mechanisms through which the seasonal variation in temperature might determine the seasonality of births. On the behavioral side, couples may desire births in spring months to improve the survival probabilities of the child or fetus. Women may desire spring births because it is uncomfortable to be pregnant in the summer (Rodgers and Udry, 1988). In addition, couples may choose to engage in intercourse more often in certain seasons if the temperature is too cold or hot.

Previous literature on the seasonality of births provides some evidence supporting temperature based explanations. Indirect evidence is provided by comparisons of seasonal patterns across socio-economic groups. Greater seasonality for lower income classes could be interpreted as evidence supporting temperature-induced seasonality, since high income groups may control their environment more completely than low income groups. The most recent analysis of socio-economic differentials in seasonality is by Kestenbaum (1987), who uses the 1980 U.S. census to identify the quarterly pattern in births in years immediately prior to the census, controlling for parental income. Consistent with a number of previous studies for the U.S., Kestenbaum finds somewhat greater seasonality among lower income groups. Kestenbaum's results are subject to the criticism that income and education in his sample will be highly correlated with residence in northern versus southern states. His results may therefore simply reflect the large north-south differentials in seasonality (see Figure 1 below), rather than provide independent evidence of socioeconomic effects.

Earlier studies of socioeconomic differentials in birth seasonality are inconclusive, being based on small geographical areas and/or short time periods. Pasamanick et al. (1959, 1960) report more seasonality in low income groups than in high income groups in Baltimore, although reexamination by Zelnik (1969) indicates no significant differences between income groups. A replication of the Baltimore study by Chaudhury (1972) supports the finding of greater seasonality among lower income groups, and Warren and Tyler (1979) find similar differences based on birth records for a county in Georgia. In contrast to these findings of greater seasonality among low income groups in the United

---

[2]See also the paper by Becker (1990) in this volume.

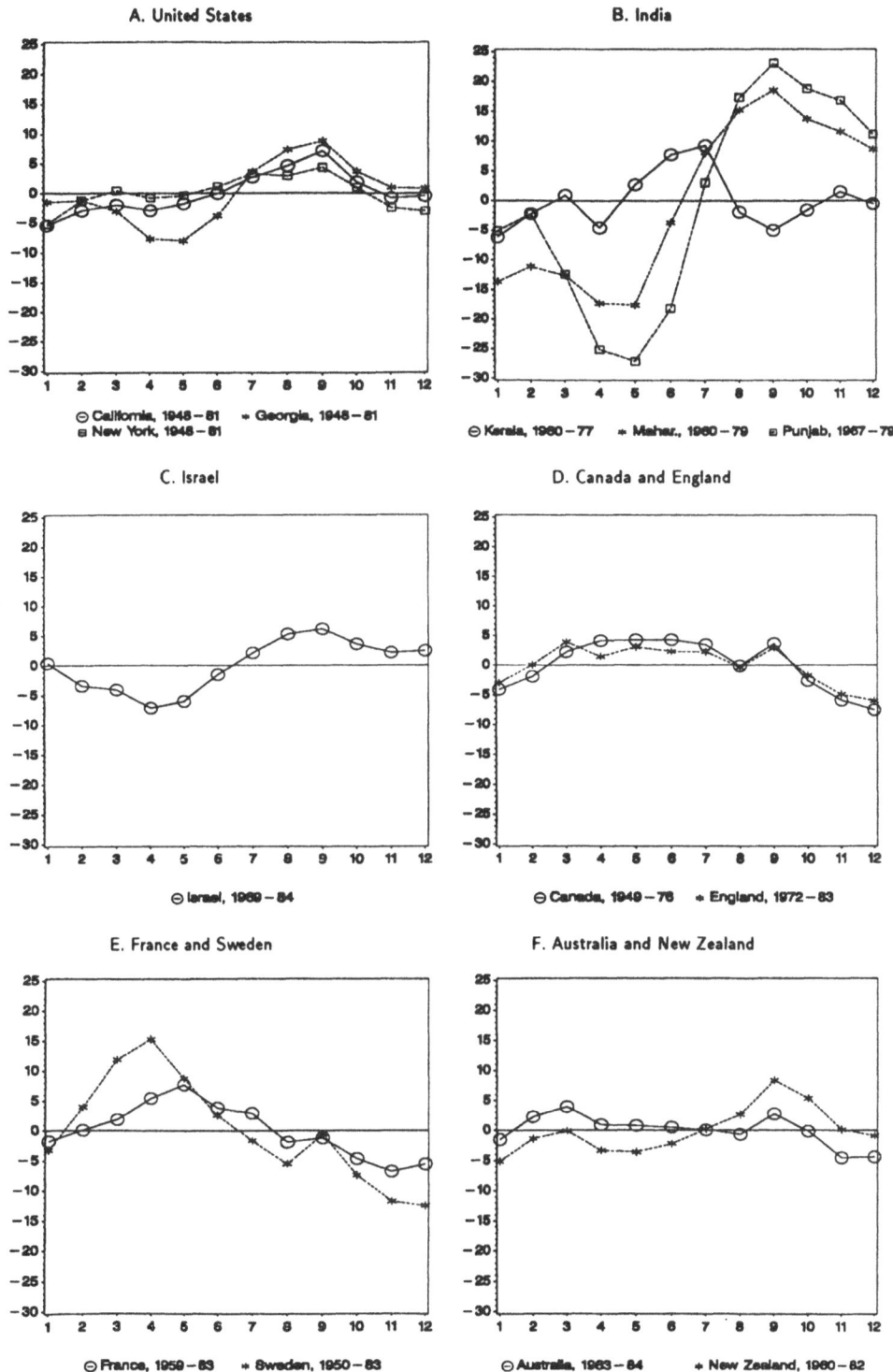

Figure 1. Seasonality of Births
Percentage Deviations from Trend by Month of Birth.

States, however, James (1971), comparing birth seasonality across social classes in England, finds *less* seasonality among low income groups. His results are based on comparisons for a single year's vital statistics data, however, a sample size too short for reliable conclusions.

The role of temperature in seasonality in the United States is examined indirectly by Seiver (1985). His test of the hypothesis that summer heat explains the April-May trough in births is based on a regression of the time trend in the magnitude of the May seasonal effect for each state between 1960 and 1970 on the increase in air conditioning in that state over the same period. Seiver's interesting results provide support for the hypothesis that summer heat reduces conceptions and thus partially explain the April-May trough in births. Seiver's results do not explain the August-September peak in births, the other dominant feature of the U.S. seasonality pattern. Seiver (1989) tests the role of temperature more directly by estimating the effect of summer heat extremes on births nine months later. His results indicate that increases in summer temperature have a direct negative effect on births.

Recent work by Becker, Chowdhury, and Leridon (1986) may suggest an important effect of temperature on birth seasonality in Matlab, Bangladesh[3]. This study finds that the large seasonal variation in fecundability can be only partially explained by the seasonal variation in the frequency of intercourse. In particular, there is a peak in frequency of intercourse in September, the period of lowest fecundability. Since September is the month with highest mean monthly temperature, an effect of temperature on spermatogenesis or the foetal loss rates appears likely. As emphasized by Becker, Chowdhury, and Leridon, however, further investigation is required to determine the precise mechanisms through which temperature operates.

The biomedical literature contains some evidence that temperature affects the biological probability of conception, as well as some evidence of seasonality in this probability, which may reflect temperature. Levine *et al.* (1988) show that semen quality in a sample of 1235 men deteriorates significantly from winter to summer in New Orleans, with the size of the deterioration greatest in those occupations most exposed to external temperature. Zorgniotti and Sealfon (1988) find significantly higher scrotal temperatures in subfertile than normospermic men. Studies of small numbers of subjects suggest monthly variations in male hormones related to sexual activity (Reinberg, 1974, Reinberg *et al.*, 1978), without explaining the source of the variations directly. Research based on a sample of males in Houston, Texas indicates significant seasonal variation in sperm count, with peaks in February and March and a trough in September (Tjoa *et al.*, 1982). This pattern of peaks and troughs implies a peak in births in November and December and a trough in June, predictions that differ by one to two months from the seasonal pattern in births in Texas (Lam and Miron, 1987).

Seasonal variations in menstrual cycles and menarche have also been documented (Sundararaj *et al.*, 1978; Brundtland and Liestol, 1982), suggesting possible physiological explanations of seasonality on the female side as well. Results for a sample of women in Minnesota show some shortening of menstrual cycles during spring and summer months (Sundararaj *et al.* 1978). Again no direct explanations of the seasonal variations in menstruation are provided, although Sundararaj *et al.* speculate that the effects might result from seasonal changes in physical activity or external temperature. The temperature may also operate through effects on intrauterine mortality. Holmes (1968), MacFarlane (1970), and Tromp (1963), for example, discuss the possible relationship between temperature extremes, intrauterine mortality, and birth seasonality. The evidence is mixed on whether such an effect is empirically important[4].

To summarize, there has been significant effort devoted to demonstrating effects of temperature on fertility. This effort has produced results that are, on the whole,

---

[3]For earlier analyses of seasonality in the Matlab population, see, for example, Stoeckel and Choudhury (1972) and Becker (1981).

[4]See Slatis and DeCloux (1967), Huntington (1938), McDonald (1971), and Warren et al. (1980).

surprisingly inconclusive. The hypothesis that receives the most consistent support is that summer heat depresses conceptions. Even the evidence for this proposition is mixed, however, and existing work does not isolate the precise mechanism through which temperature affects fertility.

## THE SEASONALITY OF BIRTHS AND TEMPERATURE

In this section we illustrate the key stylized facts about birth seasonality as they relate to possible temperature effects. We do not provide an exhaustive set of estimates of birth seasonality; a more extensive set is found in our previous papers (Lam and Miron 1987,1989), which present the seasonal patterns in all forty-eight continental United States and over twenty-five countries, in many cases for long sample periods. Here we present only those patterns that are necessary to evaluate the possible role of temperature in explaining the seasonality of births. The sources of the data and a description of our methods are contained in the earlier papers.

A useful first point to make is that the seasonality in births is quantitatively important. Table 1 of Lam and Miron (1989) shows that seasonal dummies typically account for 50% of the non-trend variation in births, and in many cases as much as 80%. The peak to trough amplitudes in birth seasonals are almost always greater than 10% and are sometimes as high as 30%, even in modern, urban populations. Tests of statistical significance reject the null hypothesis of no seasonality at the 1% level in all cases. Thus, according to any reasonable measure, seasonal fluctuations are a dominant source of the variation in births.

We turn next to an examination of the actual seasonal patterns. A sample of patterns from around the world is presented in Figure 1. The coefficient for each month is interpreted as the percentage difference between births in that month and mean births per month, adjusting for the number of days in the month.

Consider first the patterns in white births in three states representing different geographic regions of the United States, shown in panel A. All of the states exhibit a September peak and an April-May trough, a pattern well established in previous studies of U.S. birth seasonality[5]. The figure exemplifies a persistent regional difference in birth seasonality in the United States. The April-May trough is noticeably more pronounced in Georgia than in California and New York, and the overall magnitude of birth seasonality is larger in Georgia. As documented in previous studies of U.S. birth seasonality, the Georgia pattern is typical of the southern United States. The systematic differences across states in the magnitude of the spring trough is the most convincing evidence in favor of a temperature based explanation of birth seasonality.

Figure 1 next shows seasonal patterns for a number of populations outside of the United States. Panel B shows the seasonal patterns for three regions of India[6]. The patterns in two cases are similar to patterns for southern regions of the United States, although the amplitude of the patterns is significantly larger in India[7]. Panel C of the figure shows the seasonal birth pattern for Israel, which is remarkably similar to that in the southern United States, exhibiting an April trough and a September peak. The patterns for both of these countries are broadly consistent with the summer heat hypothesis.

Panels D and E of Figure 1 show the seasonal patterns in Canada, England, Sweden and France. The Canada and England patterns differ from the U.S. pattern in the spring

---

[5]See, for example, Cowgill (1966), Rosenberg (1966), MacFarlane (1970), Lyster (1971), and Seiver (1985).

[6]Kosambi and Raghavachari (1951) present evidence on birth seasonality in India during the pre-WWII period.

[7]Punjab is the most northern of the Indian states included here, at roughly the same latitude as Georgia. Maharashtra is further south, at roughly the same latitude as central Mexico. Kerala is in the far south, at latitudes corresponding to Central America.

months, with no evidence of the spring trough that is so pronounced in the southern United States. They are similar to the U.S. patterns in the presence of a September peak followed by a reduction in births in October through January. The seasonal patterns for France and Sweden are typical of those found in Northern Europe. This "European pattern" consists of a global spring peak, a local September peak, and a significant trough during the late fall and early winter. The pattern seen here is markedly different from that in the United States, particularly the spring peak in Europe as opposed to the spring trough in the U.S. The patterns in Canada and the European populations obviously indicate that some factor other than summer heat is at work in producing observed seasonal patterns.

Additional evidence to this effect is shown in Panel F, which provides the seasonal birth patterns for the southern hemisphere countries of Australia and New Zealand. The pattern shown for Australia, consisting of a global March peak, a local September peak, and November-December trough, is surprisingly similar to the pattern for England[8]. The pattern for New Zealand, on the other hand, demonstrates a seasonal pattern with the key features of the U.S. pattern. It is clearly difficult to reconcile these Southern Hemisphere patterns with the simple, summer heat explanation of birth seasonality.

The difficulties involved in explaining the cross population differences in birth seasonality by means of differences in temperature seasonality are illustrated dramatically in Figure 2, which shows the seasonal patterns in temperature for all of the populations in Figure 1. Although the basic shape of the seasonal patterns in temperature are the same in most countries around the world, the differences in birth seasonality are dramatic. For example, the temperature patterns for New York and Sweden are extremely similar on a month by month basis, yet the observed birth seasonals are strikingly different. Similarly, Australia and New Zealand have temperature patterns that are identical except for the average level, yet the birth seasonals differ significantly.

It seems unlikely that any simple explanation based *solely* on temperature effects can reconcile these data. We therefore interpret these facts as leading to one of two conclusions. Either temperature operates in a more complicated manner than by simply depressing conceptions during the period of summer heat, or at least one other factor in addition to temperature is at work in producing observed seasonal patterns. In the rest of this paper, we assess the case for each of these two possibilities.

ESTIMATION STRATEGY

This section explains our strategy for examining the effects of temperature on fertility. We estimate the following two equations:

$$\ln B_t = \theta_1 t + \theta_2 t^2 + \theta_3 t^3 + \theta_4 t^4 + \sum_{s=1}^{12} \alpha_s d_t^s + \epsilon_t \qquad (1)$$

and

$$\ln B_t = \theta_1 t + \theta_2 t^2 + \theta_3 t^3 + \theta_4 t^4 + \sum_{s=1}^{12} \alpha_s d_t^s + \beta_1 T_{t-9} + \beta_2 T_{t-9}^2 + \beta_3 T_{t-10} + \beta_4 T_{t-10}^2 + \epsilon_t \quad (2)$$

where

---

[8] Analyses of birth seasonality in Australia focusing on earlier data (Cowgill, 1965, Rosenberg, 1966) identify a September global peak. Rosenberg (1966), for example, presents a seasonal pattern for South Australia for the period 1957 - 59 that is remarkably similar to the Unites States pattern in both timing and magnitude. Mathers and Harris (1983) analyze three-year periods from 1962 to 1979 and conclude that the Australian pattern has shifted from a September peak in births to a March peak in births (see also MacFarlane and Spalding (1960), Lyster (1979), and Parker (1978, 1979). The change in pattern they discuss is primarily a change in relative magnitudes. The March and September local peaks shown in our estimates appear in virtually all periods, with the September peak more predominant in earlier periods.

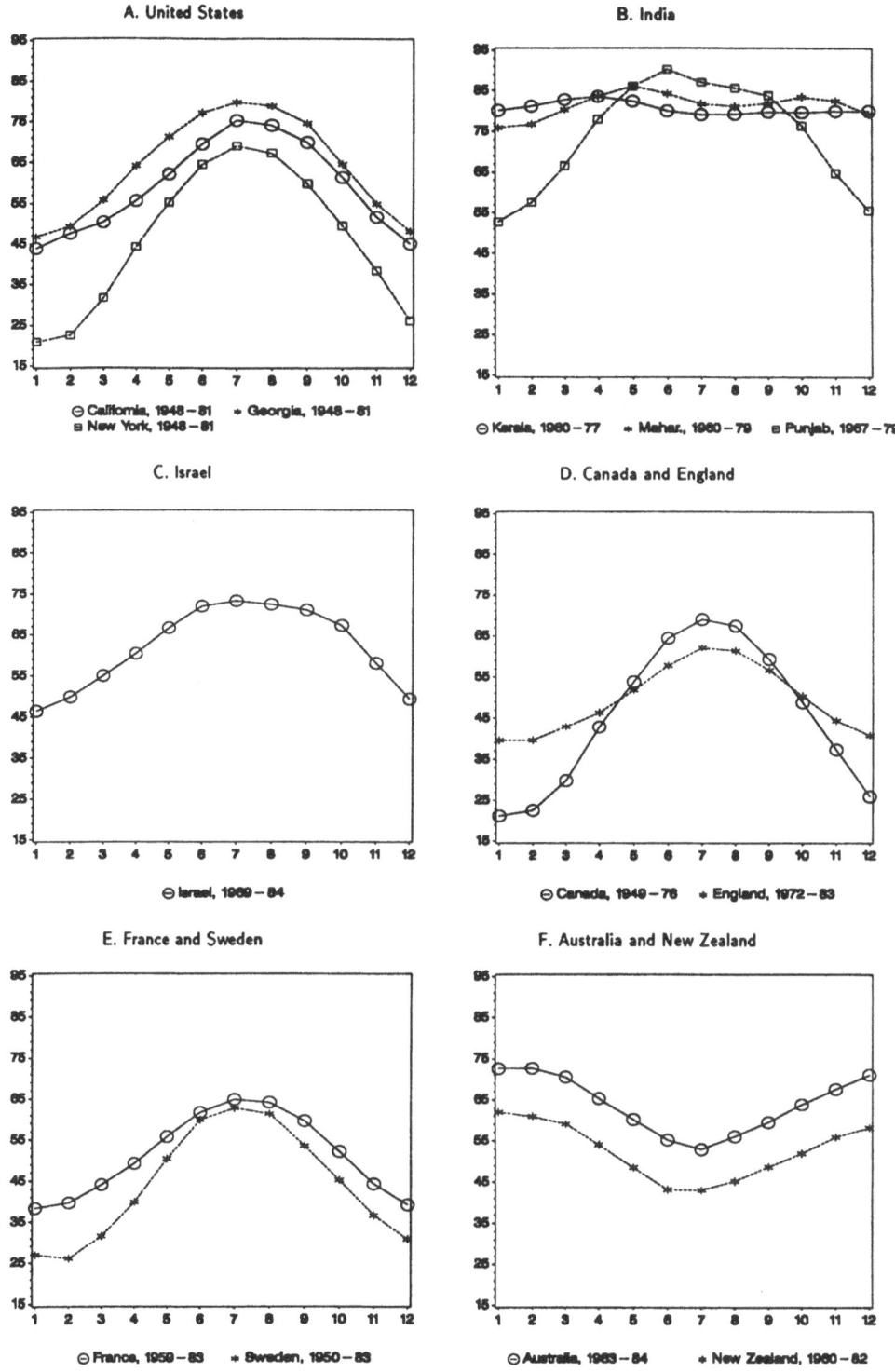

Figure 2. Seasonality in Temperature
Average Temperature (in Degrees Fahrenheit) by Month

$B_t$ is births (per day) in month $t$,

$d_t^s$ is a seasonal dummy for season $s$, and

$T_t$ is temperature in month $t$.

Several comments about this specification are in order. The equations that we estimated are deduced from relationships between temperature and births, so they combine a variety of effects. Direct physiological effects, such as an effect of heat on semen quality, will appear as effects of monthly temperature lagged nine months before births. Behavioral effects of two types will also be included in the effect of temperature. One behavioral effect is the effect of temperature on coital frequency. Another more indirect effect is an effect on the population at risk of conception. The population at risk is affected by the number of new entrants into the pool of at-risk couples in a given month, and by the number who left the pool in previous months due to a conception[9]. Our results must be interpreted with caution, since the effects of temperature on births that we identify can result from a number of behavioral and biological explanations.

The polynomial functions of time are included to capture any possible trends in births, leaving us to concentrate on the kind of short-run fluctuations that will be associated with short-run changes in temperature. The most likely source of any such long-run trends in births is new entrants into the population at risk, i.e., new couples exposing themselves to the risk of pregnancy. There might also be trends in the number of abortions or intrauterine mortalities, as well as in biological probabilities of conception, if, for example, women have been systematically trying to conceive at a later point in life.

The set of temperature variables that we include allows for a quadratic in temperature. In addition to permitting a simple linear relationship between temperature and fertility, this specification allows, for example, the possibility that temperature has no effect at moderate temperatures but has a negative effect at high temperatures, or the possibility that both extreme cold and extreme heat suppress fecundity. The extra lags of temperature (at lag 10) are needed to account for possible lagged effects of temperature on births.

## RESULTS

We implement the estimation strategy described in the previous section on the states and countries for which we examined seasonal patterns above[10]. The source for the temperature data for states is *Statewide Average Climatic History, Historical Climatology Series 6-1*, National Climatic Data Center, Ashville, NC. The data are arithmetic averages over all the reporting weather stations in that state. The source of the temperature data for countries is *Monthly Climatic Data for the World*. In each case, we use data for a city that is as close as possible to the population weighted center of the country[11]. The results are presented in Figure 3 and Table 1[12].

---

[9]See Lam and Miron (1987) for elaboration of these components of seasonal fluctuations in births.

[10]The sample periods for all regressions are as follows: CA,GA,NY (white births) 1948:1-1981:12; Kerala 1960:1-1977:12; Maharashtra 1960:1-1979:12; Punjab 1967:1-1979:12; Israel 1969:1-1984:12; Canada 1949:1-1976:12; England 1972:1-1983:12; France 1959:1-1983:12; Australia 1963:1-1984:12; New Zealand 1960:1-1982:12.

[11]The temperature data are measured in degrees Fahrenheit. For Kerala, Maharashtra, Punjab, Israel, Canada, England, France, Sweden, Australia, and New Zealand, we use, respectively, temperature data from the following cities: Trivandrum, Bombay, Amritsar, Jerusalem, Toronto, London, Paris, Stockholm, Christchurch, and Sydney. These cities were chosen, subject to data availability, to match as best as possible the geographics of the births population. For most of these locales outside the U.S. some temperature observations are unavailable ("some" ranges from 1 of 336 observations in Canada (Toronto) to 20 of 276 observations in New Zealand (Christchurch)). These are simply coded as missing values in all regressions.

[12]All test statistics are computed using the Newey and West (1987) correction, with the lag length set equal to 12 and the damp factor equal to 1.

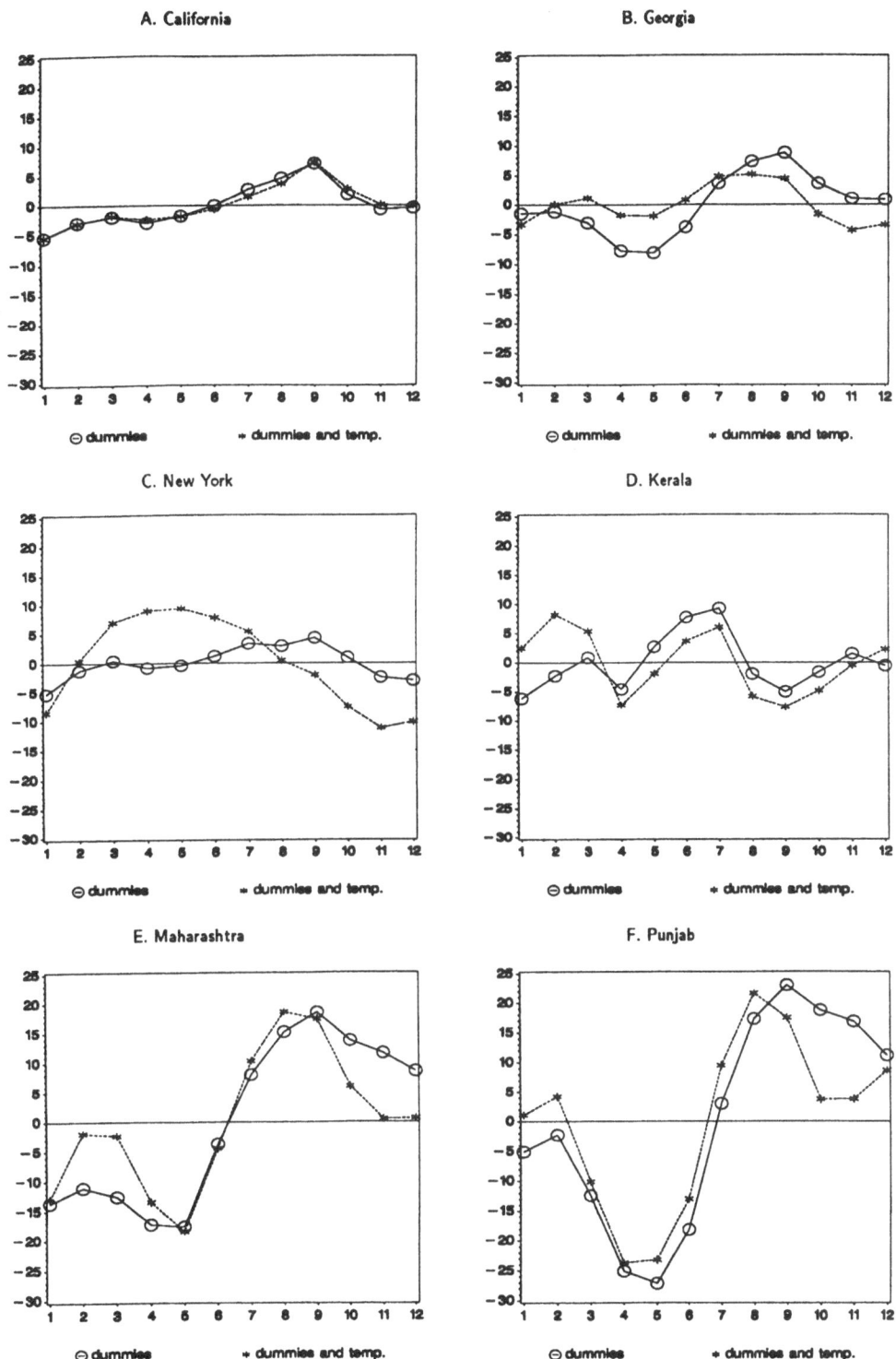

Figure 3. Temperature's Effect on Birth Seasonality
Simple Seasonals vs. Seasonals Controlling for Temperature's Effect.

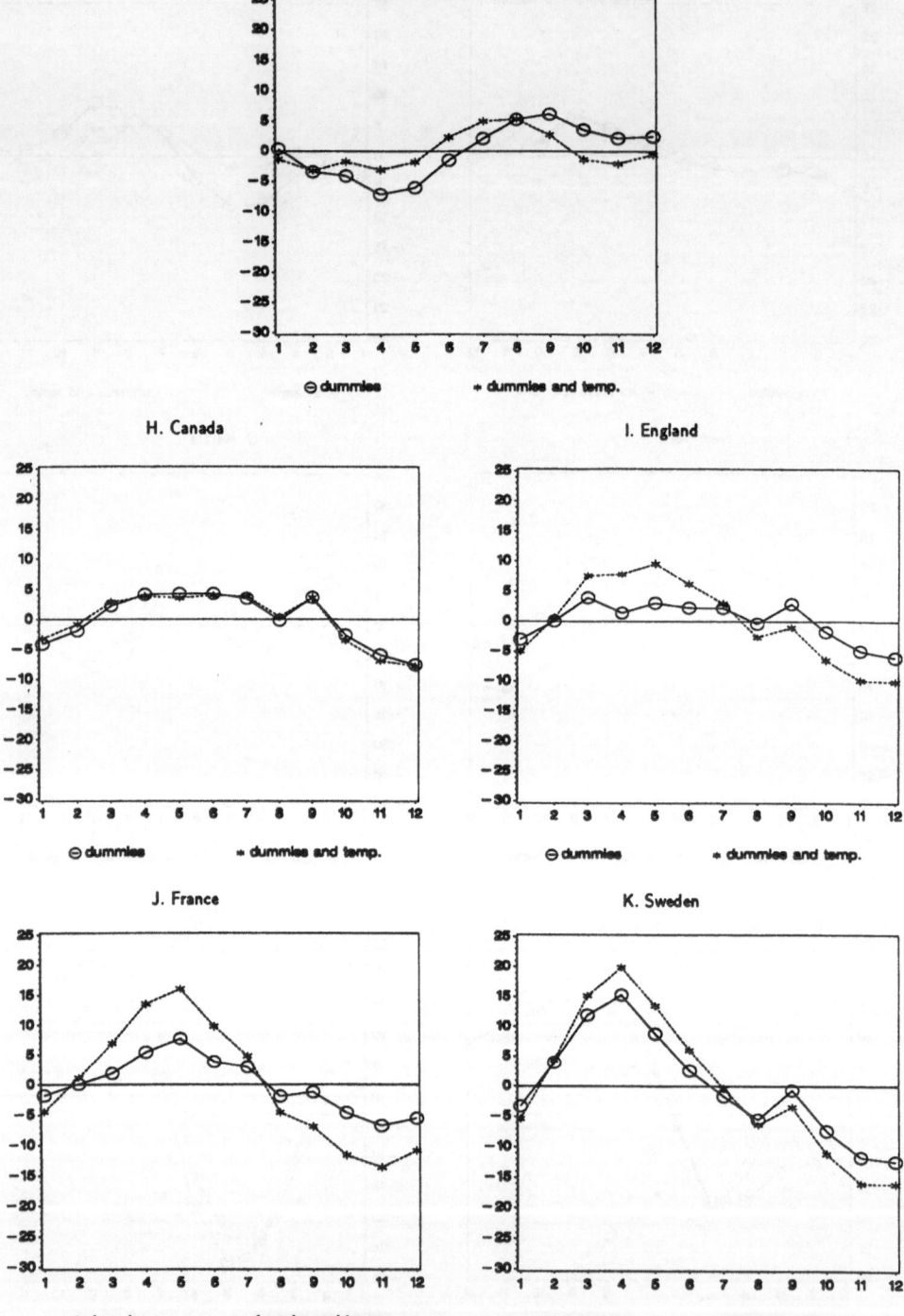

Figure 3 continued.  Temperature's Effect on Birth Seasonality
Simple Seasonals vs. Seasonals Controlling for Temperature's Effect.

The figure plots the estimated seasonal patterns in (detrended) births along with the patterns implied by our estimates of equation (2). The first set of coefficients, identified as "dummies" and marked with circles on the graphs, represents the simple unconditional seasonal pattern in births and is identical to the set of coefficients in Figure 1. The second set of coefficients, identified as "dummies and temp" and marked with stars on the graphs, represents the coefficients for each month when monthly temperature variables are included in the regression. These coefficients can be interpreted as the part of the seasonal pattern not attributable to temperature. If all seasonal fluctuations were due to the seasonality in temperature, and if that relationship could be represented as a quadratic function of monthly temperature, then this second set of coefficients would equal zero, implying no seasonality in births after the effects of temperature are removed.

Comparison of the two sets of coefficients in each panel of Figure 3 indicates that we are not in fact able to explain all of the seasonal fluctuations in births by seasonal fluctuations in temperature alone. The results for California in Panel A, for example, suggest that only a small part of the seasonal birth pattern in that state can be explained by temperature. Taken literally, the results imply that birth seasonality in California would be almost identical to the observed seasonal pattern even if there were no seasonal variations in temperature. The results for Georgia, shown in Panel B of Figure 3, show that monthly temperature explains much more of the seasonal birth pattern in that state. Including monthly temperature nine and ten months before the month of birth significantly reduces the magnitude of the April-May trough in births in Georgia. This result is exactly consistent with the hypothesis that extreme heat in July and August suppresses conceptions in those months.

Examining the unadjusted seasonal birth pattern and the seasonal birth pattern controlling for monthly temperature in the remaining panels of Figure 3, we observe a general tendency for spring births to increase when we control for the effects of temperature. This implies that there is, in fact, an identifiable world wide effect of summer heat suppressing conceptions. For several populations, notably Georgia, New York, Maharashtra, and Israel, controlling for monthly temperature removes a significant portion of the observed trough in births in the spring months. For the northern populations of England, France, and Sweden, the results imply that even though there is not an observed spring trough in births in the simple unadjusted birth pattern, spring births are in fact lower in these populations than they would be if there were no seasonal variation in temperature.

Table 1 includes results for two formal hypothesis tests on our estimated coefficients from specification (2). The first tests the null hypothesis that all of the

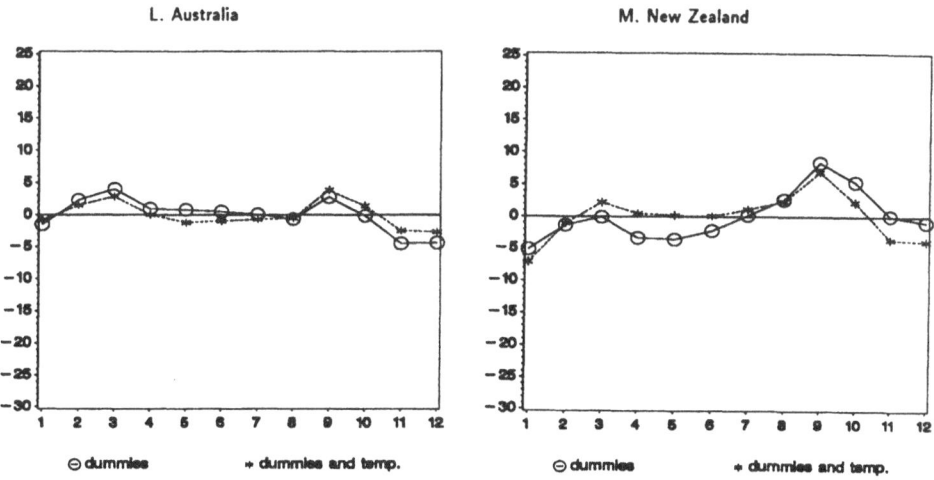

Figure 3 continued.  Temperature's Effect on Birth Seasonality
Simple Seasonals vs. Seasonals Controlling for Temperature's Effect.

## Table 1. Results of Hypothesis Tests

| Regressors | Population | | | | | | | | | | | | |
|---|---|---|---|---|---|---|---|---|---|---|---|---|---|
| | California | Georgia | New York | Kerala | Maharashtra | Punjab | Israel | Canada | England | France | Sweden | Australia | New Zealand |
| dummies | 305.10 | 256.74 | 290.15 | 97.68 | 674.41 | 544.69 | 156.52 | 335.20 | 319.66 | 216.75 | 449.18 | 199.24 | 189.17 |
| | ( .00) | ( .00) | ( .00) | ( .00) | ( .00) | ( .00) | ( .00) | ( .00) | ( .00) | ( .00) | ( .00) | ( .00) | ( .00) |
| weather | 8.13 | 10.20 | 34.12 | 11.64 | 11.27 | 5.24 | 8.00 | 4.55 | 38.73 | 29.23 | 4.51 | 2.09 | 8.31 |
| | ( .09) | ( .04) | ( .00) | ( .02) | ( .02) | ( .26) | ( .09) | ( .34) | ( .00) | ( .00) | ( .34) | ( .72) | ( .08) |

Notes: 1. The entries in the table are $\chi^2$ statistics for tests of the null hypothesis that either the dummy coefficients or the coefficients on the weather variables are jointly equal to zero. 2. Aysmptotic p-values are in parentheses.

seasonal dummy coefficients are zero, i.e., that there is no birth seasonality once the effects of monthly temperature are controlled for. These are the tests marked "dummies" in Table 1. In every population we strongly reject the null hypothesis of no remaining seasonality. The second test result shown is the test that all of the coefficients on the monthly temperature variables are zero, i.e., that monthly temperature does not explain a significant portion of the monthly variation in births. These results are more mixed. The temperature variables are significant at the 5% level in six populations, Georgia, New York, Kerala, Maharashtra, England, and France. They are significant at the 10% level in California, Israel, and New Zealand. The temperature variables are not statistically significant at conventional levels in Punjab, Canada, Sweden, and Australia.

The results in Figure 3 and Table 1 also imply, however, that there remains a significant amount of birth seasonality after the effects of temperature are removed. Especially noteworthy is the September local peak observed in many of the populations, a peak which tends to appear both before and after controlling for monthly temperature.

CONCLUSIONS

Seasonal patterns in births potentially provide important evidence regarding the effect of temperature on human reproduction. In this paper we look at the evidence suggested by comparing seasonal patterns across populations. We then attempt to directly test the effect of monthly temperature on conceptions by using monthly birth and temperature data for a variety of populations.

Our results provide mixed evidence regarding the effects of monthly temperature. Looking at seasonal birth patterns, we observe a tendency for a spring trough in births in a number of warm climate populations, suggesting that hot summers reduce conceptions. We also observe a number of patterns without obvious temperature explanations, including a persistent September local peak in births and a spring peak in births in a number of northern populations.

Estimating regressions of monthly births on monthly temperature nine and ten months prior, we find that temperature has a statistically significant effect on the seasonal birth pattern in most of the populations studied. It is in many ways surprising that temperature does not explain more of the seasonal pattern than it does, however. Significant seasonality in births remain after the effects of temperature are removed in all of the populations.

Although temperature clearly does not explain all of the observed seasonal patterns in births, the effects of temperature we identify are generally consistent with the hypothesis that high summer temperatures reduce conceptions. When we control for the effects of monthly temperature we tend to consistently increase births in the spring months in these populations, implying that summer conceptions are being systematically reduced by high temperatures.

Although our results provide evidence that high temperatures reduce conceptions in a wide variety of populations, it is important to note that our results do not indicate the mechanism causing this relationship. In particular, we cannot distinguish between biological and behavioral explanations based on these results alone. The effect could be caused, for example, by a reduction in coital frequency in hot summer months, or by an effect of heat on sperm quality. Our results do provide an important piece of the puzzle in identifying biological effects of temperature on conceptions, however. Seasonal patterns in births and temperature clearly merit further attention as researchers attempt to understand the effects of temperature on human reproductive physiology.

ACKNOWLEDGEMENTS

This paper has benefitted from discussions with Stan Becker. Excellent research assistance was provided by Todd Clark. Financial support was provided by the National Institute for Child Health and Human Development, Grant No. 1-R01-HD22141.

## REFERENCES

Becker, Stan. 1981. Seasonality of fertility in Matlab, Bangladesh. Journal of Biological Science, 13: 97-105.

Becker, Stan. 1990. Seasonal patterns of birth throughout the world. This volume.

Becker, Stan, Alaudin Chowdhury and Henri Leridon. 1986. Seasonal patterns of reproduction in Matlab, Bangladesh. Population Studies, 40: 457-472.

Brundtland, G.H. and K. Liestol. 1982. Seasonal variations in menarche in Oslo. Annals of Human Biology, 9(1): 35-43.

Chaudhury, R.H. 1972. Socioeconomic and seasonal variations in births: A replication. Social Biology, 19: 65-68.

Cowgill, U.M. 1965. Season of birth in man: contemporary situation with special reference to Europe and the Southern Hemisphere. Ecology, 47(4): 614-623.

Cowgill, U.M. 1966. The season of birth in man: The Northern New World. Kroeber Anthropological Society Paper, 1-21.

Holmes, R.L. 1968. Reproduction and environment. Norton, New York.

Huntington, E. 1938. Season of birth. Wiley, New York.

James, W.H. 1971. Social class and season of birth. Journal of Biosocial Science, 3: 309-320.

Kestenbaum, Bert. 1987. Seasonality of birth: two findings from the Decennial Census. Social Biology, 34(3-4): 244-248.

Kosambi, D.D. and S. Raghavachari. 1951. Seasonal variation in the Indian birth rate. Annals Eugenics, 16(2): 165-191.

Lam, David A. and Jeffrey A. Miron. 1987. The seasonality of births in human populations. Research Report # 87-114, Population Studies Center, University of Michigan.

Lam, David A. and Jeffrey A. Miron. 1989. The seasonality of births in human populations. Manuscript, University of Michigan.

Lee, Ronald D. 1981. Short-term variation: vital rates, prices, and weather. In: E.A. Wrigley and R. Schofield (eds), The Population History of England, 1541-1871: A Reconstitution. Harvard University Press, Cambridge.

Levine, Richard J., Brenda L. Borsdon, Ravi M. Mathew, Michelle H. Brown, Jonathan M. Stanley and Thomas B. Starr. 1988. Deterioration of semen quality during summer in New Orleans. Fertility and Sterility, 49(5): 900-907.

Lyster, W.R. 1971. Three patterns of seasonality in American births. American Journal of Obstetrics and Gynecology, 110: 1025-1028.

Lyster, W.R. 1979. New seasonal distribution of births in New South Wales. The Medical Journal of Australia, p. 150.

MacFarlane, W.V. 1970. Seasonality of conception of human populations. Biometerology, 4.

MacFarlane, W.V. and D. Spalding. 1960. Seasonal conception rates in Australia. The Medical Journal of Australia, p. 121.

Mathers, C.D. and R.S. Harris. 1983. Seasonal distribution of births in Australia. International Journal of Epidemiology, 12(3): 326-331.

McDonald, A.D. 1971. Seasonal distribution of abortions. British Journal of Preventive and Social Medicine, 25: 222-224.

Newey, Whitney and Kenneth West. 1987. A simple, positive definite, heteroskedasticity and autocorrelation consistent covariance matrix. Econometrica, 55: 703-8.

Parker, Gordon. 1978. Season of birth in New South Wales. The Medical Journal of Australia, 2: 563-566.

Parker, Gordon. 1979. New seasonal distribution of births in New South Wales. The Medical Journal of Australia, 2: 424.

Pasamanick, B., S. Dinitz and H. Knobloch. 1959. Geographic and seasonal variations in births. Public Health Reports, 74: 285-288.

Pasamanick, B., S. Dinitz and H. Knobloch. 1960. Socio-economic and seasonal variations in birth rates. Milbank Memorial Fund Quarterly, 38: 248-254.

Reinberg, A. 1974. Aspects of circannual rhythms in man. In: E.T. Pengelley (ed), Circannual Clocks. Academic Press, New York.

Reinberg, A., M. Lagoguey, F. Cesselin, Y. Touitou, J. Legrand, A. Delassalle, J. Antreassian and A. Lagoguey. 1978. Circadian and circannual rhythms in plasma hormones and other variables of five healthy young human males. Acta Endocrinologica, 88: 417-427.

Rodgers, Joseph Lee and J. Richard Udry. 1988. The Season-of-Birth Paradox. Social Biology, 35(3-4): 171-185.

Richards, Toni. 1983. Weather, nutrition and the economy: Short-run fluctuations in births, deaths and marriages, France 1740-1909. Demography, 20: 197-212.

Rosenberg, H.M. 1966. Seasonal variation of births. National Center for Health Statistics, Washington, D.C., 21:9.

Ryder, N.B. and C.F. Westoff. 1971. Reproduction in the United States, 1965. Princeton University Press, Princeton, N.J.

Seiver, Daniel. 1985. Trend and variation in the seasonality of United States fertility, 1947 to 1976. Demography, 22: 1.

Seiver, Daniel. 1989. Seasonality of fertility: New evidence. Population and Environment, forthcoming.

Slatis, H.M. and R.J. DeCloux. 1967. Seasonal variation in stillbirth frequencies. Human Biology, 39: 284-94.

Stoeckel, J. and A.K.M.A. Chowdhury. 1972. Seasonal variation in births in rural East Pakistan. Journal of Biosocial Science, 4: 107-116.

Sundararaj, N., M. Chern, L. Gatewood, L. Hickman and R. McHugh. 1978. Seasonal behavior of human menstrual cycles: A biometric investigation. Human Biology, 50(1): 15-31.

Tjoa, W.S., M. Smolensky, B. Hsi, E. Steinberger and K. Smith. 1982. Circannual rhythm in human sperm count revealed by serially independent sampling. Fertility and Sterility, 38: 454-459.

Tromp, S.W. 1963. Medical Biometerology, Elsevier, Amsterdam.

Warren, C.W. and C.W. Tyler. 1979. Social status and season of birth: A study of a metropolitan area in the Southeastern United States. Social Biology, 26: 275-288.

Warren, C.W., J. Gold, C.W. Tyler, J.C. Smith and A.L. Paris. 1980. Seasonal variation in spontaneous abortions. American Journal of Public Health, 70(12): 1297-1299.

Zelnik, M. 1969. Socioeconomic and seasonal variations in births: A replication. Milbank Memorial Fund Quarterly, 47: 159-165.

Zorgniotti, A.W. and A.I. Sealfon. 1988. Measurement of intrascrotal temperature in normal and subfertile men. Journal of Reproductive Fertility, 82: 563-66.

# SEASONAL VARIATION IN HUMAN SEMEN QUALITY

Richard J. Levine

Chemical Industry Institute of Toxicology (CIIT)
P.O.B. 12137
Research Triangle Park, NC 27709

## INTRODUCTION

In regions with warm climates the birth rate is reduced during the spring season[1-3]. Could summer heat nine months earlier have caused semen quality to deteriorate and diminish the rate of conception? Human testes and epididymides, carried in the scrotal sac, are more likely to be influenced by environmental temperatures than are female reproductive organs, located inside the pelvic cavity and protected by the full complement of the body's homeostatic mechanisms[4]. Even mild scrotal warming after two to three weeks can temporarily reduce both concentration and number of sperm in ejaculates[5]. A decrease in the frequency of sexual intercourse might account for a diminished rate of conception during summer, but the available evidence suggests, if anything, that the opposite may occur[4,6-8].

If heat sufficient to affect testes or epididymides would begin in late spring and subside by late summer, the preponderance of effects on male reproductive capacity should occur during summer. When heat subsides and semen quality returns to normal, sperm numbers may thereupon exceed baseline values for a period of time[5]. This overshoot phenomenon, which has been observed in studies of scrotal warming in human volunteers, could counterbalance any residual depression of semen quality during early fall. Seasonal variation in semen quality, therefore, is to be expected in non-Equatorial regions with warm climates if sperm concentration and total sperm per ejaculate may be affected by summer heat.

## METHODS

To evaluate seasonal variation, all four-season studies of semen quality were identified. Those whose duration exceeded one year were collapsed to a single set of 12 months. Chances of observing a seasonal pattern against background variation were improved by considering only studies with an average of $\geq 75$ specimens per month from an equal number of donors. This criterion was relaxed in cases where individuals had contributed multiple specimens since variation within persons is much less than between persons. Almost all studies reported results by month, rather than by season. Where results had been given by season, the original data were utilized to produce results by month. Since in all studies data had been obtained from geographic locations in the northern hemisphere, January, February and March are hereafter referred to as winter; April, May and June, as spring; July, August and September, as summer; and October, November and December, as fall.

*Temperature and Environmental Effects on the Testis*
Edited by A. W. Zorgniotti, Plenum Press, New York, 1991

## Table 1

## SEASONAL VARIATION IN HUMAN SEMEN QUALITY: STUDIES REVIEWED

| Reference | Location | Number* of Specimens | Donors | Study Duration (months) | Abstinence (hrs) | Donor Type | Place Specs Collected |
|---|---|---|---|---|---|---|---|
| Hotchkiss[9] | New York | 642a | 22 | 14 | >72 | Volunteers | Home |
| MacLeod[10] | New York | 591 | 8 | 10** | >72 | Students | Home |
| Tjoa[11] | Houston | 4435 | 4435 | 52 | Unspecified | Pre-vasectomy | Home |
| Mortimer[12] | Edinburgh | 1566b | 1566 | 60 | >72 | Infert. Clinic | Home |
| Spira[13] | New York | 1067c | 52 | 36 | Unspecified | Students | Lab |
| Levine[4] | New Orleans | 1155d | 900 | 37 | 48-96 | Infert. Clinic | Lab |
| Saint Pol[14] | Lille | 4169 | ?2100 | 96 | 48 | Sperm Donors | Lab |
| Politoff[15] | Basel | 2677 | ? | 12 | 72 | Infert. Clinic | Home |
| Levine[16] | Calgary | 3601 | 2191 | 46 | 72 | Infert. Clinic | Lab |

\* Substantial deviations of N of individual parameters from the numbers reported are given below.

a  Volume – 378, Total Sperm – 376

b  Percent Normal Morphology – 1452

c  Percent Motile – 831

d  Percent Normal Morphology – 1078

\*\* Lacks June and October

Table 2. Seasonal variation in human semen quality: Studies excluded

| Reference | Location | Number of | | Reason Excluded |
| | | Specimens | Donors | |
| --- | --- | --- | --- | --- |
| Baker[17] | Melbourne | 835 | 177 | Insufficient |
| | | 1227 | 114 | data reported |
| Abbaticchio[18] | Bari | 248 | 248 | Insufficient |
| | | | | numbers |
| Reinberg[19] | Houston | 355 | 260 | Insufficient |
| | | | | numbers |
| Schrader[20] | Cincinnati | 400 | 46 | Lacks March, |
| | | | | April, May |

Manual methods had been employed to measure semen quality. The precision of techniques used in the various laboratories is not known. Numerical data were taken from text or tables, where available; elsewhere, they were estimated from figures. For each study under consideration the season and month with the highest or lowest values of semen quality parameters were identified. The statistical significance of the aggregation of highest (or lowest) values in a particular season or month was assessed, using binomial probabilities. The probability of occurrence of the high (or low) value in a particular season or month of a given study was assumed to be 1/4 or 1/12, respectively. A two-sided p-value for the significance of clustering across studies was computed by summing the binomial probabilities of aggregations at least as extreme as the one observed.

RESULTS

In all, thirteen studies of seasonal variation in semen quality were identified. Information on the nine which met review criteria is given in Table 1[4,9-16]. Four of these were of infertility patients; five were of presumably normal or fertile populations. The four studies that were excluded are listed in Table 2 together with reasons for their exclusion[17-20].

Seasonal variation in ejaculate volume was examined in seven of the nine studies reviewed[4,9,10,12-14,16]. There was no significant clustering of studies reporting a particular season with the highest (or lowest) mean ejaculate volume. The largest aggregation was for the occurrence of highest volume in three studies during fall[9,14,16].

Six of the studies which met review criteria reported the results of seasonal variation in percent motile sperm[4,9,12-14,16]. In one such study the parameter examined was motility grade, a score derived from percent motile sperm, motility type, and duration of activity[9]. Highest values were noted in four of six studies during summer $(p=0.04)$[9,12-14], but the importance of this finding is uncertain. In two of these studies, both with highest values in summer, specimens had been collected at home[9,12]. It is possible that they may have been affected by exposure to ambient temperatures while being transported to the laboratory for analysis. One of the two studies reported lowest values in winter; the other, in fall.

Five studies recorded the percent of morphologically normal sperm in different seasons[4,9,12,13,16]. Although highest values were found during spring in three studies[4,9,16], there was no significant clustering either of highest or lowest values within a particular season.

Total sperm per ejaculate was given in seven studies[4,9-13,16]. Three studies reported highest values in winter[10-12]; and three, in spring[9,13,16]. Neither of these aggregations was

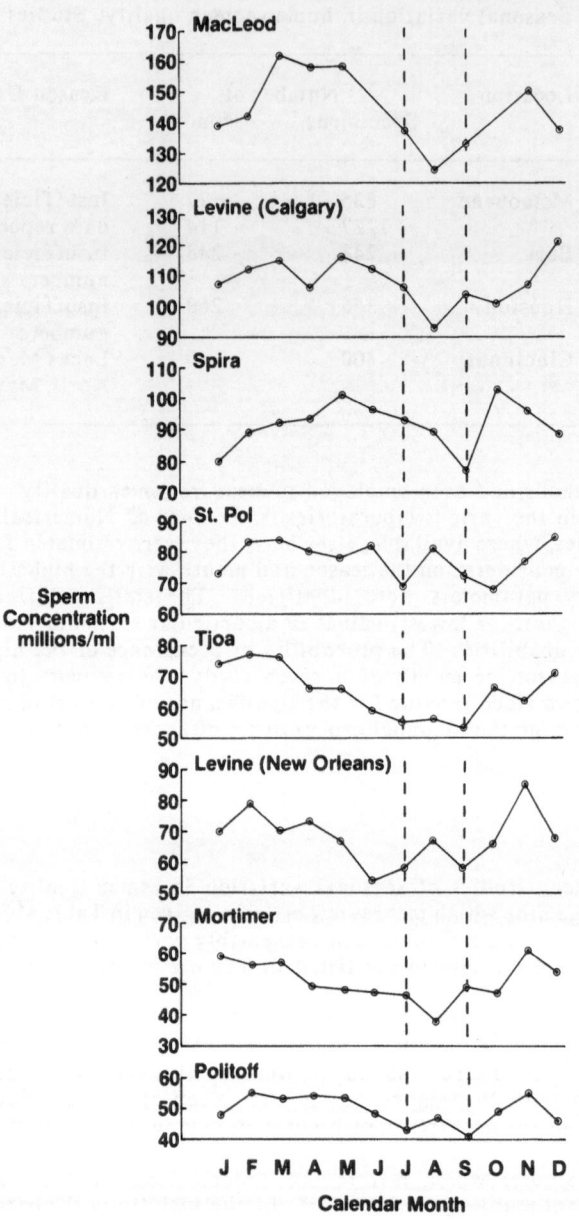

Figure 1. Sperm concentration by calendar month in each of eight studies which met review criteria. Values of summer months can be identified within the broken vertical lines. Data were obtained from MacLeod[10], Table 3, p. 477 (lacks June and October); Levine[16], unpublished; Spira[13], Figure 1b, p. 25, St. Pol[14], Table 1, p. 1031; Tjoa[11], Figure 3, p. 458; Levine[4], unpublished; Mortimer[12], Figure 2a, p. 3; and Politoff[15], Figure 1, p. 487.

significant. However, in five of seven studies lowest values were found during summer[4,11-13,16]. This clustering was statistically significant (p=0.01).

Eight studies examined sperm concentration[4,10-16]. Monthly concentration levels in each are plotted individually in Figure 1. Highest values in three studies occurred during winter[11,12,15]; in four, during spring[10,13,14,16]. Neither aggregation was statistically

Table 3. Mean of monthly sperm concentrations and percent of this value by month

Percent of Mean Concentration by Month

| Reference | Mean Conc. $(10^6/ml)$ | Jan | Feb | Mar | Apr | May | Jun | Jul | Aug | Sep | Oct | Nov | Dec |
|---|---|---|---|---|---|---|---|---|---|---|---|---|---|
| MacLeod[10] | 144 | 97 | 99 | 113 | 110 | 110 | --- | 95 | 86 | 92 | --- | 104 | 96 |
| Tjoa[11] | 65 | 114 | 118 | 117 | 102 | 102 | 91 | 85 | 86 | 82 | 102 | 95 | 109 |
| Mortimer[12] | 51 | 116 | 110 | 112 | 96 | 94 | 92 | 90 | 75 | 96 | 92 | 120 | 106 |
| Spira[13] | 92 | 87 | 97 | 100 | 101 | 110 | 104 | 101 | 97 | 84 | 112 | 104 | 97 |
| Levine[4] | 68 | 103 | 116 | 103 | 107 | 99 | 79 | 85 | 99 | 85 | 97 | 125 | 100 |
| Saint Pol[14] | 78 | 94 | 106 | 108 | 103 | 100 | 105 | 88 | 104 | 92 | 87 | 99 | 109 |
| Politoff[15] | 49 | 98 | 112 | 108 | 110 | 108 | 98 | 88 | 96 | 84 | 100 | 112 | 94 |
| Levine[16] | 108 | 99 | 104 | 106 | 98 | 107 | 104 | 98 | 86 | 96 | 94 | 99 | 112 |
| Overall Mean -- | | 101 | 108 | 108 | 103 | 104 | 96 | 91 | 91 | 89 | 98 | 107 | 103 |

significant. Remarkably, among all eight studies sperm concentration was lowest during summer (p=0.00002). Lowest monthly concentrations occurred during June[4] (one study), August[10,12,16] and September[11,13,15] (three studies each, p=0.0004), and during October[14] (one study). Lowest values, therefore, appeared to cluster within the second half of the summer season and possibly, to a lesser degree, during early fall.

Table 3 gives the mean of monthly sperm concentrations by study (an unweighted average of sperm concentration in each month) together with sperm concentration in each month expressed as a percentage of this mean value. Overall mean percentages by month across all eight studies are also indicated. This information is depicted graphically in Figure 2. It can be seen that sperm concentration is highest in February and March and lowest during the summer season. The magnitude of the average reduction during summer is about 10% from the mean of all studies and 17% from the February-March peak. The difference in extent of summer reduction between studies performed in northern Europe or Canada and those at more southerly latitudes, appears to be minimal or nonexistent (16% vs. 17%, respectively, from the February-March peak).

DISCUSSION

Evidence for seasonal variation in ejaculate volume, percent motile sperm, percent morphologically normal sperm, total sperm per ejaculate, and sperm concentration has been reviewed above. Except for highest values of percent motile sperm, which resulted in a summer cluster of borderline statistical significance and questionable biological relevance, seasonal variation could not be detected among the first three parameters. Assessment of sperm motility and morphology, however, was performed by manual methods of analysis which depended to a considerable extent upon subjective judgment. Only ejaculate volume and sperm concentration (and their product, total sperm per ejaculate) were ascertained objectively, using pipettes or graduated cylinders and hemocytometers. These parameters of semen quality had a greater likelihood for consistent evaluation within a study over the study duration. Ability to detect a seasonal pattern against background variation, therefore, would be enhanced.

Lowest values of sperm concentration and the concentration-related parameter, total sperm per ejaculate, clustered significantly during the summer season. Sperm concentration was lowest during summer in all eight studies which evaluated this parameter, including fertile as well as infertile populations. Summer depression of sperm concentration may at least partially explain the deficit of spring births in warm climates by increasing time to pregnancy as well as the proportion of men at greater risk of infertility (sperm concentration <20 million/ml).

Reports from Edinburgh[12], Lille[14], Basel[15], and Calgary[16] suggest that summer sperm concentration may be reduced also in cooler climates. Indeed, there appears to be little or no difference in extent of summer reduction between these studies and others performed at more southerly latitudes. As a consequence, it is less likely that heat alone can explain the entire phenomenon and suggests that the etiology may be multifactorial, perhaps also involving light. Furthermore, since the birthrate is highest during spring in northern Europe and Canada[2,21,22], not lowest as it is in the southern U.S., the relationship there of diminished summer sperm concentration to spring births is not clear. Seasonal variation in social or psychological factors which influence coital frequency[23,24] may predominate in these regions and obscure the effect of reduced summer sperm concentration on spring births.

Sperm concentration, ejaculate volume, and total sperm per ejaculate vary directly with the length of the preceding period of abstinence[25]. Could summer reduction in sperm concentration and total sperm per ejaculate have resulted from an increased frequency of ejaculation? Since ejaculate volume is at least as sensitive as sperm concentration to the effects of varying lengths of abstinence[25], the apparent absence of seasonal variation in ejaculate volume (even among the four studies in cooler climates) suggests that duration of abstinence did not vary with season. Lack of a summer reduction in ejaculate volume, therefore, is evidence against a role for diminished length of abstinence in the etiology of summer sperm concentration depression. The abstinence period was constrained by clinical requirements in most, if not all studies. It may not, therefore, truly reflect either abstinence in the general population or coital frequency.

There can be little doubt that sperm concentration in non-Equatorial regions is in fact reduced during summer. Although it is by no means well established, in warm latitudes, where seasonal variation in daylength (photoperiod) is minimal, heat is the most likely causative agent. The importance of an environmental factor such as heat is supported by an analysis of seasonal variation in semen quality at a New Orleans fertility clinic according to job classification. Substantial and significant deterioration during summer was likely to occur only among men with little or no access to air conditioning during the hottest part of the day[4].

Additional investigation is needed before research on seasonal variation in semen quality can lead to recommendations for the treatment of infertility. The contribution

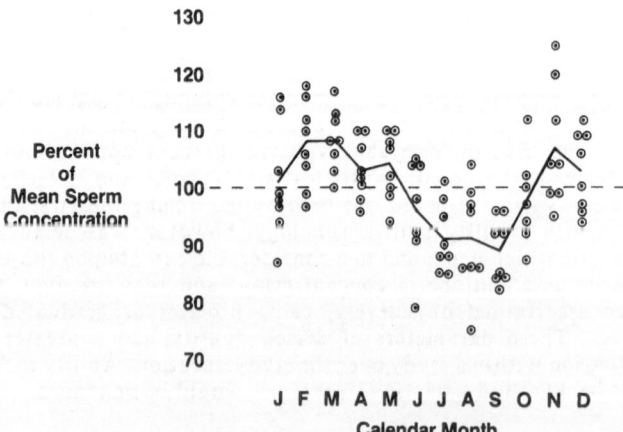

Figure 2. Percent of the mean of monthly sperm concentrations by calendar month in the eight studies which met review criteria (open circles) and the average percent of this value across all studies (solid line).

of heat must be distinguished from light, and the possible influence of photoperiod, spectrum, and intensity of illumination must be ascertained. The evidence is sufficient, however, for researchers to be cautioned now to consider the effects of season when planning and analyzing clinical or epidemiological studies.

## ACKNOWLEDGMENT

The assistance of Ravi Mathew and Michelle Brown of CIIT is gratefully acknowledged. The author thanks Dr. Stan Becker for reviewing the draft manuscript.

## REFERENCES

1. Huntington, E. Season of Birth: Its Relation to Human Abilities. John Wiley & Sons, New York (1938)
2. Rosenberg, H.M. Seasonal variation of births: United States, 1933-63. Vital and Health Statistics, Series 21, Number 9, National Center for Health Statistics. U.S. Department of Health, Education, and Welfare, Washington, DC, (May 1966)
3. Becker, S. Seasonality of fertility in Matlab, Bangladesh. J. Biosoc. Sci., 13:97-105 (1981)
4. Levine, R.J., Bordson, B.L., Mathew, R.M., Brown, M.H., Stanley, J.M. and Starr, T.B. Deterioration of semen quality during summer in New Orleans. Fert. Steril., 49:900-07 (1988)
5. Robinson, D. and Rock, J. Intrascrotal hyperthermia induced by scrotal insulation: effect on spermatogenesis. Obstet. Gynecol., 29:217-23 (1967)
6. Becker, S., Chowdhury, A., Leridon, H. Seasonal patterns of reproduction in Matlab, Bangladesh. Popul. Stud., 40:457-72 (1986)
7. Udry, J.R. and Morris, N.M. Seasonality of coitus and seasonality of birth. Demography, 4:673-79 (1967)
8. Ehrenkranz, J.R.L. Seasonal breeding in humans: birth records of the Labrador Eskimo. Fert. Steril., 40:485-89 (1983)
9. Hotchkiss, R.S. Factors in stability and variability of semen specimens: observations on 640 successive samples from 23 men. J. Urol., 45:875-88 (1941)
10. MacLeod, J. and Heim, L.M. Characteristics and variations in semen specimens in 100 normal young men. J. Urol., 54:474-482 (1945)
11. Tjoa, W.S., Smolensky, M.H., Hsi, B.P., Steinberger, E. and Smith, K.D. Circannual rhythm in human sperm count revealed by serially independent sampling. Fert. Steril., 38:454-59 (1982)
12. Mortimer, D., Templeton, A.A., Lenton, E.A. and Coleman, R.A. Annual patterns of human sperm production and semen quality. Arch. Androl., 10:1-5 (1983)
13. Spira, A. Seasonal variations of sperm characteristics. Arch. Androl., 12(Suppl):23-28 (1984)
14. Saint Pol, P., Beuscart, R., Leroy-Martin, B., Hermand, E. and Jablonski, W. Circannual rhythms of sperm parameters of fertile men. Fert. Steril., 51:1030-33 (1989)
15. Politoff, L., Birkhauser, M., Almendral, A. and Zorn, A. New data confirming a circannual rhythm in spermatogenesis. Fert. Steril., 52:486-89 (1989)
16. Levine, R.J., Brown, M.H., Mathew, R.M., Pattinson, H.A. and Mortimer, D. Seasonal variation in semen quality at a Calgary fertility clinic (in preparation, 1990).
17. Baker, H.W.G., Burger, H.G., de Kretser, D.M., Lording, D.W., McGowan, P. and Rennie, G.C. Factors affecting the variability of semen analysis results in infertile men. Int. J. Androl., 4:609-22 (1981)
18. Abbaticchio, G., de Fini, M., Giagulli, V.A., Santoro, G., Vendola, G. and Giorgino, R. Circannual rhythms in reproductive functions of human males: correlations among hormones and hormone-dependent parameters. Andrologia, 19:353-61 (1987)
19. Reinberg, A., Smolensky, M.H., Hallek, M., Smith, K.D. and Steinberger, E. Annual variation in semen characteristics and plasma hormone levels in men undergoing vasectomy. Fert. Steril., 49:309-15 (1988)
20. Schrader, S.M., Turner, T.W., Breitenstein, M.J. and Simon, S.D. Longitudinal study of semen quality of unexposed workers: I. Study Overview. Repro. Toxicol., 2:183-90 (1988)
21. Cowgill, U.M. Season of birth in man: contemporary situation with special reference to Europe and the southern hemisphere. Ecology, 47:614-623 (1965)

22. Ashley, M.J. Season of birth: stability of the pattern in Canada. Can. J. Pub. Health, 79:101-03 (1988)

23. Ehrenkranz, J.R.L. Seasonal breeding in humans: birth records of the Labrador Eskimo. Fert. Steril., 40:485-89 (1983)

24. Abas, W. and Murphy, D. Seasonal affective disorder: the miseries of long dark nights? Br. Med. J., 295:1504-05 (1987)

25. Heuchel, V., Schwartz, D. and Price, W. Within-subject variability and the importance of abstinence period for sperm count, semen volume and pre-freeze and post-thaw motility. Andrologia, 13:479-85 (1981)

# SECTION 3

## TESTIS THERMOMETRY

# THEORETICAL AND PRACTICAL CONSIDERATIONS IN

## SCROTAL TEMPERATURE MEASUREMENT

Andrew I. Sealfon

Repro-Med Systems, Inc.
Middletown, NY 10940

## INTRODUCTION

Scrotal or testicular temperature measurement serves as a clinical and research tool to study temperature effects on spermatogenesis and its relation to semen quality. While accurate temperature determination may appear simple, in reality anatomic and thermodynamic considerations may make readings unrepeatable and therefore unreliable. A better understanding of the potential problems will facilitate the work of researchers and clinicians.

One concept which causes difficulty is the difference between heat and temperature. Temperature is defined as hotness or coldness of an object or environment while heat is defined as thermal energy stored, given up or absorbed by an object or environment. Such simplistic definitions do not provide adequate insight and understanding. A more technical definition for temperature will include such concepts as proportionality of vibrational energy in solids or translational kinetic energy in gases (gases have different properties from solids) as related to absolute temperature, but this also fails to provide a clear distinction between heat and temperature. Heat may be defined as energy transferred from one system to another solely by reason of a temperature difference or gradient between the systems.

Suppose a 100 gram block of ice with an imbedded thermometer is placed into a calorimeter reading 0°C. Heat in the amount of 8000 calories is permitted to be absorbed by the ice. The result? Water at 0°C. Heat was added, the ice changed state from ice to water, but the temperature remained constant. Consider the example of a candle and a radiator. The candle is hotter, that is, has a much higher temperature. But to heat a room in winter, the radiator would provide much more heat, albeit at a significantly lower temperature than the candle. The hotter candle does not provide enough heat to heat the living room. An analogy using water might be that temperature is equivalent to water pressure, while heat is the quantity of water transferred. The greater the temperature difference or gradient, the greater the amount of heat which can be transferred or "pushed" across the gradient.

Temperature in the testis can therefore be considered as a reading of the energy state derived from the heat input which is primarily arterial blood flow and heat losses from conduction (primarily venous return blood), surface skin convection, sweat evaporation, and radiation. In any clinical temperature measurement, the complete thermodynamic system in which the temperature measurement is performed, must be considered.

*Temperature and Environmental Effects on the Testis*
Edited by A. W. Zorgniotti, Plenum Press, New York, 1991

Most of the heat from the scrotal area is removed by venous return blood. The scrotal skin serves to provide direct dissipation into ambient through sweat evaporation, convection and to a lesser extent radiation and conduction. Convection is the process by which heat is transferred from a solid, such as the scrotum, to the surrounding air. Radiation is defined as the process by which energy is transferred from a higher temperature body to a lower temperature by means of electromagnetic waves and is termed "radiant heat". Conduction is energy transfer from a region of high temperature to low temperature within a medium or between media in direct contact.

## SYSTEM CONSIDERATIONS

When the testes and scrotum are analyzed as a system, the measurement technique may be considered as a variable added to the overall thermodynamic system. Any protocol involving temperature measurement, as well as the instrument itself, will influence the measurement and alter the reading. This is the thermodynamic equivalent to the Heisenberg Uncertainty Principle where the act of measurement changes the measurement itself. Unacceptable measurement errors may be introduced unless the effects of the procedure on the thermodynamic system are considered and the protocol carefully followed. With the following specific system considerations in mind, useful and repeatable clinical and research data can be obtained.

In analyzing a thermodynamic system, the importance between transient or steady state heat transfer cannot be minimized. In the analysis of heat transfer in the testis and scrotum, the modes of heat transfer must be considered as well as determining whether this process is steady or unsteady. Under steady state conditions, the rate of heat into the scrotal system must equal the heat efflux and the temperature <u>does not change</u> with time. Although this simplifies calculations considerably, it is unrealistic in the living organism. If the thermal environment changes due to internal effects (increase in blood flow or core temperature) or environment (exposure to cold, swimming), then the temperatures will be changing with time and transient analysis becomes necessary. The rate of change must include the masses involved as well as specific changes in heat flow. System analysis of the testis and scrotum include the following factors:

(1) Core Temperature: the amount of heat available to the testes comes primarily from arterial blood flow. If metabolic heat is ignored, the total heat available to be dissipated by the testes is the arterial heat brought in minus the venous heat return. Total heat input can be calculated from the temperature of the incoming blood, by the amount of blood flow and specific heat of blood. Clearly the amount of heat available is greatly influenced by the temperature of the spermatic artery blood, which is at core temperature. Thus, elevated core temperature will place a higher thermal load on the testis which, in order to maintain cooling, must dissipate the additional heat energy.

(2) Time Factors: The mass of the testis represents a storage element which cannot change temperature instantly but requires time to reach equilibrium. If that time factor, which in the case of the testis may be many minutes if not hours, is not considered then significant error is a likely result.

(3) Ambient Conditions: The testis must dissipate its heat energy into the ambient. The ability to perform this function is dependent on the temperature gradient between the testis and ambient. A large gradient means greater heat dissipation and cooler testis temperature, while high ambient temperature will result in a small gradient, less heat dissipation and elevated testis temperature. Ambient must also be considered along with time factors as in #2 above, since prior ambient exposure to cold, for example, may result in prolonged lower testis temperature, which may cause a clinician to miss a pathological temperature elevation.

(4) Use of Skin Preparation before Measurements: The use of any evaporative fluid placed on the scrotum before measurement would invalidate the measurement. Heat of vaporization of water, which at body temperature is about 520 calories/gram, would remove copious amounts of heat and this would take the body/testis system several hours

to restore. It is doubtful that any useful short term data could be obtained with any skin prep.

## OVERVIEW OF THERMOMETRIC TECHNIQUES

There appears to be no generally accepted procedure for measuring testis temperature which has achieved widespread acceptance. This may have discouraged testis temperature surveys of any but the smallest clinical populations. Based on thermodynamic principles, indirect measurement through the scrotal wall appears to offer the most accurate, acceptable and reasonably repeatable technique.

### Thermistors and Thermocouples

Direct measurement by placing an admittedly accurate thermistor or thermocouple directly into the substance of the testis is an unacceptable inconvenience and discomfort to the patient. The invasive nature of this procedure can alter the measurement. The patient must be anaesthetized and this has been shown to interfere with temperature readings (Waites, 1970). Required skin prep to the puncture site causes large heat exchange (losses) with the environment. As will be demonstrated, such heat losses take minutes if not hours to return to equilibration. Any local vascular interference caused by placing the sensor within the testis can also result in erroneous data. Thermistor placement is also crucial since the temperature at the anterior testis is lower than that at the mediastinum (posterior). Robinson (1968) partially solved the problem by inserting the tip of the thermistor bearing needle, adjacent to the upper pole of the testis, measuring intrascrotal temperature.

Measurement of the testis by direct contact of a thermocouple applied to the scrotal skin surface is not satisfactory. The measurement site is covered with the thermocouple and the adhesive. Furthermore, there is movement of the sensor with slight subject movement and intrinsic skin movement. These variables can give inaccurate readings (Phillips, 1934).

### Infrared Thermometers

Infrared (IR) thermometers and thermography reflect the temperature of the underlying organ (testis) by indirect infrared spectrum emission but such measurements have drawbacks. The spectral parameter being measured is a function of emissivity of the surface of the skin. Although currently available infrared thermometers can detect differences in temperatures of ±0.1°C, absolute temperature, evaluation is rendered inaccurate without a known skin temperature reference. We found that absolute measurements of intrascrotal temperature by a mercury immersion thermometer (vide infra) are substantially different from infrared surface emissivity readings. As different spectra are used, different emissivities require new baseline data. Relative readings are acceptable provided that new baseline data are taken both for the normal and pathologic state. These cannot be compared directly with readings taken on different IR equipment or with the mercury thermometer method. In addition to the above, electronic devices are subject to inaccuracies owing to drift. The most satisfactory device we have tested is the Barnes MT3 (no longer marketed) which is chopper stabilized against a reference, resulting in the accuracy and repeatability required for these measurements.

### Liquid Crystals

Liquid crystal films change color depending on contact temperature. The accuracy of this technique is subject to interpretation and at the present time (±0.5°C) is not sufficient for the measurement of small temperature differences (<0.5°C). These may prove useful as a screening technique in the future.

### Mercury Thermometer Invagination Technique

A method for non-invasive measurement of intrascrotal temperature evolved by Zorgniotti et al. (1973) appears to offer an acceptable and rapid method to obtain

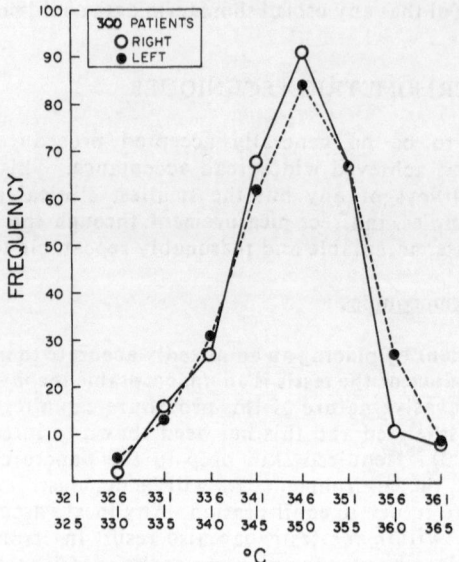

Figure 1. Temperature/frequency distribution curve of readings in 300 consecutive infertile patients.

thermometric data especially on a large scale. This procedure calls for the disrobed patient to equilibrate supine in a room at 21-23°C for six minutes. A mercury immersion thermometer is prewarmed to slightly above the temperature expected and then placed against the most prominent part of the anterior testicle and the loose scrotal skin gently drawn about it with the fingers so as to "immerse" the bulb. When the falling mercury column stops (in < 8 sec), indicating equilibration, the reading is recorded.

The method was checked in the operating room on patients undergoing scrotal surgery, where the bulb was inserted through a small incision, and also placed between the anterior testis and scrotal wall. The mean temperature was +0.13°C higher than mean temperature obtained by the external technique. We have made several thousand readings by this method. Figure 1 shows a temperature/frequency distribution curve made of readings in 300 consecutive infertile patients. The curves are symmetrical suggesting that the method and the cohort produce homogeneous results.

Deep Body Thermometry

A new measurement technique shows promise for testis temperature measurement, and is called DBT or Deep Body Temperature thermometry. Specifically, two plates are separated by a thermal insulator. Electric heat is then applied to the outer plate and the temperature difference between the two plates is measured by a built-in thermistor. As long as there is a difference in temperature between the two plates, a thermal gradient exists and heat continues to flow from the outer plate to the inner plate. Once the two plates reach the same temperature, there is no longer any temperature gradient between the plates and the temperature measurement below the skin surface (possibly 1.1 - 1.5 cm) is obtained.

This technique should make non-invasive deep body measurements possible and would appear to be an ideal instrument for testis measurements. The caution is that the instrument must create the thermal gradient by placing heat into the tissues until a null gradient is achieved. There is concern that the amount of heat energy entered into the system (small mass) may be sufficient to introduce an (positive) error into the measurement.

102

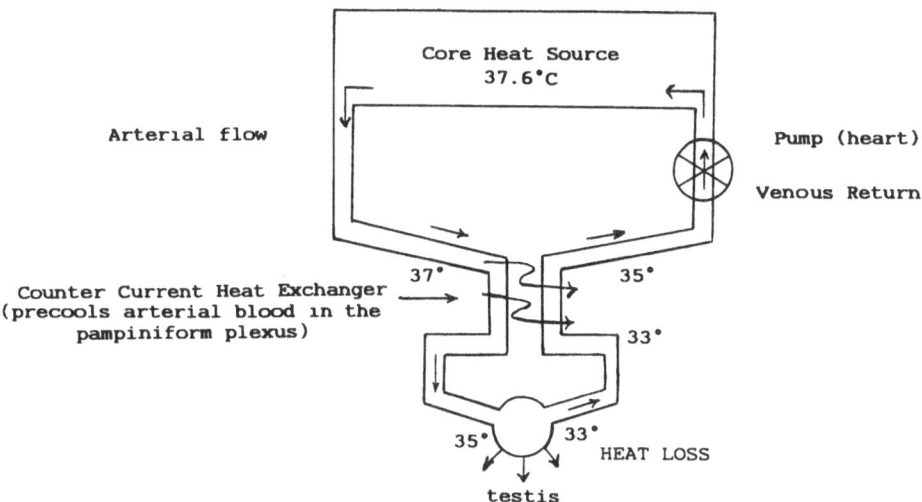

Figure 2. Fixed dissipator (scrotum). Heat loss is by radiation, evaporation and conduction.

## ANALYSIS OF VALIDITY OF NON-INVASIVE INTRASCROTAL TEMPERATURE METHOD

This is the system analysis used to determine the validity of the mercury bulb invagination technique for intrascrotal temperature measurement. To analyze this, a thermodynamic model of the testis must be considered and the relationship between heat flow, temperature and time determined. A thermodynamic model of the human testis is shown (Figure 2). This depiction represents the heat flows in and out of the testis including a countercurrent heat exchanger (explained later in the text). Figure 3 is an electrical equivalent of the thermodynamic model shown (Figure 2). Testis temperature is a function of the heat entering the testis by arterial flow, heat loss to the testis by venous return and heat loss through scrotal skin from conduction, radiation, convection

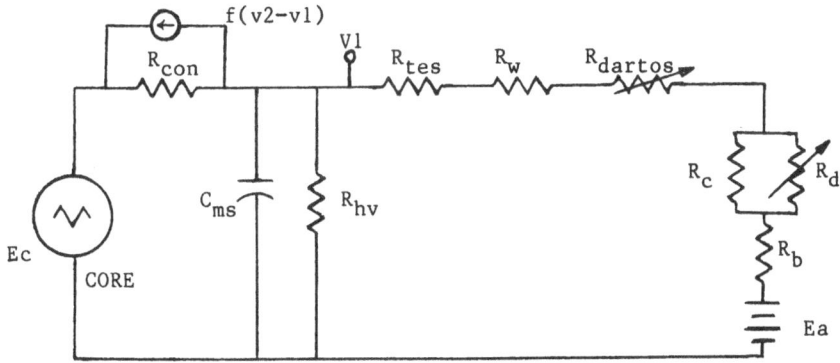

Figure 3. Open loop electronic equivalent system.
$V_1$ = Internal testicular temperature
$R_1$ = Countercurrent heat exchanger
$R_2$ = Complete network which includes $R_{tes}$ (gradient to surface of testis), $R_w$ (gradient through walls), $R_{dar}$ (gradient change with dartos) and $R_{ds}$ (convective and conductive losses at skin surface)
C = Heat storage of testis mass

and evaporation. *Based on the masses involved and subsequent small heat values, metabolic heat production in the testes is ignored in these calculations.* Heat flow takes the mathematical form of:

$$q = @A(t_1 - t_2) \qquad \text{(Eqn 1)}$$

This is a simple statement that heat runs downhill (i.e., from hot to cold) and is basically a statement of the second law of thermodynamics. The amount of heat gained or lost q, equals the differences of the temperatures (the driving force) times the surface area normal to the direction of heat flow, times the coefficient of heat of the material involved.

The testis itself represents a mass which takes time to gain or lose enough heat to effect a temperature change. This time factor takes the general form:

$$\frac{dT}{dt} = \frac{1}{tau}(T_0 - T_r) \qquad \text{(Eqn 2)}$$

This linear differential equation can be integrated to the form:

$$T = T_r + (T_0 - T_r)e^{-t/tau} \qquad \text{(Eqn 3)}$$

where $T_0$ is the temperature at the beginning, $T_r$ is the steady state temperature and tau is the time factor for the tissue element to change temperature. For example, if a subject's testes are exposed to a cold ambient for one hour, then placed into a 20°C room where testis temperature is measured over the next hour, how will the temperature increase? This is an analytical question which involves how much heat energy is available to the testis, what mass of the testis is, and other thermal losses as mentioned.

The model of the testis (Figure 2) can now be modified to include the mercury bulb thermometer and t = 0 after the six-minute equilibration period. The mass of the mercury bulb and the latent heat required to make the measurement is of several orders of magnitude below testis latent heat and can, therefore, be ignored. Similarly the natural heat generation of the cells of the testis can also be ignored since this energy is insignificant when compared to the core heat available from testis arterial flow. The effects of this temperature measurement can be analyzed by writing standard thermal equations around the testis model.

The objective is to find the solution to the equation of this network. To simplify the solution, the effects of the heat exchanger are simulated by a linear transfer function in which heat flow across the heat exchanger is a function of the mean temperature difference (from hot to cold). In electrical values, this is represented by a current flow which is linearly dependent on the voltage difference. For initial calculations, this linear transfer function is omitted until the end.

First, we must write the equation for the solution of the internal testicular temperature. To do this we must first recognize that the heat process can be represented by an electrical network. This network is shown in Figure 3. We apply Kirchoff's Current Law, which states that the algebraic sum of all currents leaving a node is zero at all instants of time. This is obviously a result of the law regarding the conservation of charge, which states that current cannot be created or sunk at a nodal point. The same concept can be applied to heat; heat cannot be generated or absorbed at a boundary or point. The heat that enters must leave and once the heat leaves, no more can come out.

Since we are interested in the nodal point, which represents testicular temperature, we will try to solve for $V_1$.

The current entering $V_1$ is: $(E_1 - V_1)/R_1$.

The currents leaving are: $C\frac{dV_1}{dt}$ and $\frac{V_1}{R_2}$.

The sum of these currents is zero:

$$\frac{E_1 - V_1}{R_1} - C\frac{dV_1}{dt} - \frac{V_1}{R_2} = 0$$

Grouping terms together:

$$C\frac{dV_1}{dt} = \frac{E_1}{R_1} - V_1\frac{(R_1 + R_2)}{R_1 R_2}$$

This yields:

$$\frac{dV_1}{dt} + V_1\frac{1}{C}\frac{(R_1 + R_2)}{R_1 R_2} = \frac{E_1}{CR_1}$$

This represents a nonhomogeneous linear differential equation with constant coefficients of the form:

$$\frac{dE}{dt} + PE = Q.$$

We can multiply every term in this equation by an integrating factor $e^{Pt}$ without changing its value.

This results in: $e^{Pt}\frac{dE}{dt} + PEe^{Pt} = Qe^{Pt}$,

the solution of which can be seen by recalling the derivative of a product:

$$d(xy) = xdy + ydx.$$

Relating this to our equation with $P = \frac{1}{C}\frac{(R_1 + R_2)}{R_1 R_2}$ and $Q = \frac{E_1}{CR_1}$,

this results in the following:

$$V_1 = e^{-(1/C)[(R_1 + R_2)/R_1 R_2]}\int\frac{E_1}{CR_1}\cdot e^{(1/C)[(R_1 + R_2)/R_1 R_2]} + Ke^{-(1/C)[(R_1 + R_2)/R_1 R_2]}$$

Recognizing that $e^{ax}dx = \frac{e^{ax}}{a}$

and by reducing terms, we get:

$$V_1 = \frac{E_1 R_2}{R_1 + R_2} + Ke^{-(1/C)[(R_1 + R_2)/R_1 R_2]t}.$$

The constant K must be solved by determining the initial conditions at t = 0. In an electrical circuit generally there are no voltages prior to turning on the system. However, in our thermal situation, there will exist a (non-zero) value of temperature before we begin to change things. Thus, our solution will look like this:

$$V_1 = T_i e^{-(1/C)[(R_1 + R_2)/R_1 R_2]t} + \frac{E_1 R_2}{R_1 + R_2}(1 - e^{-(1/C)[(R_1 + R_2)/R_1 R_2]t})$$

where $T_i$ = the initial starting temperature.

What does this equation mean at the initial conditions?

By setting t = 0, creates an "e" to the zero power which equals one. Thus, the second term becomes (1 -1) and is zero. The first term $e^0$ also becomes one and therefore sets $V_1$ to $T_i$, the initial starting temperature.

What happens at large values of t? (In other words, we do something and wait a long time for the system to settle down.) In this case "e" raised to a large negative number limits toward zero and the first term disappears. In the second term, the natural logarithm also disappears and we are left with:

$$V_1 = \frac{E_1 R_2}{R_1 + R_2}$$

which means the storage effect of the mass of the testis no longer affects the system and the resulting temperature is a function of the core temperature and the losses through the system. I believe this is in accord with common sense.

Unfortunately, $R_2$ is a twice removed simplification of a complex network, whose actual values would be a function of the other resistances in the network.

$$R_2 = \frac{R_3 (R_2 + R_D + R_4)}{R_3 + R_2 + R_D + R_4}$$

At this point a computer program was written to evaluate this expression. As a result of this analysis, the gradient between the testis and the skin is relatively small due to the small thermal mass of the skin, its lower blood flow rate, and its inability to substantially alter the heat flow as in Equation 1. The temperature *at the scrotal surface* is solely due to the starting temperature and subsequent heat flows. The equations also demonstrate that it is unlikely that other effects, such as nervous tension or local skin anomalies could achieve significant alterations in heat flow, with resulting temperature errors. In addition, since the measurement is performed within less than 8 sec, there is insufficient time for drastic errors to occur assuming such gradients could be shown to exist (see Equations 2 and 3). In other words, if one were to introduce some sort of error in applying the thermometer to the testis, there is not enough time for the heat to flow and alter the temperature being measured in < 8 seconds. (One could test this theory by holding the thermometer for a full five minutes and see if it changes from the original reading obtained at equilibration.)

## DIGRESSION TO EXPLAIN AN EXAMPLE TO SHOW HOW THE THERMODYNAMIC EQUATIONS RELATE TO THE ABOVE ANALYSIS

Assume that the testis has equilibrated for six minutes at some temperature, $T_1$. The mass of both testes is measured at 150 grams. This mass is primarily water with a coefficient of heat of 1 calorie per gram per degree Celsius. In other words, for the sake of a simple analysis, it would take 150 extra heat calories to change the testes 1.0°C.

$$Q = mc (T_2 - T_1) \qquad \text{(Eqn 4)}$$

This equation states that heat (Q) is a function of the mass (m) times the specific heat of the material times the difference in temperature rise or fall. Equation 4 may be written for any homogeneous system in the absence of work other than a change in volume ($W = {}_{v1}\int^{v2} pdV$). Thus if, for example, the combined mass of both testes is 150 grams and the specific heat of the testes is essentially that of water and we are looking for a 1.0°C change, Equation 4 becomes Q = 150 x 1 x 1 = 150 calories. In previous unpublished work, we have calculated that <u>total</u> net calories available to <u>both</u> testes is around 900 calories per hour. Thus, if the testes were suddenly cooled by placing them in an infinite water bath at, say 24.0°C, then the amount of heat required to get them back to 34.0°C would be:

$$Q = 150 \times 1 \times 10 = 1500 \text{ calories}$$

assuming a constant heat flow of 900 calories per hour (actually as delta T decreases, so does heat flow; see Equation 3), a return to "normal" (34.0°C) would require a minimum of 1.6 hours.

We have reported a mean difference of one degree (or greater) between normospermic and subfertile patients. Since intrascrotal temperature measurement by the immersion thermometer method is performed in < 8 sec, what heat gradient would be required to elevate the normal temperature 1.0°C? To simplify the calculations, let us assume a glass thermometer with a thermal conductivity "K" of 25 x 10$^{-4}$ cal/cm/°C (time rate of transfer of heat by conduction through unit thickness across unit area for unit temperature difference, i.e., calories per second per square centimeter with a delta T of 1.0°C). Going back to Equation 1 and modifying for thermal conductivity, we get:

$$Q = \frac{K (t_2 - t_1) aT}{d} \qquad \text{(Eqn 5)}$$

where a is the area, d is the thickness and T is the time.

Let us assume we place a glass thermometer with a glass thickness of 1 cm and one square inch of contact (a large surface area which equals 6.4516 cm$^2$). This thermometer is placed against the testis for 10 sec and placement of the thermometer causes a temperature error in the testis of 1.0°C, which requires the testis to absorb an additional 150 calories. Note that 10 sec is much longer than it actually takes to contact the testis for this measurement. With these assumptions the result is:

$$150 = \frac{25 \times 10^{-4} \times (t_2 - t_1) \times 6.4516 \times 10}{1}$$

thus, $t_2 - t_1 = 930$. If the testes were at 32°C to start, then in order to get both testes up one degree Celsius with a glass bulb mercury thermometer, you must have a contact size of 1 square inch, hold contact for 10 sec, and must maintain 962°C on the other side of the thermometer. If one were to argue only one testicle was changing, a reduction of the mass to 75 grams would require 481°C to create the error. (*One side cannot be heated independently* because of thermal gradients across the scrotal septum.) Therefore, one who argues that some "local" skin factor is causing a temperature increase, or the glass thermometer causes an elevated error, is not considering the thermodynamic laws which govern heat transfer!

## CONSIDERATIONS OF CORE TEMPERATURE AND SCROTAL RECTAL GRADIENTS IN THE STUDY OF TESTIS TEMPERATURE

Gradients based upon rectal temperature add another variable and, in our experience, tend to blur significance between mean temperatures of normospermic subjects and subfertile patients.

As seen in the thermodynamic model of the testis, core temperature is the initial value which through thermal losses and resulting temperature gradients produce a temperature less than core value at the testis. The value of core measurement is minimal since the apparent optimum temperature for spermatogenesis seems to be some absolute value and not related to core temperature. Analysis of the biochemistry of testis tissue slices and homogenates, often require reaction optima at an absolute temperature which is several degrees cooler than core temperature, and represent scrotal temperatures. Thus, while the core temperature value ultimately determines at what value the testes will equilibrate, the core value and resulting gradient serve no useful function since it appears that the absolute value is what determines chemical equilibrium that permits effective spermatogenesis. A patient with normal core temperature and elevated testis temperature has a small gradient, while a patient with elevated core and elevated testis temperature would have a "normal" gradient and both are likely to have subfertile semen owing to this temperature elevation.

The estimation of core temperature is not trivial. Rectal measurements were taken on several subjects, and their value as related to core is ambiguous. The depth of insertion, core equilibrium, and transient effects serve to obfuscate these numbers. Oral temperatures are worse and do not even correlate to rectal values at the higher temperatures. Core values were correlated to tympanic membrane temperature values to obtain reasonable core accuracy. This is beyond the scope of clinical evaluations and would serve no useful purpose other than to confirm that core temperature, even accurately determined, probably is irrelevant to the study of subfertile semen.

## USING SYSTEM ANALYSIS

The implications of system thermodynamic analysis as applied to scrotal temperature measurements can be demonstrated in a critique of a recent article by Kurz and Goldstein in which a correlation between scrotal skin temperature and intratesticular temperature was claimed. Far more interesting was their observation that scrotal temperature was apparently measured as being higher than intratesticular temperature, a finding that disagrees with previous investigators.

Since a difference of the skin temperature and testicle temperature or gradient exists, then a heat flow is developed as shown in the equations above. In the data presented by the authors, this difference defines a heat flow from the surface of the skin (in the direction of) into the testicle, which is contrary to logic.

Since the scrotal wall is said to be warmer than the testicle, heat flowing towards the testicle must be dissipated. The only means to remove heat in the center of the thermal mass would be via venous return flow. For heat to flow into the venous system it must be hotter and, therefore, must somehow be derived from the arterial circulation. Since previous understanding places the arterial flow posterior by the mediastinum, this does not agree with the anatomy of the arteries to the testis. Since ambient temperature is lower than scrotal, heat would have to flow in two directions, into ambient and into the testicle, which is not possible.

If the measurement of temperature reflects a steady state or if the system is dynamic and attempting to reach equilibrium, evaluating temperature on a minute by minute basis or allowing six minutes of equilibration may not be sufficient to reach thermodynamic equilibrium. (The temperature curve as the equations above show is logarithmic; therefore, assigning equal time intervals as in minute by minute reading would make readings appear to stabilize, while logarithmic time intervals would suggest a different interpretation of equilibration.) In the steady state condition, the body would be supplying a constant flow of heat in the form of arterial blood flow (volume, temperature and venous return flow); the losses at the scrotum due to conduction, radiation, convection and sweating all reach equilibrium; and a steady state temperature results at the testis, which is reflected at the scrotum.

In the dynamic state, something has upset the steady state condition by adding or removing heat and the system logarithmically adjusts to the new condition, while attempting to reach a new steady state temperature. Kurz apparently measured scrotal skin, then after the induction of anesthesia the temperature of the unexposed unprepared skin was measured every minute until 3 consecutive identical measurements were recorded. Then a dry scrotal shave was performed and scrotal skin again measured. Finally, a povidone-iodine scrub and air-drying preceded the insertion of a 29 gauge needle thermistor to read intratesticular temperature. From the data presented, both scrotal and testis temperature continue to decrease throughout these procedures. This led to the conclusion that intratesticular temperature was warmer than scrotal temperature and that scrotal shaving decreased testis temperature by some 0.34°C (right) and 0.99°C (left). Suppose the shaved hairs could be reattached immediately after shaving and taking this measurement; would the temperature rise by this same amount?

Although obviously this suggestion is given tongue in cheek, it is meant to call attention to two important issues: the first is that the system may not have been in equilibrium in the first place and, therefore, the decrease in temperature occurred as a natural consequence of trying to reach equilibrium with the added thermal load of shaving and handling; secondly, a hysteresis factor may be in effect whereby the testicle appears not to return to its initial starting temperature after all these procedures done to it are reversed.

The failure to consider a complete thermodynamic system analysis can lead to difficulties in data interpretation. Brindley implanted a radio telemetry device within the scrotum which permitted deep scrotal temperature measurements under a wide range of different physical activities and conditions. One of his experiments consisted of running 1900 meters in 8-1/2 minutes in 5 - 6°C ambient air, then sitting for approximately an hour while recording temperature. The purpose of this experiment was to determine the effects of different sitting positions and different underwear. His results indicate that as ambient air decreases, deep scrotal temperature also decreases. He also noted that the greater the insulation of the underwear, the higher the temperature.

After running 1900 meters in 8 minutes, the core temperature has undoubtedly increased as did blood flow and pressure. By exposing the scrotum to the 6°C ambient temperature in England in February, the countercurrent heat exchanger was able to substantially precool the testes which is in accord with results from our computer model. Remaining stationary for 60 - 100 minutes after exercise (see Brindley's Fig. 5) the equilibration result becomes undefined. Core temperature and blood flow would decrease (less cardiovascular output while sitting), thus less heat inflow is possible while ambient temperature increased from the 5 - 6°C to 20°C. The time factor to reach equilibrium after these effects become steady state must be considered. Due to the dynamic nature of the data obtained with different variables changing in opposite directions, trend analysis becomes difficult and conclusions based thereupon, may be erroneous.

CONCLUSION

In the measurement of testis temperature careful consideration must be given to the entire thermal system and all measurements must be analyzed from that system perspective. Time factoring and transient analysis applied to these measurements may insure repeatable, significant data. However, if the data derived appear to violate the laws of physics as they are presently understood, the likelihood is that the data are suspect and that linear regression will not compensate for regressive analysis.

REFERENCES

Andrews, F.N. 1940. Thermo-regulatory function of rat scrotum. I. Normal development of effect of castration. Proc. Soc. Exptl. Biol. Med., 45: 867.

Bazett, H.C., Love, L., Newton, M., Eisenberg, L., Day, R. and Foster, R., II. 1948. Temperature changes in blood flowing in arteries and veins in man. J. Appl. Physiol., I: 3.

Brindley, G.S. 1982. Deep scrotal temperature and the effect of clothing, air temperature, activity, posture and paraplegia. Br. J. Urol., 54: 49.

Hammel, H.T. 1968. Regulation of internal body temperature. Ann. Revl. Physiol., 30: 641.

Hardy, J.D. 1961. Physiology of temperature regulation. Physiol. Rev., 41: 521.

Harrison, R.G. and Weiner, J.S. 1949. Vascular patterns of the mammalian testis and their functional significance. J. Exp. Biol., 26: 304.

Jequier, E. 1986. Human whole body direct calorimetry. IEEE Eng. Med. Bio., 5: 12.

Kandeel, F.R. and Swerdloff, R.W. 1988. Role of temperature in regulation of spermatogenesis and the use of heating as a method for contraception. Fertil. Steril., 49: 1.

Kurz, K.R. and Goldstein, M. 1986. Scrotal temperature reflects intratesticular temperature and is lowered by shaving. J. Urol., 135: 290.

Moule, G.R. and Knapp, B. 1950. Observations of intra-testicular temperatures of Merino rams. Aust. J. Agric. Res., 1: 456.

Phillips, R.W. and McKenzie, F.F. 1934. The thermoregulatory function and mechanism of the scrotum. Univ. Mo. Agr. Expt. Sta. Res. Bull., 217: 1.

Riemerschmid, G. and Quinlan, J. 1941. Further observations on the scrotal skin temperature of the bull, with some remarks on the intratesticular temperature. Onderstepoort J. Vet. Res., 17: 123.

Strandness, D.E., Jr. and Sumner, D.S. 1968. Venous hemodynamics and control of venous capacity in venous valvular incompetence. In: "Hemodynamics for Surgeons", Grune and Stratton, New York.

Waites, G.M.H. 1970. Temperature regulation and the testis. In: "The Testis I", A.D. Johnson (ed), Academic Press, New York.

Waites, G.M.H. and Setchell, B.P. 1964. Effect of local heating on blood flow and metabolism in the testis of the conscious ram. J. Reprod. Fertil., 8: 339.

Zorgniotti, A.W. and Sealfon, A.I. 1988. Measurement of intrascrotal temperature in normal and subfertile men. J. Reprod. Fertil., 82: 563.

# NON-INVASIVE SCROTAL THERMOMETRY

Adrian W. Zorgniotti

Department of Urology
New York University School of Medicine

A key factor in testis temperature investigation is thermometry. The insertion of thermocouples and thermistors into the substance of the testis or into the scrotum goes back to the 1920's. As this is invasive, large scale studies are difficult owing to refusal on the part of volunteers and even patients to participate. Thermistors are also not ideal for reasons which affect the accuracy of readings: e.g., use of anaesthesia (Waites, 1970), evaporation of liquid applied to the scrotum (skin preparation) (Zorgniotti et al., 1980), and temperature may vary with depth of placement since we know that temperature is higher at the mediastinum testis than peripherally.

Brindley (1982) implanted a sensor between the two testes of spinal cord injured patients. This could be externally interrogated to obtain temperature readings, a procedure which might encounter difficulty obtaining the approval of a Human Experimentation Committee in the United States. Brindley, it might be added, did not hesitate to have himself implanted.

In order to solve the problem of surveying large cohorts, a non-invasive method for intrascrotal temperature measurement was evolved by Zorgniotti and MacLeod (1973) largely because infrared thermometers, available at that time, were not sufficiently accurate to differentiate what they suspected were temperature differences of less than 1.0°C between normal and oligospermic men. This non-invasive method is readily accepted by the subject so that there is not even a need for patient consent in the diagnostic setting. Readings are made by drawing the loose anterior scrotal skin about the bulb of a mercury thermometer placed against the scrotum overlying the anterior testis. Brindley (op cit) modified this technique by placing the thermometer at the midline between the testes and called it the "invagination" method.

Instrument accuracy is not in question since the mercury thermometer measures absolute temperature and is calibrated to United States Bureau of Standards with divisions = 0.05°C. To test the validity of the method, Zorgniotti and MacLeod (op cit) measured temperatures by inserting the thermometer via a small scrotal incision and also externally, yielding a difference between the internal and external mean temperature of 0.1°C. Insertion of a mercury thermometer at operation is not new as Badenoch (1945) and Phillips and McKenzie (1937) have measured temperature by this means. Intrascrotal readings obtained by this non-invasive method reflect testis' temperature since the testis and epididymis make up the largest thermal mass in the hemiscrotum. Factors which alter the reading (scrotal skin temperature, scrotal skin circulation and heat coefficient of the glass and mercury prewarmed thermometer) are negligible by comparison.

*Temperature and Environmental Effects on the Testis*
Edited by A. W. Zorgniotti, Plenum Press, New York, 1991

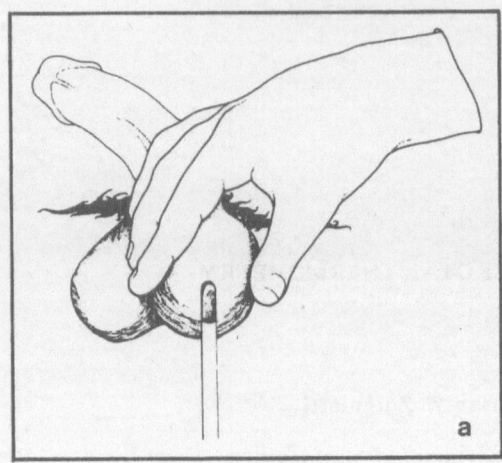

Figure 1a.  Non-invasive method for intrascrotal temperature measurement.  The calibrated laboratory thermometer (range: 32 - 40°C; divisions 0.05°C) is placed against the scrotum overlying the testis.

Three hundred consecutive men who presented for infertile marriage and abnormal semen had intrascrotal temperature estimated by A.W.Z. or by a Physician's Assistant. The subject was disrobed and placed supine for 6 min in an ambient of 21 - 23°C. Zorgniotti and MacLeod (op cit) and Brindley (op cit) found that a six minute equilibration period was sufficient to obtain readings which are suitable to the clinical setting.  The stubby thermometer has a range of 32 - 40°C with divisions of 0.05°C (Model No. C1148: Brooklyn Thermometer Company, Farmingdale, NY 11735, U.S.A.). Clinical thermometers, because of constriction in the mercury column, are less satisfactory for this purpose.

In order to speed temperature estimation, the thermometer is prewarmed to around 37.0°C by placing the bulb in contact with an electric light or immersion in warm water. The mercury bulb is then quickly placed against the scrotum over the most prominent part of the anterior testis.  The scrotal skin is then gently drawn about the bulb, invaginating it completely. Some thermometers have an "immersion line"; this should be covered by skin. When the falling mercury column reaches equilibrium, usually in about 8 sec, the intrascrotal temperature is recorded.  The procedure is then repeated for the contralateral testis (Figure 1 a,b). Measurements in men who, in cold and windy weather

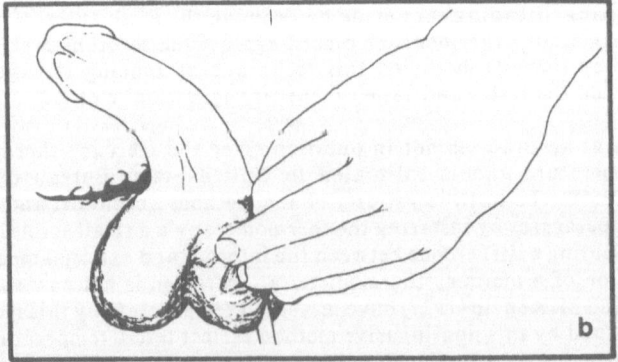

Figure 1b. The skin is wrapped around the bulb of the thermometer. When equilibrium is reached, this measures the intrascrotal temperature.

112

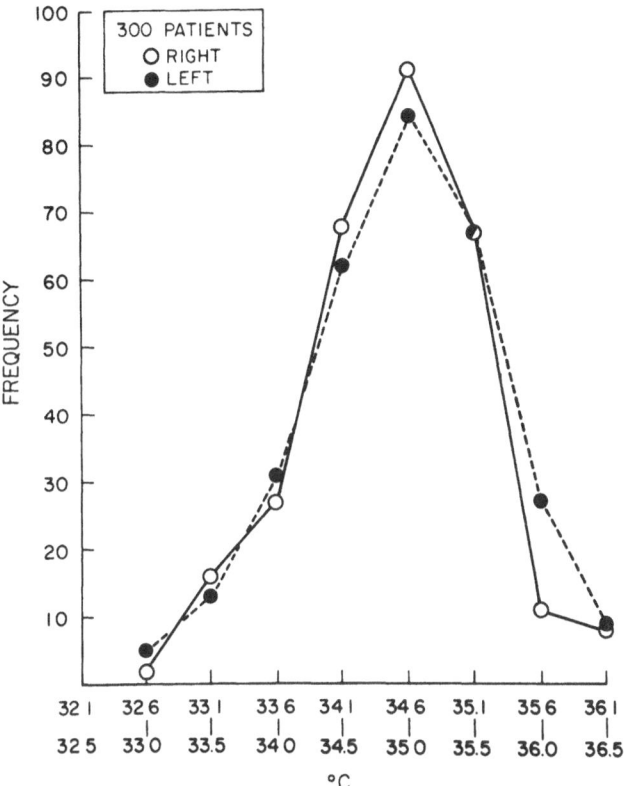

Figure 2. Symmetrical distribution curves of right and left intrascrotal temperatures from 300 infertile patients.

present wearing cotton trousers (jeans), may be inaccurate owing to chilling of the testes. A long period of equilibration to indoor temperature is necessary and it is usually easier to ask the subject to return on another milder day, wearing woolen pants.

Frequency distribution for right and left intrascrotal temperatures were plotted (Figure 2). The narrow, symmetrical peaks obtained suggests that the results are homogeneous. Mieusset (1987) used this method in his series of 187 men and concluded that the mean variation between two readings was 0.08°C. The invagination technique, using a glass thermometer is inexpensive and gives accurate intrascrotal temperature estimations (Table 1).

Table 1. Mean normal readings by scrotal invagination - °C

|                          | Mean | SD  | Range       |
|--------------------------|------|-----|-------------|
| Zorgniotti and MacLeod*  | 33.5 | 0.6 | 32.1 - 34.8 |
| Mieusset**               | 34.6 | 0.5 | 33.6 - 35.6 |
| Brindley***              | 35.3 | N/A | N/A         |

*Volunteers mean 170 mil/ml (101 -279 mil/ml)
**Fertiles mean 82 mil/ml (5 - 500 mil/ml)
***"Normals" otherwise unspecified

Table 2. Normal and pathologic temperature (°C) readings by infrared thermometry

|  |  | Mean | SD | Range |
|---|---|---|---|---|
| Normal volunteers | R | 31.8 | 1.0 | (30.7 - 32.8) |
|  | L | 32.2 | 1.7 | (31.0 - 33.6) |
| Varicocele patients | R | 33.1 | 1.4 | (30.5 - 34.3) |
|  | L | 33.2 | 1.4 | (30.7 - 34.3) |

Differences significant by t test

## INFRARED THERMOMETRY

Measuring emissivity of the scrotal skin by non-contact infrared thermometry and thermography reflects the temperature of the underlying testis (Comhaire, 1986). Because of problems inherent in electronic devices as well as variations in skin emissivity related to degree of pigmentation, none of the infrared thermometers currently on the market that were tried have the sensitivity (±0.1°C) needed to discriminate small differences. Infrared non-contact thermometers have the advantage of producing an almost instantaneous reading. Similarly, liquid crystal thermometry is not sufficiently sensitive and probably should not be used if we are trying to discriminate differences of 0.5 - 1.5°C which represent the difference between normal and elevated testicular temperature.

In spite of these difficulties, Zorgniotti et al. (1979) published data on temperatures obtained with a Thermal Master IT-4M infrared thermometer manufactured, at that time, by Barnes Engineering of Stamford, CT. The instrument is no longer available but it is still in use and possesses the requisite accuracy ±0.1°C to give reliable readings of skin surface temperature (Table 2).

## REFERENCES

Badenoch, A.W. 1945. Descent of the testis in relation to temperature. Brit. Med. J., II: 601.

Brindley, G.S. 1982. Deep scrotal temperature and the effect on it of clothing, air temperature, activity, posture and paraplegia. Brit. J. Urol., 54: 49.

Comhaire, F. 1986. Varicocele and its role in male infertility. Oxford Rev. Reprod. Biol., 8: 165.

Mieusset, R. et al. 1987. Association of scrotal hyperthermia with impaired spermatogenesis in infertile men. Fertil. Steril., 48: 1006.

Phillips, R.W. and McKenzie, F.F. 1934. The thermoregulatory function and mechanism of the scrotum. In: Missouri Univ. Agricultural Experimental Station Research Bulletin, #217: p. 12.

Waites, G.M.H. 1970. Temperature regulation and the testis. In: "The Testis", A. Johnson et al. (eds.), Academic Press, New York, vol. I, pp. 241.

Zorgniotti, A.W. and MacLeod, J. 1973. Studies in temperature, human semen quality and varicocele. Fertil. Steril., 24: 854.

Zorgniotti, A.W. et al. 1979. Infrared thermometry for testicular temperature determinations. Fertil. Steril., 32: 347.

Zorgniotti, A.W. et al. 1980. Chronic scrotal hypothermia as a treatment for poor semen quality. Lancet, 1: 904.

# DEEP BODY INTRASCROTAL THERMOMETER:

# THEORY AND METHODOLOGY

Hiroshi Takihara, Masatoshi Yamaguchi,
Yoshikazu Baba and Jisaburo Sakatoku

Department of Urology
Yamaguchi University School of Medicine
Ube, Yamaguchi, Japan

## INTRODUCTION

The deep body thermometer, which was originally devised by Fox and Solman[1,2] in England, is a new device for the measurement of body temperature. At present, the device has been improved upon and is now being evaluated for various clinical applications. In this paper, the principle of the deep body thermometer will be described and the application for the screening of varicocele will be introduced.

### The Principle of the Deep Body Thermometer

In principle, the deep body thermometer applies a method in which a certain region of the body surface is covered by a heat insulating material in order to prevent cooling from the external atmosphere. The temperature of the body surface is then equalized to that of the deeper region. That is, covering an area of the body surface with a heat insulating material leads to a reduction of heat diffusion, so that the temperature of the covered body surface becomes higher than that of the body surface exposed to the external air. It is speculated that if heat insulation is complete and the area is of sufficient size, outward heat current will reduce to almost zero even at tissues near the body surface and, consequently, the difference in temperature between the deep region and the surface will also approach zero. However, the thermal conductivities of the actual heat insulation materials are only about one-tenth of the skin thermal conductivity[3]. Thus, Fox and Solman[1,2] introduced the zero-heat-flow method to achieve an almost complete thermal insulator.

The temperature measuring probe has two thermistors separated by a thermal insulator, with an electrical heating element placed at the rear of the probe. The temperature on either side of the insulating layer is compared by a differential amplifier. The heater temperature is controlled in such a way as to achieve a situation in which no temperature gradient arises across the insulating layer and thus the heat current will be always zero regardless of the quality of the insulating material. Such control which maintains the heat current at zero, is known as the zero-heat-flow method or the thermal flow compensation method. As long as the zero-heat-flow condition is maintained, the probe is equivalent to an ideal thermal insulator. It prevents heat loss from the skin surface beneath the probe and the skin surface temperature will equilibrate with the deep tissue temperature. This could be measured by the lower thermistor in contact with the skin (Fig. 1).

*Temperature and Environmental Effects on the Testis*
Edited by A. W. Zorgniotti, Plenum Press, New York, 1991

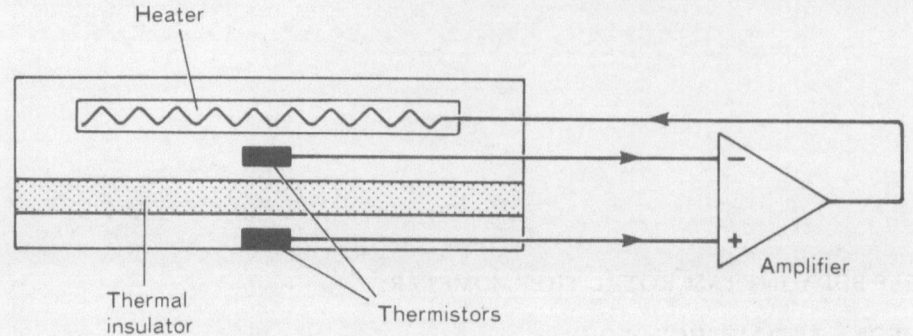

Figure 1. Schematic diagram showing a cross-section of the deep body thermometer probe, and control circuit (Fox et al., 1973).

However, it became evident that when thermometry was actually conducted with this method, the zero-heat-flow was not always completely obtained. After an investigation of this problem, it was found that heat escaped from the external edges of the probe. Thus, the temperature tended to be lower away from the center. Even though the center could be kept at zero-heat-flow, the heat current from the center toward the external edge was not prevented[4]. Thus, the heat insulating material was surrounded by a metal guard of good heat conductivity in order to maintain equal temperature between the circumference and the center.

Figure 2 shows the new model, the metal guard type probe improved by Togawa[4]. This improvement enabled the prevention of heat current toward the external circumference[5]. The efficacy of Togawa's improvement became evident through theoretical analysis and through clinical comparison between the old and new model[5].

Structure of the Deep Body Thermometer

The probe improved by Togawa is available from Terumo Corporation which is a disk-shaped probe. It is surrounded by an aluminum lining which is attached to the body surface. Rubber foam is used as heat insulating material, the elasticity of which enables the central sensor to attach closely to the skin. The thermistor detects temperature ranging from 30 to 40°C, or 20 to 45°C, with an accuracy of ±0.1°C.

Monitoring of Temperature and Accuracy

Various investigations have been conducted to evaluate whether the deep body thermometers can accurately measure temperature and rapid fluctuations of temperatures

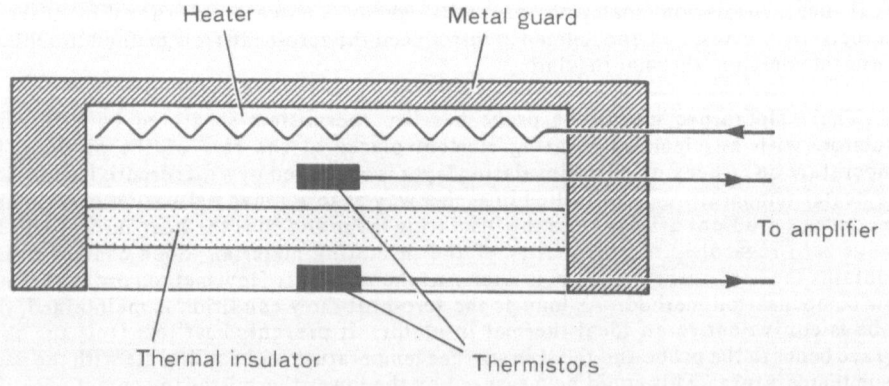

Figure 2. Schematic diagram of the probe with a guard (Togawa et al., 1973).

in the deep areas. The measurement of muscle temperature using a deep body thermometer has been studied from this point of view by Togawa[6] in 1976. A thin thermocouple wire was inserted into the biceps muscle of the subject approximately 10 mm from the skin surface, and the deep body thermometer probe was placed just above the thermocouple tip. After the temperature had stabilized, the subject was requested to support a 1.5 kg weight by the palm of the hand for about 10 min. Then, the temperature measured by the deep body thermometer rapidly followed internal temperature changes as measured by the thermocouple.

## Effect of the External Atmosphere

Theoretically, the deep body thermometer is equivalent to covering the body surface with a material of almost complete heat insulation and should in principle be unaffected by the eternal air temperature. It is an important feature of the deep body thermometer that ambient temperature changes do not affect the temperature sensor mounted on the probe. In fact, it has been confirmed that the deep body thermometer is hardly affected by external air temperature, from the results of simultaneous measurement by both methods of the skin temperature near the probe attached to the forehead. When the body was exposed to cold atmospheric air of about 14°C, although the skin temperature was greatly reduced, there was hardly any change in the deep body temperature[7].

As described, the deep body thermometer differs from the conventional body thermometer in principle and is expected to be able to obtain information about body temperature more easily. Its significant characteristics are that it can be easily attached

Room Temperature (24-25°C)

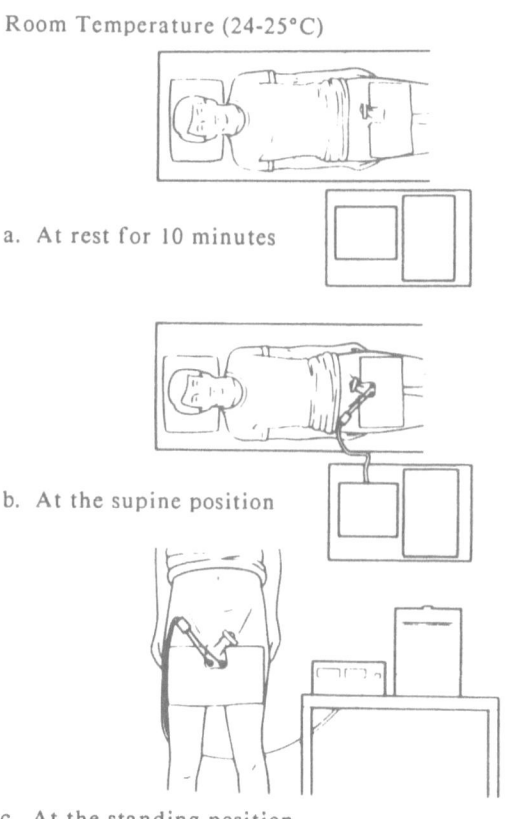

a. At rest for 10 minutes

b. At the supine position

c. At the standing position

Figure 3. Procedure of intrascrotal deep body temperature measurement.

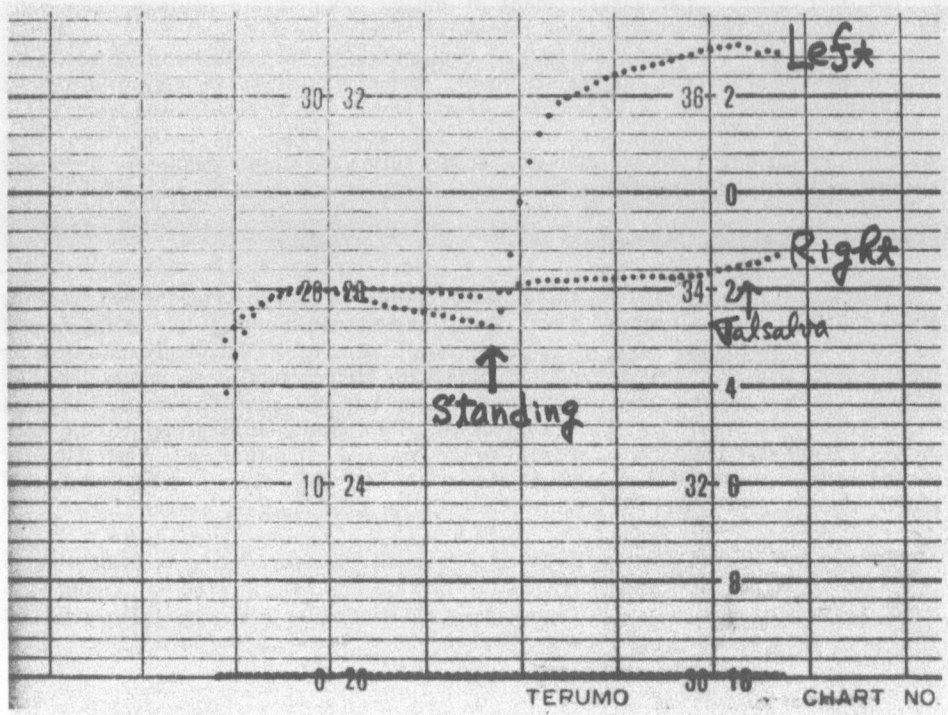

Figure 4. Intrascrotal deep body temperature in patient with left large varicocele.

to the body and is unlikely to be affected by external air temperature. It can therefore be applied to various medical fields. However, there has been no report of its application to deep body temperature for the diagnosis of varicocele.

### Methodology of Deep Body Temperature Measurement for the Non-Invasive Diagnosis of Varicocele

We attempted to use the deep body temperature measurement for the diagnosis of varicocele. The measurement was performed using Terumo DBT system. This system consists of a deep body temperature monitor Coretemp CTM 204 and Terumo Recorder TFR 102. The deep body temperature can be measured continuously by a thermal flow compensation probe PD 7, which is attached to the scrotal skin just above the testicles and fixed by adhesive tape. All patients were asked to expose their genitalia for 10 minutes while supine. Intrascrotal temperatures were recorded bilaterally in the supine position for 15 minutes and then monitored continuously with the patients standing at an environmental temperature of 24 to 25°C (Figure 3).

The temperature measured can be printed on the sheet simultaneously. In patients with left varicocele, the increase in temperature of the left side was noted (Fig. 4). Thus, the difference in temperature due to the postural change from supine to standing position of the left side was considered to be the useful index for the non-invasive diagnostic screening of varicocele.

REFERENCES

1. R.H. Fox and A.J. Solman. 1970. A new technique for monitoring the deep body temperature in man from the intact skin surface. J. Physiol., 218: 8.

2. R.H. Fox, A.J. Solman, R. Issacs, A.J. Fry and I.C. McDonald. 1973. A new method for monitoring deep body temperature from the skin surface. Clinical Science, 44: 81.

3. M. Lipkin and J.D. Hardy. 1954. Measurement of some thermal properties of human tissue. J. Appl. Physiol., 7: 212.

4. T. Togawa. 1973. Medical thermometer making use of zero heat flow method. Rep. Inst. Med. Dent. Engng., 7: 75.

5. T. Kobayashi, T. Nemoto, A. Kamiya and T. Togawa. 1975. Improvement of deep body thermometer for man. Ann. Biomed. Engng., 3: 181.

6. T. Togawa, T. Nemoto, T. Yamazaki and T. Kobayashi. 1976. A modified internal temperature measurement device. Med. Biol. Engng., 14: 361.

7. T. Togawa, T. Nemoto, T. Tsuji and K. Suma. 1976. Circulatory monitoring by an improved deep body thermometer. Digest 11th Int. Conf. Med. Biol. Engng., 40.

SECTION 4

TEMPERATURE PATHOPHYSIOLOGY OF THE TESTIS

# A THEORETICAL MODEL FOR TESTIS THERMOREGULATION

Andrew I. Sealfon and Adrian W. Zorgniotti

Repro-Med Systems, Inc.
Middletown, NY 10940

## ABSTRACT

Studies going back as far as the early 1920's show that there is a clear relationship between testis temperature and semen quality. The most intriguing question is whether there is a mechanism of thermoregulation which, in the human, maintains testis temperature within certain limits that permit euspermia. Thermoregulation is defined as maintaining some specified (optimum?) temperature plus or minus an error over internal and ambient loss factors. A computer Model has been evolved which contains no regulation or feedback. It appears to predict human testis temperature data gathered in earlier studies. The Model accounts for countercurrent heat exchange in the pampiniform plexus and predicts, with an open loop analysis, that there is no feedback or regulation. As far as thermoregulation is concerned, there appear to be no first-order effects taking place in the human testis. This suggests that ambient temperature changes cause corresponding changes in testis temperature. Also any internal changes in thermal properties such as core temperature variations or variability of the countercurrent heat exchanger will also cause temperature change.

Testis temperature as predicted by this Model is the result of the heat energy entering the testis from arterial inflow minus the venous outflow and heat loss from the scrotum. The Model predicts that the heat exchanger will function to provide precooling of arterial blood as external temperatures drop but will fail to precool effectively as temperature rises. This is predicated on the fact that the countercurrent heat exchanger becomes less effective as the temperature gradient across the exchanger becomes smaller and less heat energy is able to be transferred from arterial flow to venous flow. The Model also predicts that any diminution of the heat exchanger mechanism either from reduced venous flow or restricted scrotal heat loss will result in higher testis temperature. Lastly, the Model correctly predicted that febrile patients would experience elevated testis temperature during periods of elevated core temperature. Since more heat energy is available in the arterial blood, more heat energy is delivered to the testis under these conditions. In the human it appears that any internal or external factor causing a temperature change will not trigger or activate a feedback mechanism to control the resulting testis temperature. A major factor in subfertile semen may be the inability to check excessive temperature of the testis which impairs the ability to produce and mature fertile spermatozoa.

## STEADY STATE TESTIS TEMPERATURE

Testis temperature is a function of arterial heat entering the testis modulated by countercurrent heat exchange in the pampiniform plexus; heat loss by venous return and

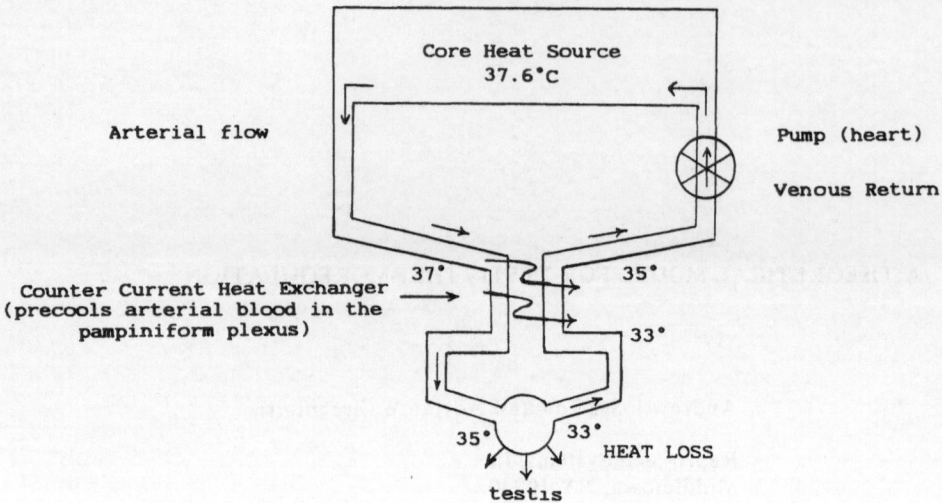

Figure 1. Fixed dissipator (scrotum). Heat loss is by radiation, evaporation and conduction.

heat loss through scrotal skin from conduction, radiation, convection and evaporation (Figs. 1 and 2). There is an equivalence between thermodynamics and electronics which permit the writing and solution of basic equations by electronic circuit analytic techniques. Thermal loss can be compared to current loss from electrical resistances. Heat storage follows the same equations as electron (current) storage in capacitance. An electronic network which simulates the thermodynamic properties of the testis can be used to determine transient and steady state temperatures of the testis.

OPEN LOOP (NON-FEEDBACK) MODEL (Figure 2)

In this example core heat is considered as a voltage or constant (core) temperature source. Heat losses are resistances, heat flow is conductance, heat storage is capacitance and the countercurrent heat losses are considered as an active linear network whose current (quantity of heat) is inversely proportional to the nodal voltage (temperature) for mathematical convenience. The assumption in this representation is that the resulting testis temperature is primarily the result of heat brought in by arterial flow, with heat loss via the venous return and loss through the scrotal skin being essentially fixed.

Basic electronic equations can be written which are directly related to the heat flow characteristics of the Model. The mathematical result is a nonhomogeneous linear differential equation with constant coefficients whose solution is seen of the form:

$$de/dt + PE = Q$$

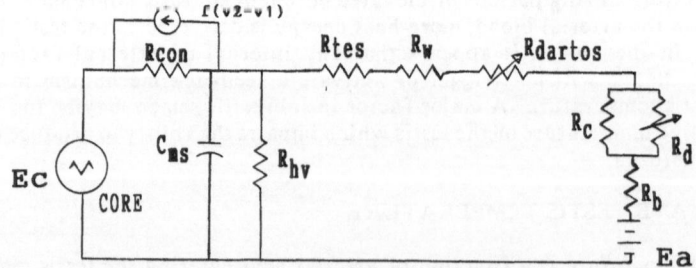

Figure 2. Open loop electronic equivalent system.

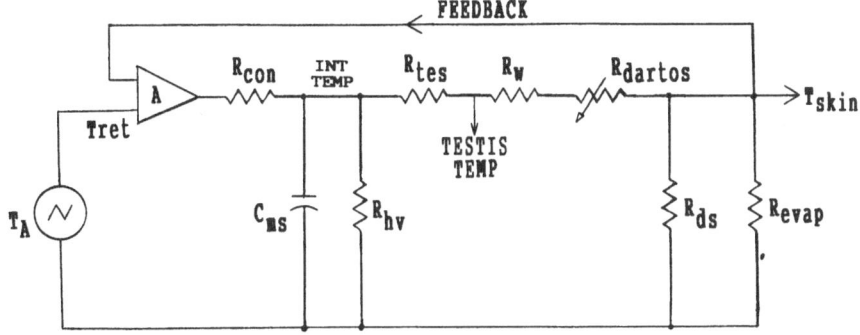

Figure 3. Closed loop model.

These mathematics are covered in a previous chapter. This equation can be programmed on the computer to obtain heat flow characteristics and final temperatures. The computer can be asked the following:

(1) How does temperature rise after discontinuance of scrotal cooling?

(2) What is the rate of scrotal cooling during equilibration to ambient upon disrobing?

(3) What happens when core temperature rises and all else remains constant?

CLOSED LOOP MODEL (Figure 3)

This Model which provides significant thermoregulatory capability would be able to stabilize testis temperature at some desired value (plus or minus an error). In order to achieve temperature stability, some means of altering the main heat flow in or out of the scrotum would be required. The difference between the desired output and the actual output represented by the system input is termed the "error". A closed loop system is actuated by such an error signal:

Error = desired output - actual output

A block diagram is shown (Figure 4). To gain thermodynamic regulation, the body would have to sense testicular temperature and using feedback, control the heat into or out of the testis. The mass of the testis would influence the rate of change. In a testis autoregulatory model, this would be accomplished by temperature sensing and feedback control of the heat going in or out of the testis. From the point of view of purely thermodynamic control, this regulation might best be accomplished in the countercurrent heat exchanger where small controlling inputs would influence large amounts of heat exchange. Other possibilities could include shunts between arterial and venous systems, controlled by some type of valve mechanism. Control of scrotal heat output by alterations in the scrotal surface by the dartos or cremaster muscles do not appear to be very effective thermodynamically since the total difference between maximum and minimum heat dissipation would be many times less than heat changes created by

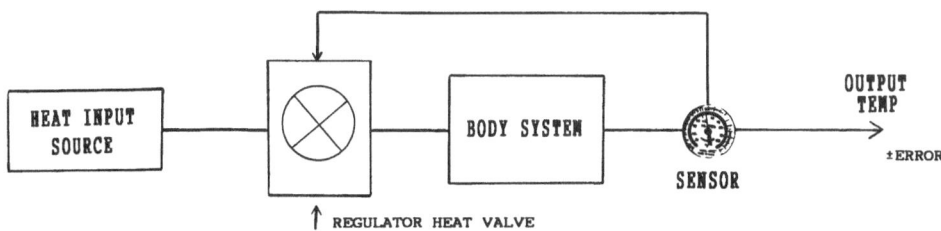

Figure 4. Block diagram of autoregulatory thermoregulator (closed loop) system.

ambient conditions or the large changes in heat availability in arterial blood, the venous return and the effects of the countercurrent heat exchanger.

## RESULTS

Open loop computer simulation closely agrees with data obtained from humans. Although much has been written about physiological mechanisms which control testicular temperature, this is not borne out by the open loop Model or data in humans. If such mechanisms exist, then their objective would be to stabilize testicular temperature at some optimum temperature value suitable for spermatogenesis. This implies that some specific temperature is maintained regardless of wide fluctuations in ambient or core temperature.

## DISCUSSION OF THE MODEL

Once the system equations were solved, a computer program was written to calculate testis and scrotal temperatures under various internal and environmental thermal conditions. The open loop thermodynamic Model demonstrates that testis temperature and scrotal temperature are closely interrelated. The circuit diagram (Fig. 2) indicates that the gradient between testis temperature and scrotal temperature is simulated by a relatively small resistance which implies good thermal conductance, and that the temperatures would remain very close to each other, within a few tenths of a degree Celsius. If this Model proves to be a close simulation of humans, this would provide additional confirmation for the belief that scrotal temperature measurement reflects the underlying testis temperature adequate for clinical measurements.

The most striking aspect of this Model is the interrelationship between scrotal heat dissipation and the countercurrent heat exchanger which form the basis of testis temperature. The Model predicts that only when the scrotum is able to dissipate heat will the heat exchanger be effective to precool arterial blood. The countercurrent heat exchanger is the dominant factor. The Model predicts that temperature of the testis is not at all stabilized but changes with ambient conditions. Specifically, as the ambient temperature rises, heat flow is reduced and the scrotum is not able to dissipate sufficient heat. This results in a rise in venous temperature and loss of precooling of the arterial inflow. (The electrical equivalent is a rise in resistance producing less current flow and therefore less "voltage" drop across the countercurrent heat exchange "Rcon".) As the ambient temperature decreases, increased dissipation by the scrotum will occur and will be reflected by greater precooling in the heat exchanger. Thus, testicular temperature will decline sharply, not indicative of temperature stability. This Model result is in complete accord with data taken on the ram by Waites and Moule (1961): "...the temperature gradient between the body and the scrotum is not in any way autoregulatory. Indeed, vascular heat exchange only serves to cool the testis when the returning venous blood is cooler than the arterial inflow and this relationship can be maintained only if heat is being lost through the scrotum." This effect is easily explained by a closer look at the heat exchanger mechanism.

If the scrotum is considered as a fixed heat dissipator with the ability to dissipate some 1500 calories/hour without a heat exchanger present, the resulting scrotal temperature would be slightly less than but close to core temperature. With the heat exchanger present, no significant change in thermal dissipation would occur at the scrotum. However, there would be measurable lowering of scrotal temperature. In other words, the core heating system would be almost totally unaware of the effects of the heat exchanger.

Human data correlates with this Model. Lazarus and Zorgniotti (1975) noted lack of regulation occurring in patients with elevated core temperature caused by fever. As the core temperature rose, so did the testis temperature. The Model predicts this effect as a result of the additional heat energy entering the testis. (Note that heat available is related to temperature and blood flow.) A second study by Zorgniotti and Reiss (1982) on normal and subfertile men, showed that clothing increased mean testis temperature equally in all subjects (≈1.5°C), again in agreement with the Model. In this case the fixed

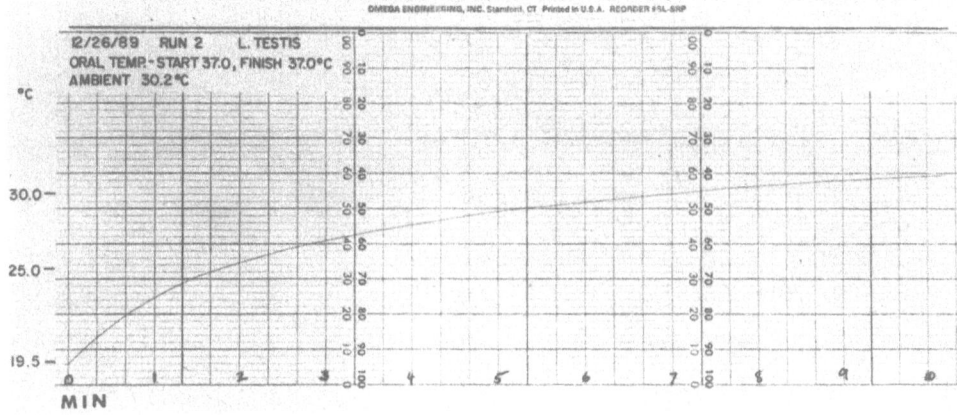

Figure 5. DBT measurement of precooled testis returning to steady state.

dissipator is not allowed to function at its preset value. The result is that less calories are transported away and intrascrotal temperature rises.

Not in agreement with the Model is the fact that men with normal and elevated temperatures had the same increase. The Model predicts that normal men would show a slightly higher increase in temperature with clothing than those with varicocele. This effect could be the result of temperature asymptotically approaching the core temperature. Obviously, scrotal temperature cannot exceed core and varicocele patients

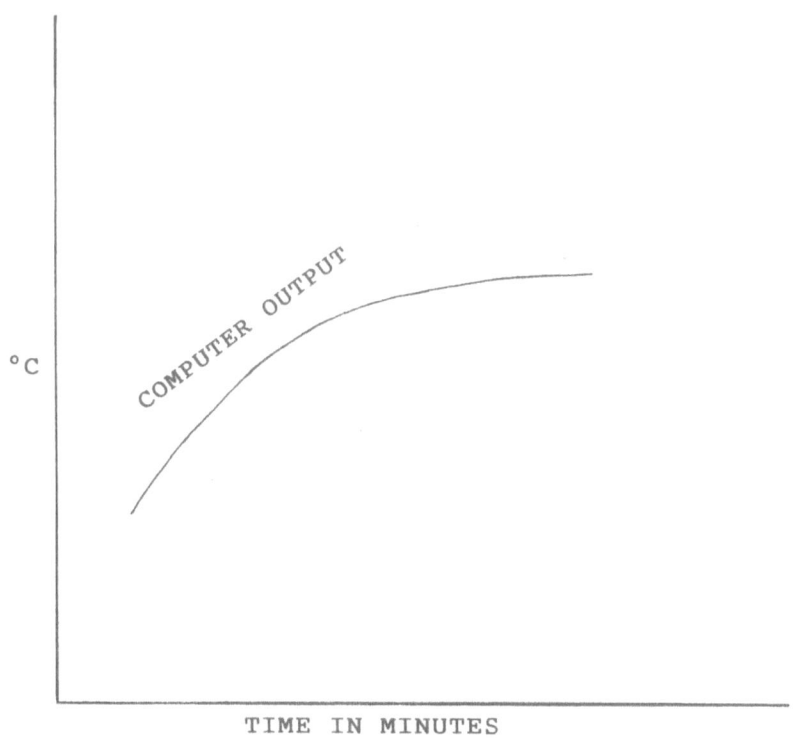

Figure 6. Computer curve for precooled testis returning to steady state.

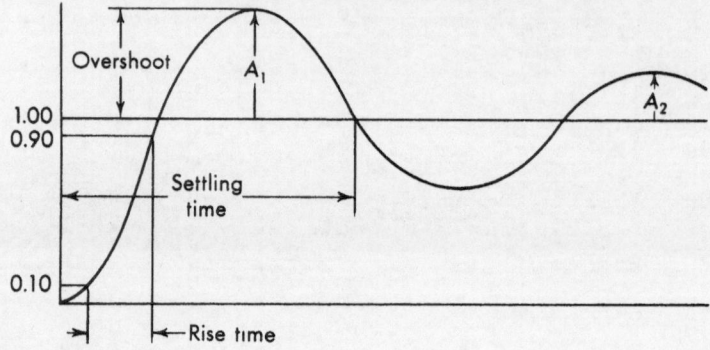

Figure 7. Time-domain specifications shown on a typical response curve.

start much closer to the core value and, therefore, should experience less temperature change. The important consideration is that the temperature does not appear to be stabilized in any way; a cold ambient produces a cold testis and a warm ambient produces a warm testis.

The significance of the capacitor in the circuit is that it represents the thermal storage of the testis' mass and is the result of the specific heat of the tissue and the amount of mass. The Model predicts that if, for example, the testis is cooled by immersion, it will take some period of time for temperature to return to its steady state value and the curve it follows will conform to the time constant equivalent of a charging

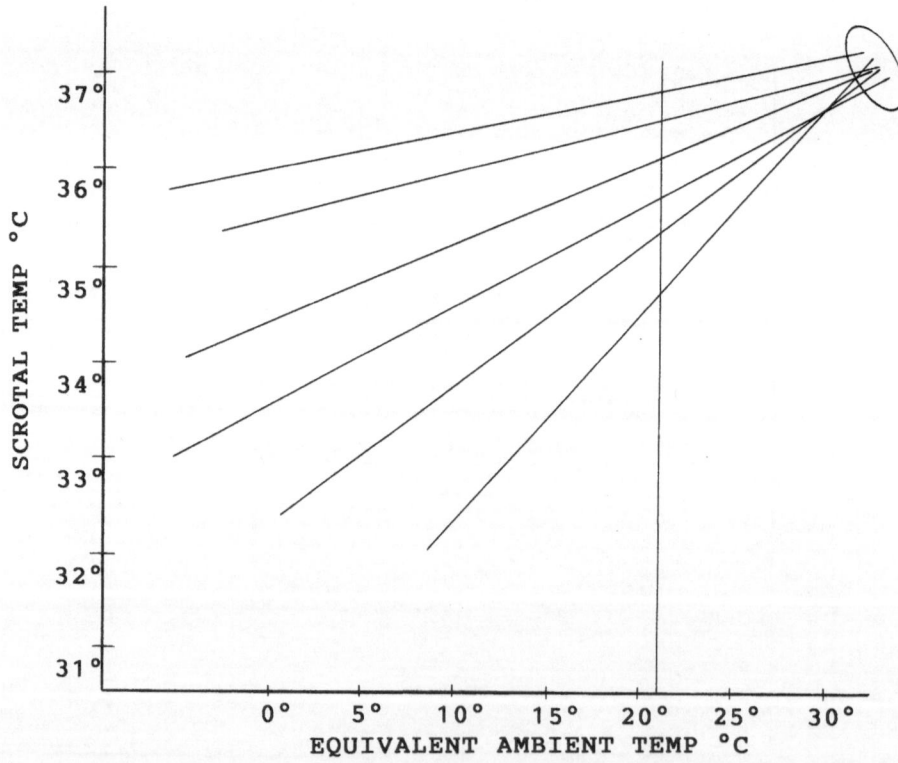

Figure 8. Family of theoretical curves showing scrotal temperature decreasing with ambient in agreement with Brindley's data.

capacitor which is of the form dE/dt + PE = Q. If any thermoregulation is present, an error signal would cause some mechanism to release a copious quantity of core heat, resulting in a much more rapid return (with an overshoot or dampened sine wave or any other curve besides the anticipated logarithmic one) to its optimum temperature.

This was tested in an experiment with the precooled testis. The subject was seated with scrotum bare. A Deep Body Temperature (DBT-Terumo) probe was affixed with tape to the scrotal skin overlying the left testis. The testis was then cooled by contact with melting ice in a plastic bag. Contact with the ice was discontinued when temperature fell 10°C below initial temperature. The testis was allowed to warm up to its steady state temperature which is shown (Fig. 5). The computer output for this comparable situation is shown (Fig. 6). The open loop simulation closely follows the human data.

If thermoregulation were present, this curve would follow a much more rapid return to equilibrium than the curve generated by a single ordered differential equation. A feedback system generally is represented mathematically by a multi-ordered differential equation which generally produces an output (as a function of time, f(t)) which contains an overshoot or dampened sine wave which is readily distinguishable from the nonregulated curves from the Model and the experiment. An example of a response containing an overshoot is shown (Figure 7).

In a recent article, Brindley measured deep scrotal temperature under varying conditions of activity, clothing and posture. This was accomplished by a radio-interrogated intrascrotal implant which permitted temperature measurement. His results show decreasing temperature as a function of air temperature. Specifically, he noted that testis temperature drops 1.0°C for every ten degree drop of air temperature. This agrees closely with the open loop computer simulation which shows decreasing testis temperature with decreasing ambient temperature. This cooling effect of the ambient is enhanced by the precooling effect of the countercurrent heat exchanger in the pampiniform plexus (Figure 8).

## DISCUSSION

The data presented point to the absence of any autoregulatory effect in the regulation of testis temperature. This is a direct consequence of an open loop system and implies that the resulting temperature of the testis varies with external thermal load, internal (core) temperature, and the effect of the countercurrent heat exchanger. The Model predicts that, beyond these, there is no significant mechanism to stabilize or control testis temperature.

What then is the role of cremaster and dartos activity? Under conditions of extreme heat and cold, these are observed to change the configuration of the scrotum. The Model does not indicate a regulatory function for these in the normal temperature range. Cremaster and dartos activity does not appear to provide temperature regulation of the scrotum and their actual function will require further elucidation.

## THERMODYNAMIC ROLE FOR THE SCROTUM

The scrotum in our Model (Figure 1) is considered to be a fixed passive dissipator that is incapable of regulating heat outflow to the surroundings. Temperature at the scrotum is a result not of intrinsic activity but of a passive thermodynamic network consisting of heat from arterial inflow, venous outflow, the countercurrent heat exchanger, and scrotal dissipation.

## ROLE OF THE PAMPINIFORM PLEXUS

It has been long recognized that the pampiniform plexus is a countercurrent heat exchanger (Harrison and Weiner, 1949; Dahl and Herrick, 1959; Waites and Moule, 1961) which serves to precool the arterial blood by venous blood. The heat exchange in the pampiniform plexus is a function of the temperature gradient and the amount of direct

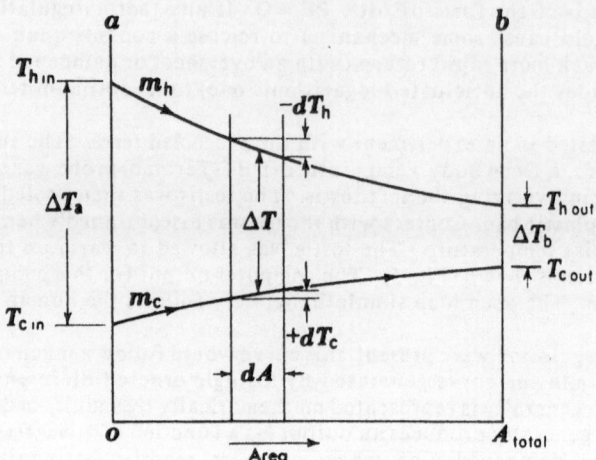

Figure 9. Temperature distribution in single-pass parallel-flow heat exchanger.

contact between arterial flow and venous flow. Nature has selected a highly effective means of heat exchange by using countercurrent flow.

## COUNTERCURRENT HEAT EXCHANGE VS. CONCURRENT HEAT EXCHANGE

Countercurrent heat exchange is widely employed in industry for heat recovery and added efficiency during steam power generation, distillation, liquefaction of gases, etc. A concurrent heat exchanger (i.e., parallel flow or both fluids flowing in the same direction) suffers from an inherent inefficiency since both fluids are approaching temperature equilibrium asymptotically as seen (Figure 9). In parallel flow systems the final outlet temperature of both fluids approach a temperature which is in between inlet temperature (Figure 10). The second law of thermodynamics will not permit the cold fluid to become hotter than the hot fluid.*

In the countercurrent heat exchanger, a temperature gradient is maintained along the length of the exchanger (Figure 1), in which hot fluid (arterial in our case) is able to be cooled to below the outflow temperature (venous), making the exchanger extremely effective over shorter lengths.

To analyze the heat flow in the countercurrent heat exchanger, the basic law of cooling is again applied:

$$q = @A(T_2 - T_1) \qquad \text{(Eqn 1)}$$

where @ = coefficient of heat transfer, A = area of contact, q = flow rate of heat.

This states that the flow rate across a boundary is a function of the contact area and the difference of the temperatures (gradient). This may be modified to include fluid flow:

$$q = W_a C_{pa}(T_{a1} - T_{a2}) = W_v C_{pv}(T_{v2} - T_{v1}) \qquad \text{(Eqn 2)}$$

---

*As a point of fact, the second law of thermodynamics has never been proved but is considered a fundamental law of nature which states: it is impossible for a heat engine to produce net work in a complete cycle if it exchanges heat only with bodies at a single fixed temperature. This is a basic statement prohibiting the perpetual motion machine. Either account for all the heat in and out of a system, or find that work was done in the system. In physics, work is defined as a force on an object undergoing displacement.

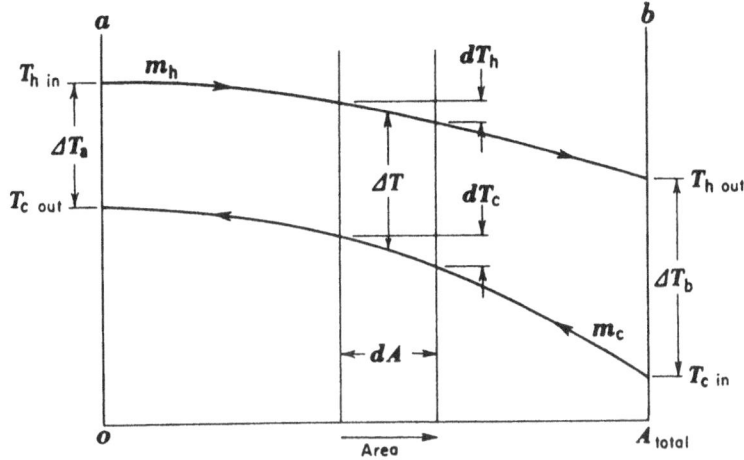

Figure 10. Temperature distribution in single-pass countercurrent heat exchanger.

where q = heat transferred, $C_{pa} = C_{pv}$ = specific heat of blood, $W_a$ = flow of arterial blood, $W_v$ = flow of venous blood, $T_{a1}$ = temperature initial of arterial blood, $T_{v1}$ = temperature initial of venous blood.

Let us perform a hypothetical calculation with the assumption that the appendage receiving the blood flow is thermally huge and whose outlet temperature is not affected by inlet temperature (Figure 11).

$$q \text{ cal/sec } °C = 1 \text{ ca/gm } °C \times 0.175 \text{ gm/sec } (37.0°C - 35.0°C)$$
$$= 1 \text{ ca/gm } °C \times 0.175 \text{ gm/sec } (35.0°C - 33.0°C)$$

$$q = 0.35 \text{ cal/sec}$$

Now suppose the flow rate in the venous system were reduced by 30% in this hypothetical example. What would be the return arterial temperature flow to this limb (assuming the venous return stays at 33.0°C)?

Note that the venous flow was absorbing 0.35 cal/sec at full flow. By reducing the flow 30% and assuming equal temperature, the venous system is capable of sinking or removing only 0.245 cal/sec. (q = 0.1225 x 1 x (35.0°C - 33.0°C). Assuming the same arterial flow, then the arterial outflow ($T_{a2}$) to the limb would become:

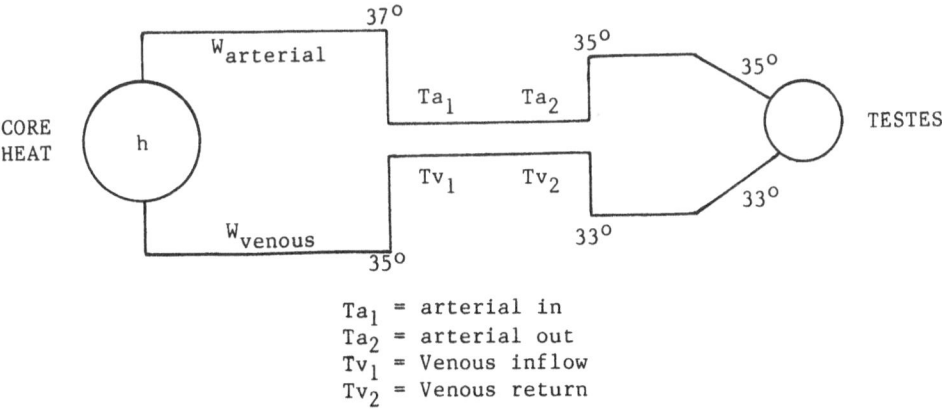

Figure 11. Countercurrent heat exchanger to huge limb.

$0.245 = 0.175 (37.0°C - T_{a2})$; thus $T_{a2} = 35.6°C$

Now consider the fact that the limb is not thermally stable but is affected by inlet arterial temperature, as the testis would be. If dissipation in the testis is assumed to be constant using the same equation as the heat exchanger, it also dissipates some 0.35 calories per second. If the new inlet temperature goes from 35.0°C to 35.6°C, then the venous return to the heat exchanger $(T_{v1})$ becomes:

$$Q = MC (T_{a2} - T_{v1}) \qquad \text{(Eqn 3)}$$

$Q = 150 \times 1 \times 2 = 300$ calories $= 150 (35.6°C - T_{v1})$, $T_{v1} - 33.6°C$ showing if the scrotum dissipates the same heat and the mass is constant, the gradient is maintained.

But now $T_{v1}$ is not 33.0°C but 33.6°C. Will the heat exchanger be able to exchange as effectively with the smaller gradient? As the gradient across the exchanger decreases, the ability to transfer heat also decreases as seen in Equation 1. Thus, a likely scenario is that the exchanger cannot increase the temperature of venous return (this may have to be confirmed experimentally) with the result that the venous system performs as follows:

$q = 0.1225 (T_{v2} - T_{v1})$ the reduced flow rate and the new $T_{v1}$

$q = 0.1225 (35.0°C - 33.6°C) = 0.1715$ cal/sec

This is a reduction from 0.245 based on constant temperatures. That forces a new calculation for $T_{a2}$.

$T_{a2}$ becomes: $0.1715 = 0.175 (37.0°C - T_{a2}) = 36.2°C$

The cycle of calculations can be repeated for each new value of $T_{a2}$. As the countercurrent heat exchanger is thrown out of balance, $T_{a2}$ will rise. As $T_{a2}$ rises, the venous flow becomes unable to pick up as much heat since it is assumed the testis continued to provide a constant two degree decrease with its dissipation. This is unrealistic as the testis would become unable to handle the additional thermal load imposed by the increasing $T_{a2}$. As a result, the temperature drop in the testis will decrease, which will further remove the countercurrent exchanger from having any effect. For completeness the extra cycle looks like this.

The heat picked up by the venous side of the exchanger:

$q_v = 0.1225 (35.0°C - 34.2°C) = 0.98$ (The 0.1225 is the reduced flow rate; the 34.2°C is the new $T_{v1}$ which is $T_{a2}$ minus the 2 degrees assumed dissipated in the testis.)

Now if that is the heat picked up by the venous system, then the arterial side must have given up a similar amount of heat. Thus, the new temperature at the outlet of the arterial side, or the temperature of the arterial blood going into the testis, $T_{a2}$.

$q_a = 0.98 = 0.175 (37.0°C - T_{a2}) ==> T_{a2} = 36.44°C$ (The 0.175 is the arterial flow rate which is not reduced; 37.0°C is the core temperature; $T_{a2}$ is the arterial outflow temperature, 36.4°C.)

This calculation is converging and can be repeated but it appears that the countercurrent heat exchanger no longer influences the testis temperature when the gradients across the exchanger fall below about 1°C. The assumption that the testis will maintain its losses at a constant value and hold its temperature gradient at a 2°C constant is an oversimplification. In reality the higher arterial temperature would cause additional heat energy to enter the testis following the idea that the quantity of heat is a function of the temperature, mass and specific heat of the material. If the testis finds itself unable to dissipate the extra heat, then its gradient would decrease as indicated above.

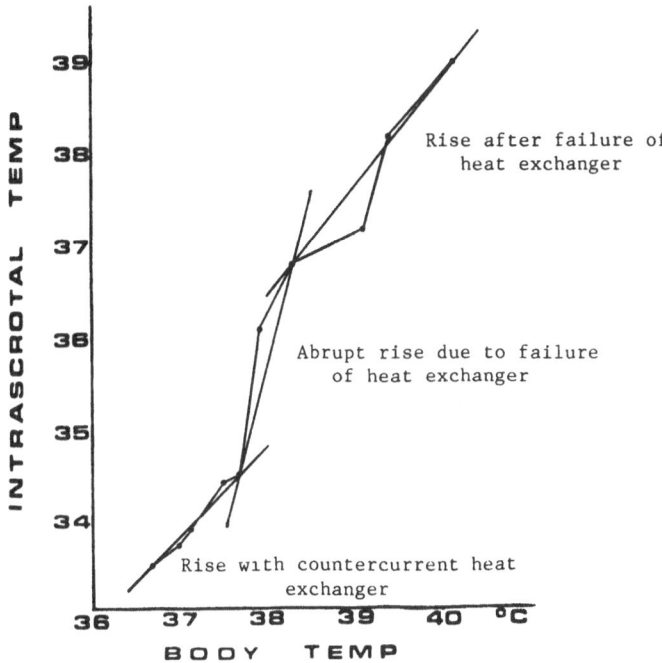

Figure 12. Plot of intrascrotal temperature versus body temperature in a febrile patient.

Although the reiterative solution lacks elegance, the trend becomes obvious. This has several implications for testis temperature. If the venous return is in any way slowed, or lies outside the heat exchanger, temperature will rise in the scrotum. A rather small decrease of the flow will cause a significant rise in testis temperature. Although it is unclear from the indirect calculations exactly at which point of equilibration it will occur in this admittedly arbitrary example, a rise of 1.5°C appears likely. (This is caused by the rise of $T_{a2}$ -- the precooled arterial blood flow -- going from 35.0°C to 36.4°C during this hypothetical 30% reduction of venous return flow.)

Lazarus and Zorgniotti noted that a 1.5°C rise occurs abruptly in the testis temperature of febrile patients. Zorgniotti and others have measured an approximate +1.5°C elevation in testis temperature of subfertile patients as compared to a fertile group. The above independent and theoretical calculation approaches a similar 1.5°C, which represents the failure of the countercurrent heat exchanger in the pampiniform plexus. Without feedback, as external temperature rise, testis temperature also rises until the countercurrent heat exchanger fails. Then the testis temperature rises abruptly 1.5°C and then continues to rise as ambient temperature continues to rise. Subfertile semen may be the result of the failure of the heat exchanger mechanism which results in a sudden rise in testis temperature of some 1.5°C and to which any additional temperature rise must be added (Figure 12).

When retrograde flow is observed, the countercurrent heat exchanger has long since failed and cannot produce any measurable cooling effects after venous flow had dropped by 40%.

There are environmental implications as well. As ambient temperature increases, either directly or indirectly, or as a result of well-insulated clothes, the ability of the scrotum to dissipate 900 to 1200 calories per hour is decreased and will result in a rapid rise in incoming arterial temperature because of failure of the heat exchanger. In fact, if the temperature internally or externally were to increase, the added load on the heat exchanger would render it ineffective. This explains the failure of the countercurrent heat exchanger in febrile patients. In this example, the value $T_{a1}$ rises along with the core temperature. The scrotum is unable to dissipate additional heat and venous return temperature rises causing failure of the exchanger. It is the nature of a heat exchanger that, as a working gradient appears across it, and it becomes effective, dramatic temperature swings are achieved with relatively small changes in heat dissipation. The converse is equally true: as the heat exchanger becomes ineffective, i.e., the gradient across it becomes too small for effective heat exchange; a small increase of heat into the system causes the heat exchanger to fail and results in large increases in temperature of arterial blood, which is no longer precooled, and testis temperature rises. This results in a flip-flop temperature swing of about 1.5°C in this region.

What happens when the testis is exposed to cold? The heat exchanger becomes more efficient since core temperature is maintained but the venous return is now much cooler. The exchanger moves more heat and additional cooling results. Thus, exposure to cold makes the testis proportionally even colder. Our hypothesis is that this cycle is altered by the cremaster muscle which then limits scrotal heat dissipation by contracting and pulling the testes up against the warmer trunk.

Additional confirmations for the Model come from papers from D.S. Pal and J. Steketee which are concerned with the thermal recovery of skin after cooling. Although the latter paper calculated temperature recovery on the forehead, the results are similar to our Model predictions for the scrotum. The equations derived for general skin recovery bear a striking similarity to our results, and since thermoregulation of the forehead or arm are unlikely, the obvious conclusion is that the scrotum reacts to heat transfer in the same unregulated manner as any other section of human skin and subcutaneous tissues.

CONCLUSION

In the manner in which thermoregulation is defined by engineers, there is no autoregulation of the human testis in any of the evidence so far. In order to have regulation, a feedback mechanism must regulate the heat flow either into or out of the testis to maintain constant temperature. In case of variation of thermal load, either hot or cold, the temperature of the testis will not stabilize at a value suggesting regulation. In the event that core temperature increases, this causes a thermal rise to occur indicating a lack of (sufficient) feedback to obtain temperature stabilization. Since no stabilization occurs, slight or no regulation is available by definition. As measured in humans and as simulated in a computer model, there appears to be no thermoregulatory effect to maintain physiological testis temperatures in the fertility range.

REFERENCES

Brindley, G.S. 1982. Deep scrotal temperature and the effect of clothing, air temperature, activity, posture and paraplegia. Br. J. Urol., 54: 49.

Bazett, H.C., Love, L., Newton, M., Eisenberg, L., Day, R. and Foster, R., II. 1948. Temperature changes in blood flowing in arteries and veins in man. J. Appl. Physiol., I: 3.

Dahl, E.V. and Herrick, J.F. 1959. A vascular mechanism for maintaining testicular temperature by counter-current exchange. Surg. Gynecol. Obstet., 108: 697.

Ehrenberg, L. and Von Ehrenstein, G. 1957. Gonadal temperatures and spontaneous mutation rate. Nature, 180: 1433.

Hammel, H.T. 1968. Regulation of internal body temperature. Ann. Rev. Physiol., 30: 641.

Harrison, R.G. and Weiner, J.S. 1949. Vascular patterns of the mammalian testis and their functional significance. J. Exp. Biol., 26: 304.

Jequier, E. 1986. Human whole body direct calorimetry. IEEE Eng. Med. Bio., 5: 12.

Lazarus, B.A. and Zorgniotti, A.W. 1975. Thermoregulation of the human testis. Fertil. Steril., 26: 757.

Minard, D. and Copman, L. 1972. Elevation of body temperature in health. In: "Temperature: its Measurement and Control in Science and Industry", J.D. Hardy (ed), Robert E. Krieger Publishing Company, Huntington.

Mitchell, J.W. 1976. Heat transfer from spheres and other animal forms. Biophysical J., 16: 561.

Mitchell, J.W. and Myers, G.E. 1968. An analytical model of the counter-current heat exchange phenomena. Biophysical J., 8: 897.

Pal, D.S. and Pal, S. 1990. Prediction of temperature profiles in the human skin and subcutaneous tissues. J. Math. Biol. 28: 355.

Phillips, R.W. and McKenzie, F.F. 1934. The thermoregulatory function and mechanism of the scrotum. Univ. Mo. Agr. Expt. Sta. Res. Bull., 217: 1.

Robertshaw, D. and Vercoe, J.E. 1979. Scrotal thermoregulation of the bull. (*Bos* sp.). J. Physiol.

Scholander, P.F. and Krog, J. 1957. Countercurrent heat exchanger and vascular bundles in sloths. J. Appl. Physiol., 10: 405.

Scholander, P.F. and Schevill, W.E. 1955. Counter-current vascular heat exchange in the fins of whales. J. Appl. Physiol., 8: 279.

Steketee, J. and Van Der Hoek, M.J. 1979. Thermal recovery of the skin cooling. Phys. Med. Biol. 24(3): 583.

Waites, G.M.H. 1961. Relation of vascular heat exchange to temperature regulation in the testis of the ram. J. Reprod. Fertil., 2: 213.

Waites, G.M.H. 1970. Temperature regulation and the testis. In: "The Testis I", A.D. Johnson (ed), Academic Press, New York.

Zorgniotti, A.W. 1982. Elevated intrascrotal temperature, II. Indirect testis and intrascrotal temperature measurement for clinical and research use. Bull. NY Acad. Med., 58: 541.

Zorgniotti, A.W., Cohen, M.S. and Sealfon, A.I. 1986. Chronic scrotal hypothermia: Results in 90 couples. J. Urol., 135: 944.

Zorgniotti, A., Reiss, H., Toth, A. and Sealfon, A. 1982. Effect of clothing on scrotal temperature in normal men and patients with poor semen. Urol., XIX: 176.

Zorgniotti, A.W. and Scalfon, A.I. 1988. Measurement of intrascrotal temperature in normal and subfertile men. J. Reprod. Fert., 82: 563.

# THE MULTI-LEVEL COMPARTMENTATION OF THE SIMULATION MODELS OF THE

# COUNTER-CURRENT HEAT EXCHANGE (CCHE) MECHANISM OF THE TESTIS

Giuseppe Tritto

Service d'Urologie (Hôpital Saint-Antoine)
Clinique Hartmann (Neuilly)
Paris, France

## INTRODUCTION

Extrinsic and/or intrinsic thermoregulation of the testis represents a controversial area in the regulation of spermatogenesis and fertility[1]. The application of scrotal hypothermia devices (improperly called testicular hypothermia devices) in clinical field[2,3] and the introduction of simulation models of the counter-current heat exchange mechanism at the pampiniform plexus in an artificial domain, presented for the first time in Berlin (7th Int. Symposium of Operative Andrology, March 19, 1988)[4,5], indicate practical and theoretical approaches to interfere with and to analyze testis thermoregulation, with the aim to ameliorate sperm and fertility parameters.

## MODELLING AND SIMULATION

Modelling and simulation represent the artificial transfer, supported by computer technology, to obtain graphic and geometric models of human organs in reconstructive surgery[6]. In the andrological field the application of this new methodology sustains the possibility to evaluate the known and to discover unknown parameters of the mechanisms involved in the thermoregulation of the testis[7,8].

## MULTI-LEVEL COMPARTMENTATION

The multi-level compartmentation of the morpho-functional organization of human organs is a new concept on which the realization of complex anatomical models at multiple and hierarchical levels is accomplished for simulation experiments[9]. Since the relative importance of the different systems involved in the thermoregulation of the testis is not well understood, a multi-level compartmentation of simulation models of these systems, on structural (morphological) and functional (physiological) bases, is provided.

In relation to the presently known morpho-functional systems that act on testis temperature, a basic scheme of possible simulation models of the anatomo-functional mechanisms, intrinsic and/or extrinsic to the testis, is proposed, putting into the core the testis as a multi-faced T-sensitive and/or T-dependent organ (Fig. 1).

## CCHE SIMULATION MODELS

### Physical and Anatomical Background

The anatomo-functional correlations on bioengineering bases represent a new methodology to the understanding of physiological processes[10]. The counter-current heat

*Temperature and Environmental Effects on the Testis*
Edited by A. W. Zorgniotti, Plenum Press, New York, 1991

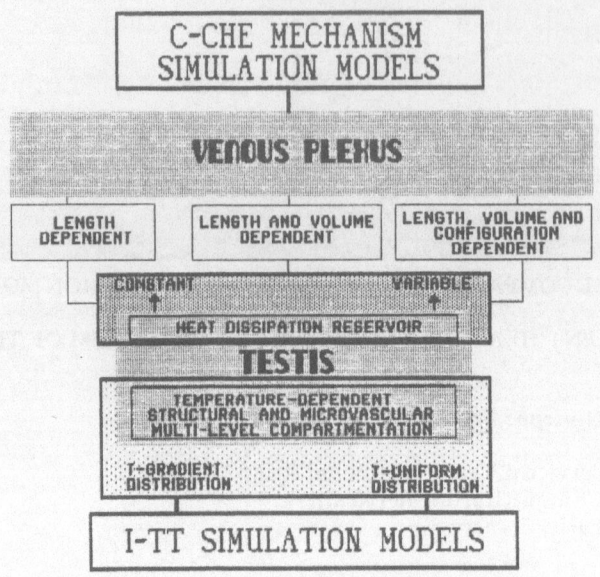

Figure 1. In CCHE mechanism simulation models, the testis is considered as a heat reservoir, depending on the geometric parameters of the venous plexus for its thermoregulation at constant or variable dissipation. In I-TT simulation models, the complex and hierarchical organization of the vascular microarchitecture of the testis sustains the flow and gradient of thermic diffusion, with uniform or non-uniform distribution.

exchange mechanism, that is, the process of transfer of heat between fluids of different temperatures flowing in opposite directions in adjacent conduits, is present in many species, including higher mammals and man. Normally, it is sustained by specific networks of intermingled arteries and veins, named "Wundernetze" by German anatomists, "wonderful networks" in English terminology, and "retia mirabilia" in Latin.

The efficiency of such systems in physical terms depends on:

(1)  the net surface across which heat can be conducted and exchanged (that is, the intimacy of the conduits);
(2) the relative rates of flow in the associated conduits (that is, the length of time that the streaming fluids are in association); and
(3)  the length of associated conduits (that is, the sectorial length of the conduits that participates in the heat exchange).

An anatomical system classically defined as a counter-current heat exchange model is the pampiniform venous plexus, arranged around the spermatic (testicular) artery. The arrangement and the architecture of the blood vascular system serving the testis-epididymis (studied with corrosion resin casting techniques combined with scanning electron microscopy in different species) seem to respond to the same identical criteria: a spermatic artery that coils many times as it runs to the upper pole of the testis; a highly branched and voluminous pampiniform plexus that completely surrounds the artery with a conical shape[11,12,13].

## CCHE - PAMP I and II, CCHE - PAMP-FEM Models

On these bases simulation models of the counter-current heat exchange mechanism are realized in computer-graphics in order to evaluate the role and burden of the length, volume and configuration of the pampiniform plexus in the process of heat transfer. The aim is to demonstrate the possibility of modifying the geometric parameters of the plexus with microsurgical corrections and, consequently, to improve the efficiency of the system[14]. The core of the model is represented by a closed system in which the thermic variation of the fluid in the conduits depends exclusively on the difference $\Delta T$ at every level in a single instant in paired conduits. The generation of the arterial and venous temperature gradients is based on the fact that the closed system, submitted to Newton's equation for heat exchange, is a dynamic system in relation to the flow of fluid in the paired vessels. The constant values are represented by the intrinsic coefficient of heat exchange k, by the density of the fluid $\rho$, and by the specific heat of the liquid c. The variables at the input in the system are organized in two sets of determinants:

(1) in the first set the (re)setting variables are introduced:
   - $T_o$ is the temperature of the spermatic artery at the origin from the aorta;
   - $\Delta T_w$ is the loss of temperature (in °C) at testicular level, between the level of temperature of the arterial flow in entrance and the level of temperature of the venous flow in exit. This factorial number expresses the function of testis as a thermic reservoir at constant heat dissipation, independent of the volume and of the flow into.

(2) in the second set the geometric and rheological variables are introduced:
   - arterial and venous lengths (aL and vL); vL/aL = length ratio
   - arterial radius at the origin ($ar_1$) and at the entry in the testis ($ar_2$)
   - venous radius at the origin from the testis ($vr_2$) and at the entry in the venous system ($vr_1$); $vr_2/vr_1$ = venous radius ratio
   The organization of the arterial-venous system is consistent with the 100% heat exchange surface, represented by the arterial tube completely surrounded by the venous tube for the length L. The variation of the vL expresses the length of the wrapped part of the venous tube, which is the real efficient compartment of the heat exchange mechanism based on the complete intimacy of the surfaces of the conduits.
   - arterial and venous flows (av and vv); av/vv = flow ratio.


## CCHE - PAMP I and II Simulation Experiments

Setting basic simplified conditions ($T_o$ = 37°C; $\Delta T_w$ = 1°C; vL = aL, (1/2)aL, (1/4)aL, (1/8)aL; $ar_1$ = $ar_2$; $vr_1$ = $2ar_1$; $vr_2$ = $2vr_1$; av = 8vv), in relation to the length variation of the system in scalar series, the venous gradient excursion difference (VGED), that is, the difference between the venous temperature at the origin from the testis and the venous temperature at the entry in the venous system, and the arterial gradient excursion difference (AGED), that is, the difference between the arterial temperature at the origin and the arterial temperature at the entry in the testis, are evaluated in four series of simulation procedures:

· In the first series of simulation experiments, the length of the venous conduit (vL) is progressively reduced from 1:1 ratio to 1:4 ratio, considering the venous channel as a cylinder (A) or as a conical conduit (B) (Fig. 2A,B). In both configurations the reduction of the venous length gives rise to Fig. 3A,B.

On the venous side to the progressive elevation of the exit temperature from the testis and the gradual reduction of the distal temperature in the vein, the net result is the progressive narrowing of the gradient with a higher setting of entry. On the arterial side, the only effect is the progressive elevation of the entry temperature of the artery in the testis.

Comparing cylindrical and conical venous conduits at the same rate of flow, the only change is represented by the output, which is the quantity of fluid that flows through the unitary section area in the unitary instant at that rate. Under these conditions the rate of flow in a conical venous conduit is greater than in a cylindrical conduit. The obvious effect on the venous gradient is that with high venous length, the gradient is larger in conical than in cylindrical conduit; with low venous length the gradient is larger in cylindrical than in conical conduit. On the arterial channel the conical configuration always sustains a larger gradient than the cylindrical configuration. On the whole, the conical venous configuration assures a large gradient on both venous and arterial

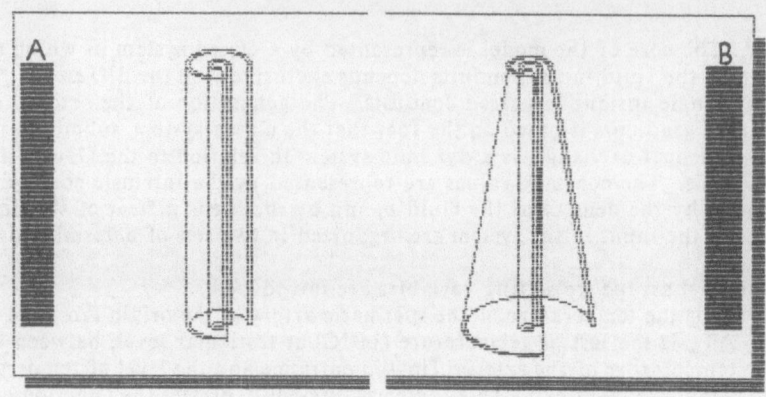

Figure 2A,B. In computer-graphic procedure, the telescopic view of the venous and arterial conduits is an oversimplification to indicate the complete heat exchange between the paired surfaces.

compartments, but the venous gradient is critically sensitive to length reduction of the conical venous tube.

·In the second series of simulation experiments, the venous flow is reduced quasi to zero, simulating the experimental situation of interruption or ligature of the venous channel, as a cylinder (A) or as a conical conduit (B) (Fig. 4A,B).

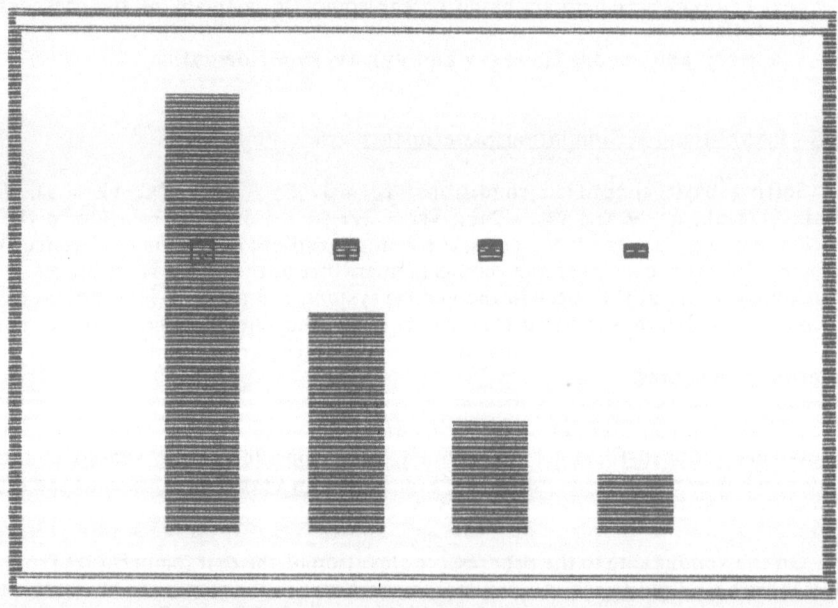

Figure 3A. Cylindrical configuration. Gray column: venous plexus length; Black column: venous gradient excursion difference; Asterisk: mean testis temperature

| Venous length (vL) | 200 | 100 | 50 | 25 |
|---|---|---|---|---|
| VGED | 1.818 | 1.386 | 0.896 | 0.420 |
| AGED | 0.910 | 0.695 | 0.451 | 0.213 |

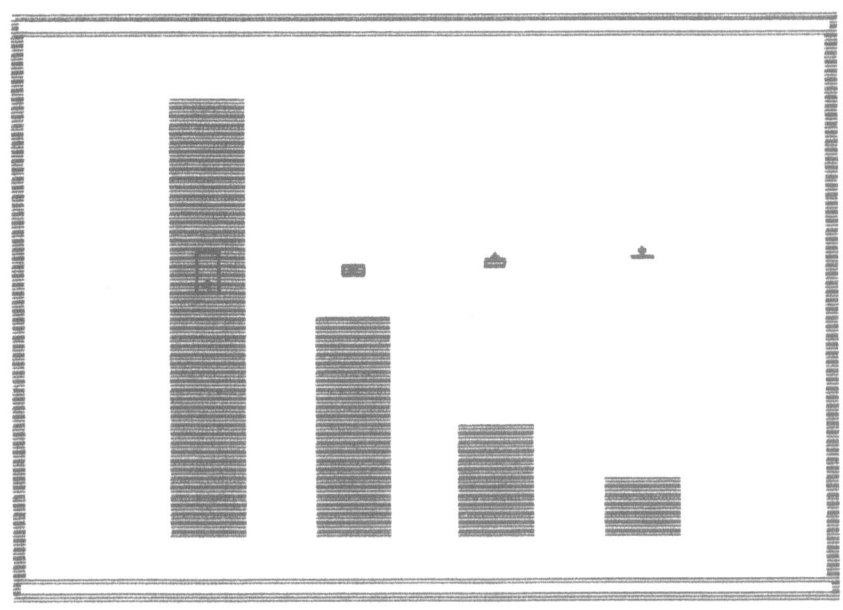

Figure 3B. Conical configuration.

| Venous length (vL) | 200 | 100 | 50 | 25 |
|---|---|---|---|---|
| VGED | 3.156 | 0.624 | 0.077 | 0.007 |
| AGED | 3.248 | 1.862 | 0.793 | 0.269 |

The comparison between intervenous and interarterial gradients in both cylindrical and conical channels in the presence or absence of venous flow demonstrates:

(a) that without venous flow the gradient is prefixed by the thermostat and is independent of the length of the system. The $\triangle T$ venous (and arterial) gradient excursion difference is also length-independent and practically constant.
(b) that in the presence of venous outflow the gain is evident, permitting to realize a large gradient both in the artery and vein, equally dependent of the length of the system.

· In the third series of simulation experiments, the venous conduit is positioned in proximal, middle or distal position in relation to the testis around the artery, both in cylindrical or in conical configuration, in normal flow condition or in the absence of flow. The effect of the length and position of the venous conduit is evaluated in relation to the mean testis temperature. No effect is pointed out from different locations of the venous conduit around the arterial conduit on the temperature gradient in this model: this is perfectly in accord with the expectancy in the physical simulation model.

From the above three series of simulation experiments regarding length, variation of flow and position, the final resultant gradient is unbiased by the position of the venous conduit, while the length confirms its role. Great evidence is available on the effect of venous tube length and shape on the core-testis temperature. It has been demonstrated that the conical configuration permits lower levels of mean testis temperature. Stopping the venous outflow would reset the core-testis temperature at constantly high levels for the cylindrical configuration. These predictive values demonstrate that the interruption of the venous flow at the distal end of the venous channel raises the reservoir core-temperature, representing a further wasting effect on the testis thermoregulation.

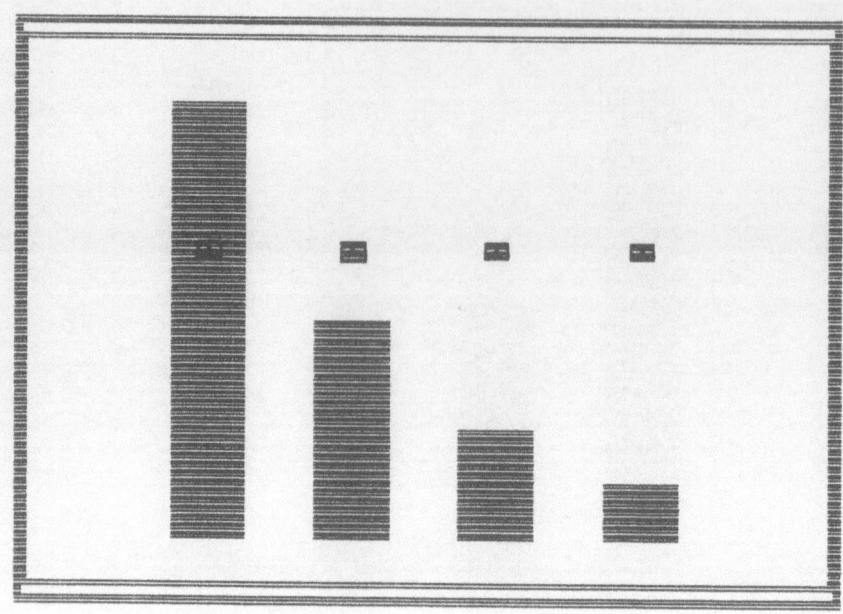

Figure 4A.  Cylindrical configuration.

| Venous length (vL) | | 200 | 100 | 50 | 25 |
|---|---|---|---|---|---|
| VGED | flow 1 | 1.818 | 1.386 | 0.896 | 0.420 |
| | flow 0.1 | 1.053 | 1.053 | 1.053 | 1.041 |
| AGED | flow 1 | 0.910 | 0.695 | 0.451 | 0.213 |
| | flow 0.1 | 0.053 | 0.053 | 0.053 | 0.053 |

In the fourth series of simulation experiments, subtraction of or sliding a sector of the venous channel at a preselected level are evaluated in relation to the mean testis temperature. In the subtraction procedure a biplanar section is obtained starting from the middle of the common length:  proximal and distal sectors are lost, leaving <u>in situ</u> the middle sector (Fig. 5). In the sliding procedure a planar section permits to lose the basal (near the testis) sector, sliding the venous cone of the same length (Fig. 6).

While in the cylindrical configuration both procedures do not modify the temperature gradient that remains univocally length-dependent, in the conical configuration the reduction process modifies the volume of the venous conduit in relation to the selected procedure and, consequently, the quantity of fluid involved in the heat exchange transfer. This is a critical factor that transforms the relationship linked to the progressive reduction of venous length from unimodal variation (length-dependent) in cylindrical configuration to bimodal variation (length-, volume-dependent) in conical configuration.

Selecting a sector in conical configuration at a constant flow rate means to change the fluid volume: the temperature gradient is not only length-dependent, but also volume-dependent.  At the same total length reduction, the subtraction procedure maintains a larger excursion gradient than that in the sliding procedure with normal or stopped flow (Fig. 7A,B).

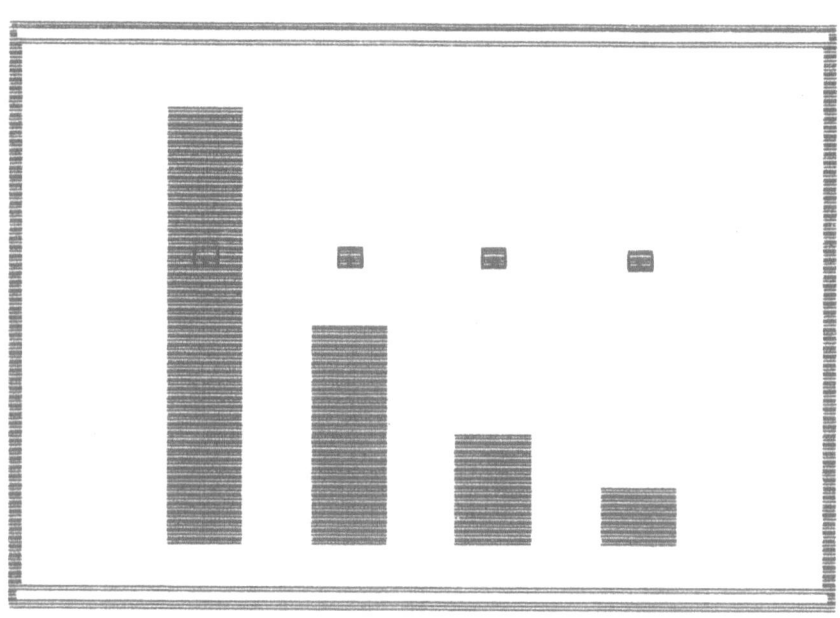

Figure 4B. Conical configuration.

| Venous length (vL) | | 200 | 100 | 50 | 25 |
|---|---|---|---|---|---|
| VGED | flow 1 | 1.982 | 1.372 | 0.635 | 0.161 |
| | flow 0.1 | 1.065 | 1.074 | 1.104 | 0.863 |
| AGED | flow 1 | 1.127 | 0.928 | 0.593 | 0.257 |
| | flow 0.1 | 0.074 | 0.074 | 0.104 | 0.135 |

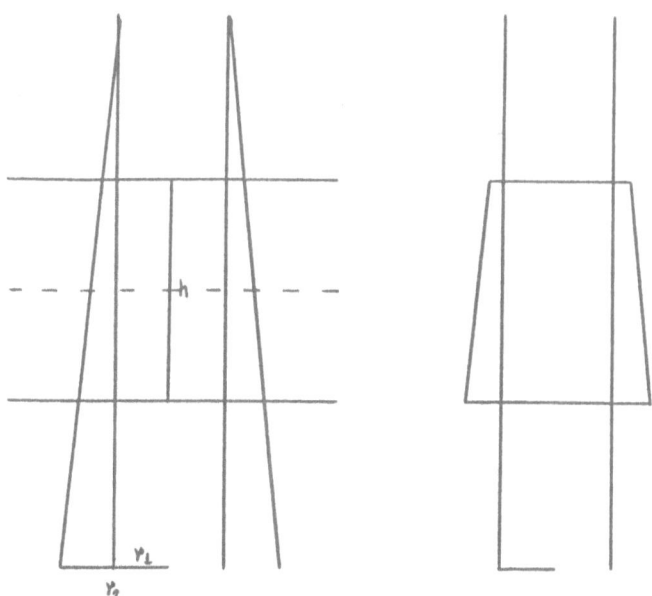

Figure 5. Subtraction procedure.

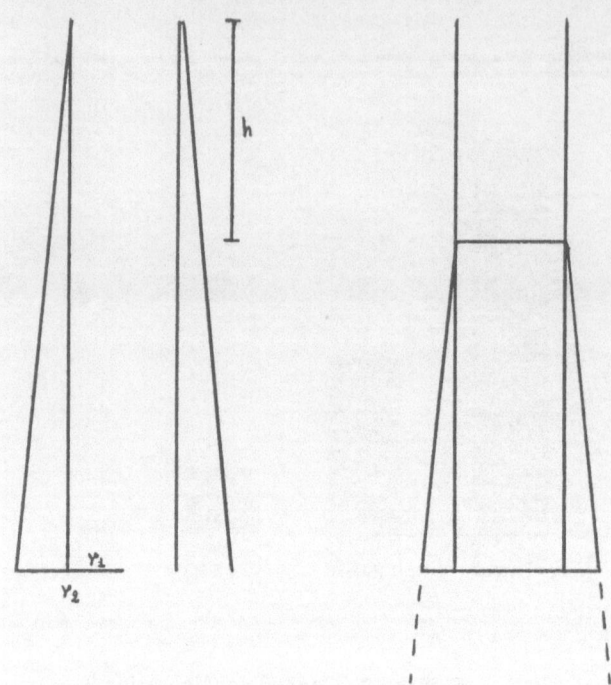

Figure 6. Sliding procedure.

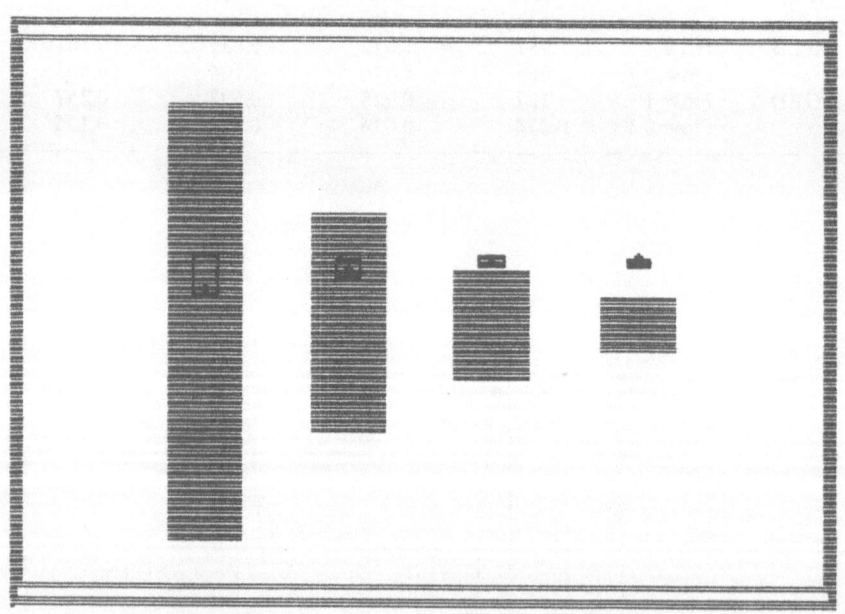

Figure 7A. Subtraction procedure: conical sector at height h.

| Venous length (vL) | 200 | 100 | 50 | 25 |
|---|---|---|---|---|
| VGED | 3.156 | 1.207 | 0.546 | 0.214 |
| AGED | 3.248 | 1.355 | 0.623 | 0.242 |

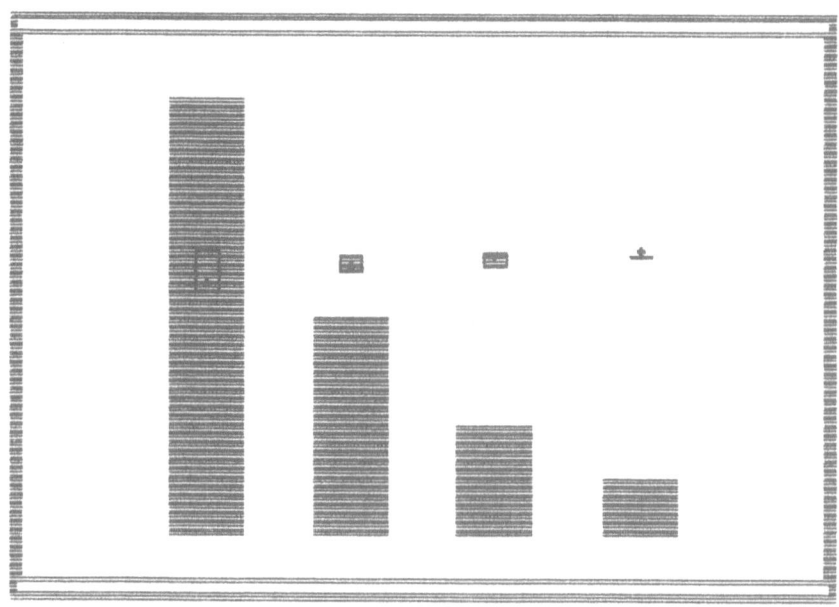

Figure 7B. Sliding procedure: conical sector at height h.

| Venous length (vL) | 200 | 100 | 50 | 25 |
|---|---|---|---|---|
| VGED | 3.156 | 1.539 | 0.385 | 0.133 |
| AGED | 3.248 | 1.016 | 0.684 | 0.257 |

### CCHE PAMP-FEM Simulation Experiments

A more sophisticated configuration of the venous plexus can be considered in front of the idealized conical configuration: the spiral organization of the venous plexus in conical configuration. Spiral cylindrical (Fig. 8A) and spiral conical (Fig. 8B) configurations are provided in simplified graphics for running simulations.

Complex three-dimensional reconstructions, applying the Finite Element Method, are realized for biomechanical evaluation (Fig. 9A,B). The vascular wall is structurally built, using three-dimensional solid elements to simulate smooth muscle cells, and spring and truss elements to simulate elastic and collagen fibers. The objective is to analyze the non-linear wall motion of the artery in relation to the pulsatile blood flow governed by the Bessel equations and in accordance to Womersley's theory, and taking into consideration the compliance of both normal and dilated veins of the plexus.

Changing the geometric parameters (radius, length and thickness) of the vessels and the three-dimensional pattern of the arterial-venous heat-exchange system within the anatomical range, the finite element-based structure analysis program calculates and displays with a step-wise procedure the new three-dimensional geometry of the venous system in relation to different situations (inflow or outflow variations of temperature oscillations).

In spiral cylindrical configuration the number of coils and the length of the pitch are selected to obtain the desired basic pattern. In spiral conical configuration the number of plates, the number of coils in the plate, the length and the configuration of the pitch are selected. Comparing the spirals in the scalar series of venous length variations, only

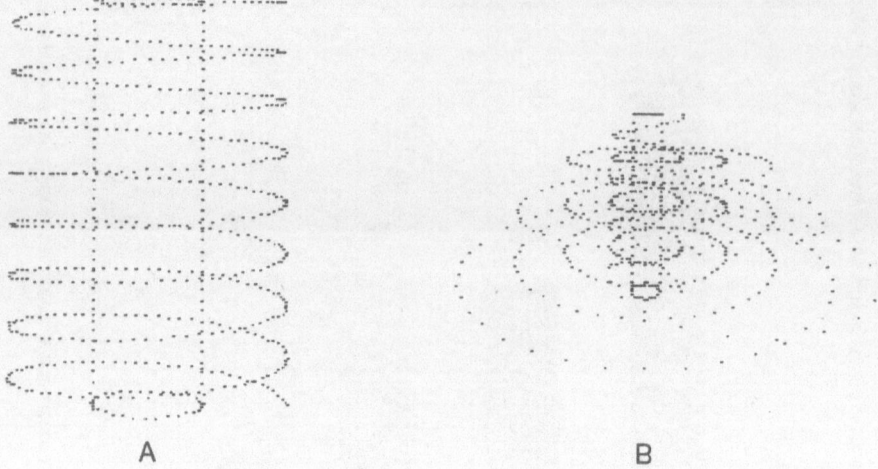

A B

Figure 8A,B. Spiral cylindrical and spiral conical configurations.

a minimal change of the venous gradient is obtained in cylindrical configuration. The $\triangle T$ venous gradient excursion difference is also length-independent, practically constant but narrow (Fig. 10A). In conical configuration a significant change of the venous gradient is obtained, especially with high venous length. The $\triangle T$ venous gradient excursion difference is length-dependent but constantly narrow (Fig. 10B).

The effect of the plates in spiral conical configuration is evident in the drop-jumping configuration of the venous curve temperature (Fig. 11). With the reduction of the venous flow quasi to zero, the gradient is prefixed by the thermostat in both spiral configurations and is independent of the length and shape of the system.

## I-TT SIMULATION MODEL

### Physical and Anatomical Background

Vascular microarchitecture, three-dimensional organization of the seminiferous tubules and of the interstitial tissues and lymphatic spaces in the testis present complex patterns in different species[15,16]. The differences in the vascular architecture express the differences in the form and arrangement of seminiferous tubules and their relationship to the blood vessels. The variations in the organization of interstitial tissues and, consequently, in the volume of interstitial lymphatic spaces can be correlated to the physiological volumetric variations of the testis and to the dumping power of its capacity-container.

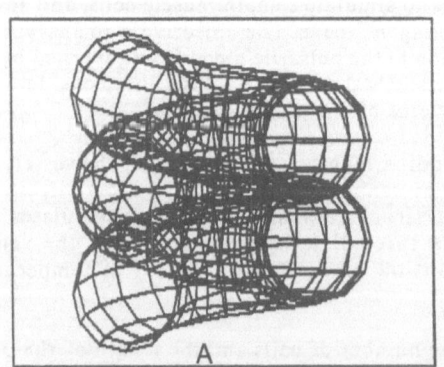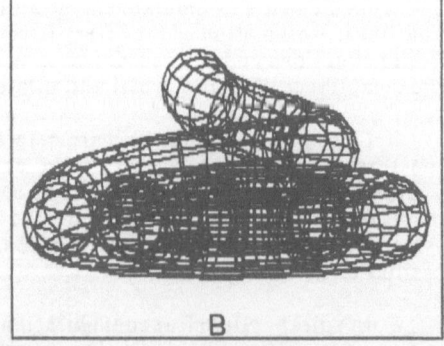

Figure 9A,B. Application of the Finite Element Method to the biomechanical evaluation of complex three-dimensional reconstructions.

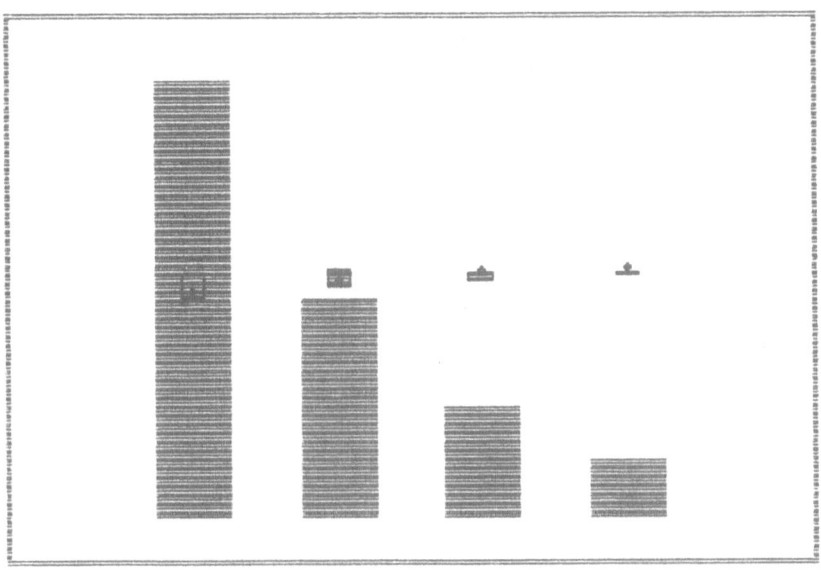

Figure 10A. Spiral cylindrical configuration showing length-independence of the $\triangle T$ VGED.

The relative anatomical positions of entry and exit of both arterial and venous ports seem important because they indicate the physical trigger points of initial and final temperatures during thermic flow diffusion.

I-TT Models

The testis can be considered as a homogeneous medium with equal input and output flow rates, embedded into an insulator shell (vaginalis, scrotum). The T-value in a single point of the testis parenchyma is related to the arterial port input and to the venous port

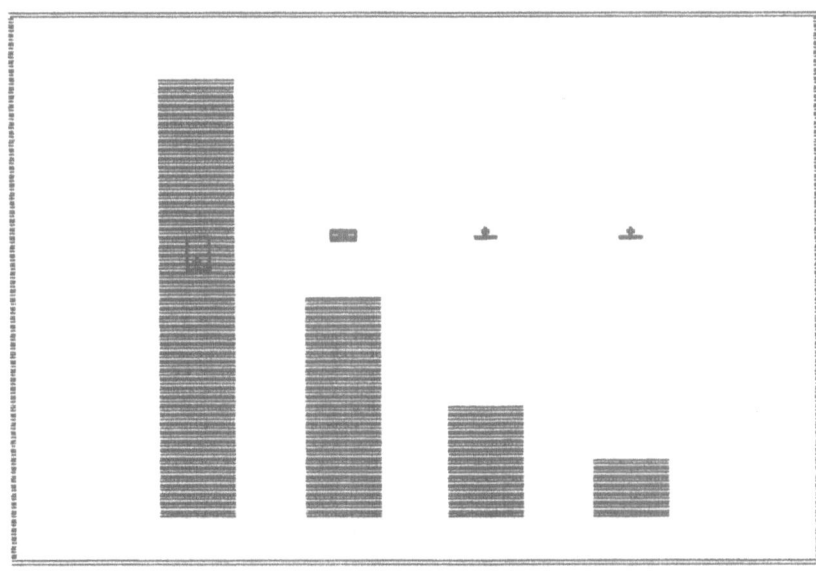

Figure 10B. Spiral conical configuration showing length-dependence of the $\triangle T$ VGED.

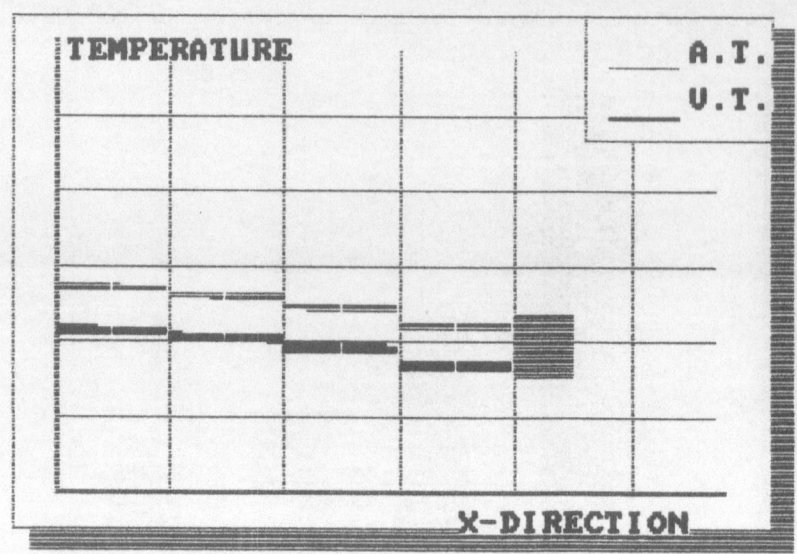

Figure 11. The drop-jumping configuration of the venous curve temperature in the spiral conical configuration.

output positions, the direction of flow (linked to the microvascular geometry and architecture), and the distance from the shell. In this configuration a T-gradient pattern is always generated.

Simulation Experiments

(1) Effect of varying the position of entry and exit ports from opposite bipolar to parallel unipolar thermic flow directions:

In the configuration of opposite bipolar ports, a regular distribution of isothermic bands or zones with a progressive gradient is obtained (Fig. 12). In the configuration of

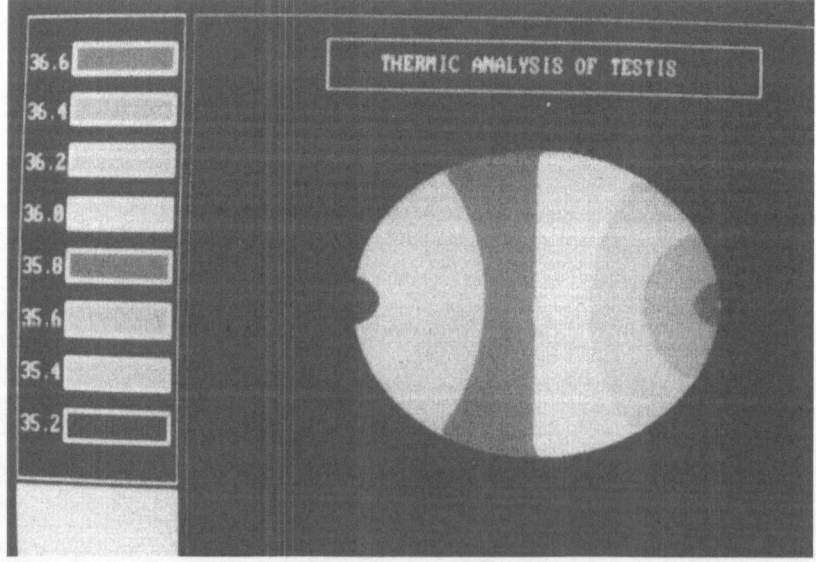

Figure 12. Configuration of opposite bipolar ports. Right pole: arterial port; left pole: venous port.

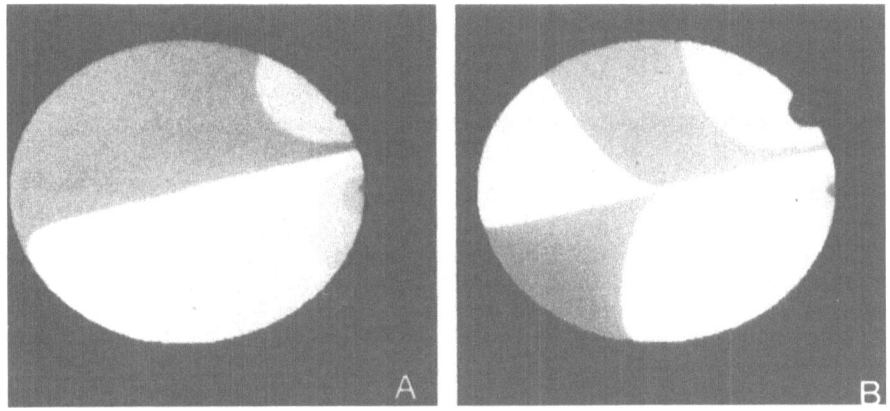

Figure 13. Configuration of unipolar parallel, convergent or divergent ports. (A) Equal low flow rates; (B) Equal high flow rates.

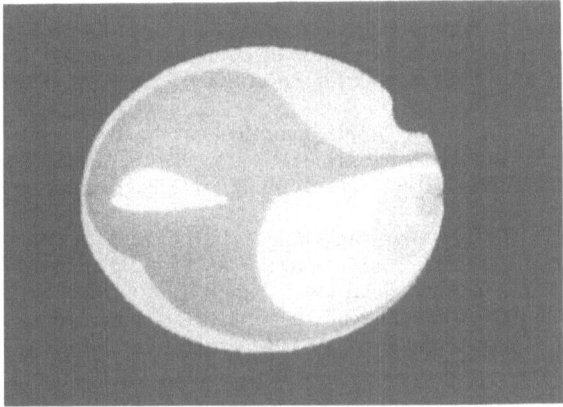

Figure 14. Simulation in high flow, unipolar divergent configuration, in the presence of shell steady-state temperature activity.

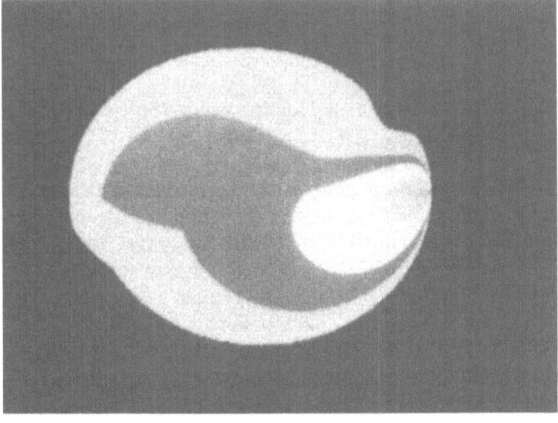

Figure 15. Cooling simulation in high flow, unipolar divergent configuration in the presence of T-shell activity.

unipolar parallel, convergent or divergent ports, isothermic zones of different extensions are obtained, depending also on the different flow rates (Fig. 13A,B).

The directional variation sustains the spotting effect in a homogeneous medium, i.e., the appearance of "hot spots" into an isothermic zone, as an effect either of the flow speed or the steady-state temperature of the shell on the particular ovoidal structure (Fig. 14).

(2) The shell cooling effect:

If the shell is artificially and externally cooled, the T-gradient distribution within the testis changes, with reduction or disappearance of the spotting areas (Fig. 15). This modulating effect, even if apparently quantitative, can also be qualitative in nature if the regional distribution of flow and zonal organization of the seminiferous tubules are considered.

DISCUSSION AND CONCLUSION

The realization (modelling) of different simple simulation models of the morpho-functional systems involved in the thermoregulation of the testis permits to test the intrinsic and specific parameters of each mechanism in relation to the testis dynamics.

In the field of Computer Sciences, each problem in a single model is evaluated separately in order to maintain the degree of complexity at a low level, considering also the absence of real physical information on the characteristics of the systems involved.

All these elaborations in simple compartmentalized simulation models offer the possibility of specifying the hierarchical levels of the different morpho-functional mechanisms and their physical (structural and functional) interdependency.

In the future the ideation and choice of different therapeutic approaches and devices in the thermomodulation of the testis can be sustained, establishing the logical hierarchy of control in the multi-level compartmentation systems (comparatively investigated in simulation models) of the testis and specifying the particular levels involved in the therapeutic regulation of heat generation and dissipation.

REFERENCES

1. Kandeel, F.R. and Swerdloff, R.S. 1988. Role of temperature in regulation of spermatogenesis and the use of heating as a method for contraception. Fertil. Steril., 49: 1.

2. Zorgniotti, A.W. and Sealfon, A.I. 1984. Scrotal hypothermia: new therapy for poor semen. Urology, 23: 439.

3. Zorgniotti, A.W., Sealfon, A.I. and Cohen, M. 1986. Chronic scrotal hypothermia: results in 90 infertile couples. J. Urol., 135: 944.

4. Tritto, G. 1988. The first computer-assisted simulation model of the counter-current heat exchange mechanism: the length-volume dependence on cylindrical and conical configuration of the pampiniform plexus. Sixth Forum of Int. Androl., Paris, (May 3-4), Abst. 100.

5. Tritto, G., Pirlo, G. and Tritto, M.C. 1989. The idealized conical configuration of the pampiniform plexus: a simplified computer-assisted simulation model of the counter-current heat exchange mechanism. Fourth Int. Congr. of Andrology, Florence, (May 14-18), Abst. 198.

6. Tritto, G. 1989. The Finite Element Method: a new tool in gross anatomy applied to reconstructive surgery. XXIVth Congress of the European Society for Surgical Research, Brussels, (May 28-31), 21/S2/, Abst. 58, 27.

7.  Tritto, G. 1989. The computer-graphics simulation modelling of human organs in reconstructive andrology: the FEM approach applied to the penis. Fourth Int. Congr. of Andrology, Florence, (May 14-18), Abst. 197.

8.  Tritto G. 1988. Computer-assisted simulation model of the counter-current heat exchange mechanism at the testicular vascular pedicle. IEEE, 10th EMBS Conference, New Orleans, (Nov. 4-7).

9.  Tritto, G., Franich, A. and Tritto, M.C. 1989. The computer-assisted simulation modelling of the multi-level compartmentation of the human penis in reconstructive andrology. 7th Int. Conf. on Math. and Computer Modelling, Chicago, (Aug. 2-5), Abst. 132.

10. Fung, Y.C., Perrone, N. and Anliker, M. 1972. Biomechanics: its foundation and objectives. Prentice-Hall, New Jersey.

11. Chubb, C. and Desjardins, C. 1982. Vasculature of the mouse, rat, and rabbit testis-epididymis. Am. J. Anat., 165: 357.

12. Ohtsuka, A. 1984. Microvascular architecture of the pampiniform plexus-testicular artery system in the rat: a scanning electron microscope study of corrosion casts. Am. J. Anat., 169: 285.

13. Weerasooriya, T.R. and Yamamoto, T. 1985. Three-dimensional organization of the vasculature of the rat spermatic cord and testis. Cell Tissue Res., 241: 317.

14. Tritto, G. 1988. New perspectives in reconstructive microsurgery of the pampiniform plexus. Int. Conf. on Reproductive Endocrinology, Beijing, (Nov. 2-6).

15. Fawcett, D.W., Neaves, W.B. and Flores, M.N. 1973. Comparative observations on intertubular lymphatics and the organization of the interstitial tissue of the mammalian testis. Biol. Reprod., 9: 500.

16. Suzuki, F. and Nagano, T. 1986. Microvasculature of the human testis and excurrent duct system. Cell Tissue Res., 243: 79.

# THE PHYSIOLOGY OF TESTICULAR THERMOREGULATION IN THE LIGHT OF

# NEW ANATOMICAL AND PATHOLOGICAL ASPECTS

Ahmed Shafik

Department of Surgery
Faculty of Medicine, Cairo University
2, Talaat Harb Street
Cairo, Egypt

Varicocele, the abnormal dilatation and tortuosity of the pampiniform plexus, is common in men; its prevalence is 15% or more. Although contradictory data have been published recently, there is substantial evidence that varicocele impairs spermatogenesis and, consequently, semen quality. Varicocele is considered the most frequent cause of male infertility.

## PHYSIOANATOMICAL CONSIDERATIONS

For better understanding of etiology and pathology of varicocele, and for proper planning of treatment, the physioanatomical aspects of the pampiniform plexus and the surrounding fasciomuscular tube need a special mention here.

### Venous Plexuses of the Spermatic Cord

A previous study [4] has shown that the cord veins are arranged into two systems: the superficial consists of the cremasteric plexus, and the deep one includes the pampiniform and vasal plexuses. The two systems are separated by the internal spermatic fascia and wrapped together by the external spermatic fascial tube (Fig. 1).

The Pampiniform Plexus. The largest plexus is the pampiniform, which lies in the anterior cord compartment. At the lower part of the cord the veins are of small size with poor muscularization. Valves are found in two to three central veins (Fig. 2). At the midcord and external inguinal ring level, the veins are larger, more muscular, and valveless (Fig. 3). The plexus at the midinguinal level consists of two large, well-muscularized and valveless veins surrounded by three to four small ones.

The Vasal Plexus. A small plexus which lies in the posterior cord compartment is the vasal plexus. It is formed of small, poorly muscularized and valveless veins arranged around the ductus deferens (Fig. 4).

The Cremasteric Plexus. The veins of the cremasteric plexus lie between the cremasteric muscle bundles; a few are found scattered in the cremasteric and external spermatic fasciae (Fig. 4). They are smaller than the pampiniform veins, poorly muscularized and valveless.

Communicating Veins. The small, poorly muscularized and valveless veins which connect the three cord plexuses are the communicating veins. They traverse the cremasteric and spermatic fasciae and transmit blood in both directions. They are especially abundant between the pampiniform and cremasteric plexuses.

*Temperature and Environmental Effects on the Testis*
Edited by A. W. Zorgniotti, Plenum Press, New York, 1991

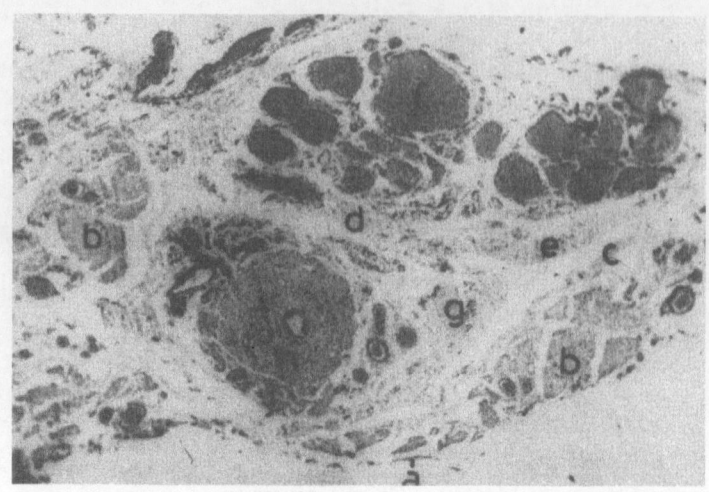

Fig. 1. Transverse section of the spermatic cord showing the cord compartments, the venous plexuses and the cremasteric internus muscle. Hematoxylin and eosin, x 28. (a) external spermatic fascia; (b) cremasteric externus muscle; (c) internal spermatic fascia; (d) transverse septum; (e) pampiniform tube; (f) pampiniform plexus; (g) vasal tube; (h) ductus deferens; (i) vasal plexus; (j) cremasteric internus muscle. (From Shafik et al. [19]).

## THE FASCIOMUSCULAR TUBE OF THE SPERMATIC CORD

The fasciomuscular tube consists of three layers which are extensions of the anterior abdominal wall musculature. These layers comprise the external and internal

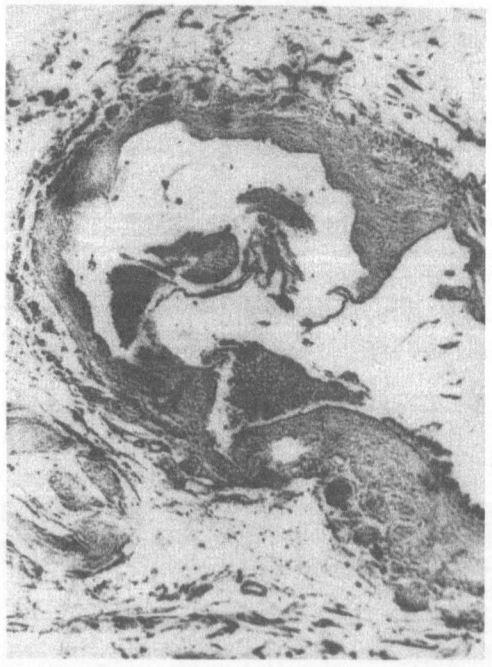

Fig. 2. Photomicrograph showing a valve in a pampiniform vein. Gomori, x 250. (From Shafik [4]).

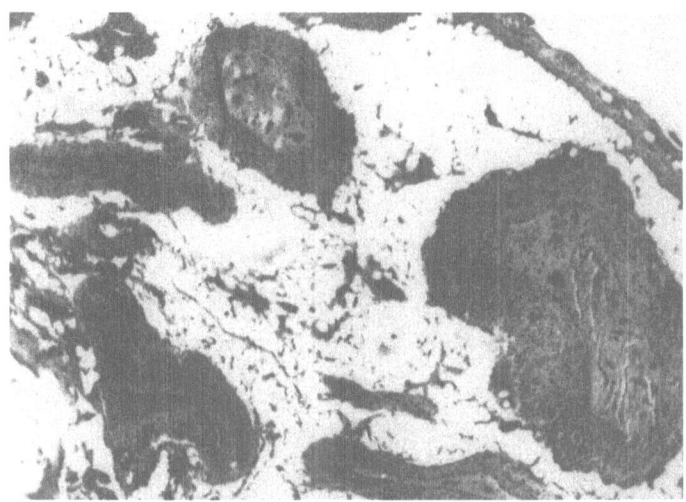

Fig. 3. Photomicrograph of the spermatic cord at the external inguinal ring level, showing that the pampiniform veins are fairly muscular. Hematoxylin and eosin, x 100. (From Shafik [4]).

spermatic fasciae, and between them, the cremasteric muscle and fascia. Microscopic studies showed the tubal fasciae to consist mainly of elastic fibers impregnated with collagen and arranged in a crisscross pattern, lending the fasciae a textile nature (Fig. 5)[19].

The fasciomuscular tube is divided into two compartments, the anterior or pampiniform, and the posterior or vasal, the two being separated by a "transverse fascial

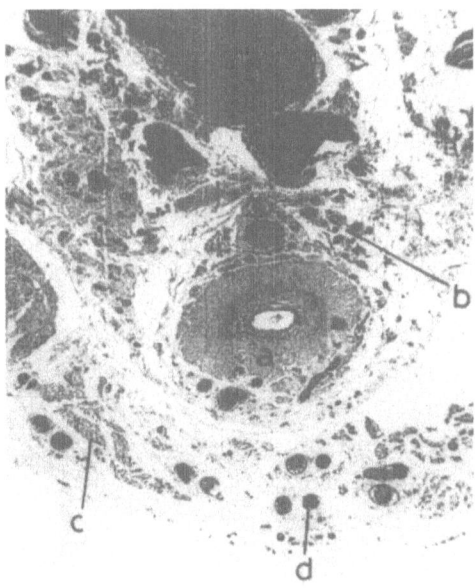

Fig. 4. Photomicrograph of the lower part of the spermatic cord showing the vasal venous plexus in the posterior cord compartment. The cremasteric plexus lies between the cremasteric muscle bundles and fascia. (a) ductus deferens; (b) vasal plexus; (c) cremasteric externus muscle; (d) cremasteric plexus. Hematoxylin and eosin, x 100. (From Shafik [4]).

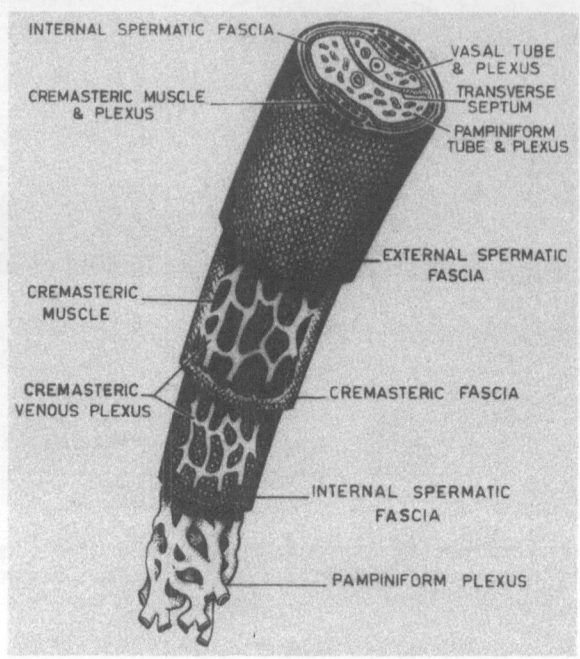

Fig. 5. The fasciomuscular tube and the venous plexuses related to it. The crisscross textile nature of the fasciae is demonstrated. (From Shafik et al. [19]).

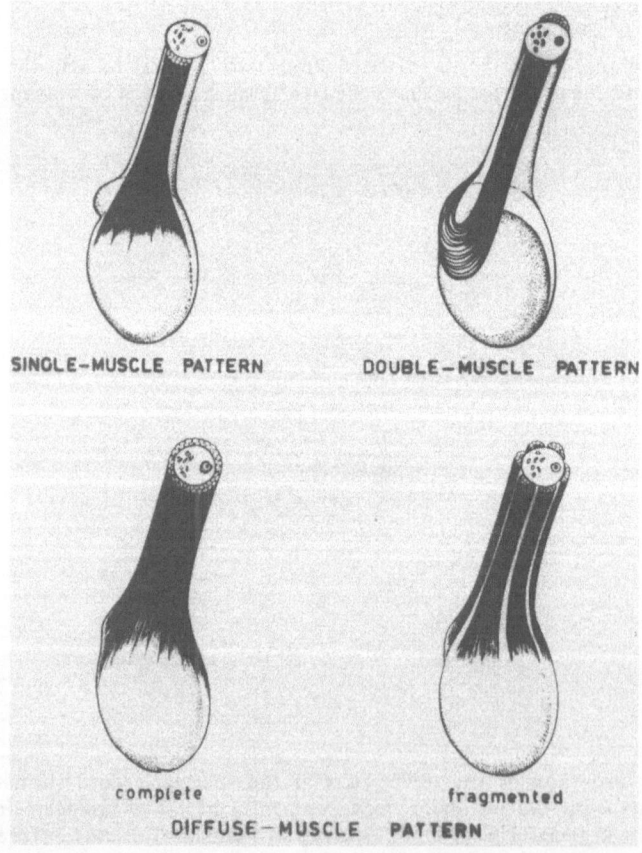

Fig. 6. Diagram of the cremasteric externus muscle patterns. (From Shafik [5]).

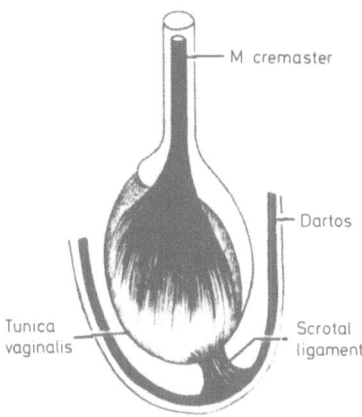

Fig. 7. Diagram of the scrotal ligament. (From Shafik [10]).

septum" deriving from the internal spermatic fascia (Fig. 1)[4, 19]. The pampiniform compartment, surrounded by a fascial tube from the internal spermatic fascia, encloses the pampiniform plexus and the testicular artery. The vasal compartment, surrounded by a similar fascial tube, ensheathes all of the ductus deferens with its artery, the vasal venous plexus and the cremasteric internus muscle [5, 9].

## The Cremasteric Muscles

There are two cremasteric muscles, the externus and internus. The cremasteric externus arises from the obliquus internus abdominis and transversus abdominis muscles, extends along the spermatic cord and testicle, and inserts into the tunica vaginalis. It exists in three patterns: single, double, and most commonly, diffuse (Fig. 6) [5].

The cremasteric internus muscle, as identified in the human (Fig. 1) [5, 9], originates from the transversus abdominis muscle and descends in the spermatic cord, commonly in the vasal compartment and rarely in the pampiniform compartment. It inserts into the tunica vaginalis. In spite of their similar origin and common termination, there is no connection between the cremasteric externus and internus since they are separated by the internal spermatic fascial tube. The cremasteric internus muscle seems to help in the draining of the ductus deferens.

## The Scrotal Ligament

At the scrotal bottom is the scrotal ligament, a small fibromuscular band which ties the dartos closely to the testicle at its lower pole (Fig. 7) [6, 10]. It consists of muscle bundles with collagen fibers scattered between them, and is more developed and muscular in children than in adults. Elsewhere, the dartos is loosely connected with the underlying fasciomuscular tube by areolar tissue.

The ligament plays a significant role in testicular thermoregulation in that it provides the mechanism which keeps the testicular movements in harmony with those of dartos. Furthermore, it constitutes a connection along the testicle between the dartos and the cremasteric muscle, thus synchronizing the action of both muscles [10]. Absence of the scrotal ligament disturbs the thermoregulatory mechanism of the testicle, and leads to the syndrome of "aligamentous testicle" which may cause male infertility [16].

## The Fasciomuscular Pump

A study of the anatomical and histological structure of the fasciomuscular tube has shown that it has more important functions than being a mere fascial covering of the spermatic cord and the testicle [5, 19]. It acts as an "autoelastic stocking" which supports

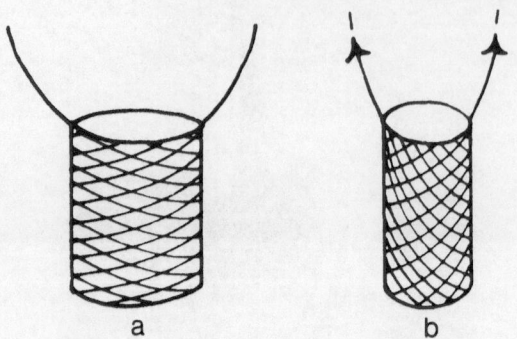

Fig. 8. Diagram of the sphincteric action of the fasciomuscular tube. (a) The tube at rest; (b) the tube during contraction. (From Shafik [11]).

the cord veins. In addition, it constitutes a "pump" by which the blood is pushed up the cord. For this purpose, the spermatic and cremasteric fasciae are provided with a very large number of elastic fibers. The crisscross pattern of fibers provides the fasciomuscular tube with maximal efficiency necessary to its functional performance. The tube possesses a sphincteric action on the cord veins which is greatly potentiated by the contraction of the cremasteric muscle (Fig. 8). This mechanism participates in pumping blood from the testicle and the cord.

The pumping effect is augmented by division of the fasciomuscular tube into two compartments by the transverse fascial septum. The spermatic cord is thus composed of two units, each representing a separate pump containing its own venous plexus, the anterior containing the pampiniform plexus and the posterior containing the vasal plexus. Each pump is surrounded by a layer of elastic fascia (internal spermatic fascia) and supported by a muscular cushion (cremasteric muscle). The two units are wrapped together by an outer common elastic fascia (external spermatic fascia). This arrangement helps to increase the fasciomuscular pump efficiency.

The impregnation of the elastic fibers of the fasciomuscular tube with collagen has an important function. Being inelastic, the collagen fibers limit excessive relaxation of the tube. This prevents tubal subluxation which might result from venous stagnation during erect posture.

### Venous Return from the Testicle

The pampiniform and vasal plexuses collect the venous blood from the testicle and epididymis, while the cremasteric muscle plexus drains the fasciomuscular tube. In the erect position, blood is returned to the heart by the pumping action of the fasciomuscular tube, the cremasteric muscle being the main component of the pump [5, 19]. Each plexus presents certain anatomical features in adaptation to its function [4]. Adequate muscularization of the pampiniform veins augments the pumping mechanism efficiency of the pampiniform plexus. The pampiniform valves provide a protective mechanism against testicular congestion. Snapping shut when the venous pressure rises during increased intra-abdominal pressure, they prevent the full force of this high venous pressure from being transmitted to the testicle. With increased venous pressure and closure of the pampiniform valves, the blood draining from the testicle escapes via the communicating veins to the other valveless plexuses from where it is ejected upwards through the fasciomuscular tube pumping action. These natural mechanisms are designed to prevent testicular congestion and its injurious effects on spermatogenesis.

The location of the cremasteric plexus between the cremasteric muscle bundles increases its venous drainage efficiency. This is important, as the plexus may have to cope with extra blood escaping from the pampiniform veins when the pampiniform valves shut upon venous pressure elevation. Further, as the cremasteric veins lie outside the strong internal spermatic fascial tube and are poorly muscularized and valveless, they yield easily in adaptation to the extra blood volume. For this reason, cremasteric plexus dilatation and varicosity occur early in varicocele [4].

The vasal veins are well supported by the ductus deferens and by their location within a narrow fascial compartment. This renders the veins less liable to dilatation and varicosity.

## Tubovalvular Antireflux Mechanism

The tubovalvular antireflux mechanism is provided by the fasciomuscular tube and the testicular vein valves to protect the testicle from the high venous tension induced by increased intra-abdominal pressure [4-6, 11, 19]. As a prolongation of the abdominal wall muscles, the tube remains in harmony with changes in them. When these muscles contract, increasing intra-abdominal pressure, there is simultaneous contraction of the cremasteric muscle (being a part of the obliquus internus and transversus muscles), as well as traction on the spermatic fasciae in a longitudinal direction (being extensions from the obliquus externus muscle and the fascia transversalis). This mechanism effects constriction of the fasciomuscular tube and tightens the elastic stocking around the cord veins and the testicle, thus preventing venous reflux from the abdominal veins to the testicle (Fig. 8). Together with the shutting of the valves of the testicular veins, this mechanism prevents the full force of the high venous pressure from being transmitted to the testicle.

The location of the valves in the testicular system of veins varies. However, whether they are located down in the pampiniform veins as demonstrated by Shafik [4] or high up in the testicular vein as recorded by Revington [2] and Kohler [1], the valves function to prevent venous reflux to the testicle. Nevertheless, the scarcity and inconsistent position of the valves emphasize the importance of the tubal antireflux mechanism as compared with the valvular one [11].

## Role of Valves in Reflux

A recent study [21] has shown that absent or incompetent testicular valves alone do not lead to reflux under normal physiologic conditions. The pressure gradient and the negative pressure created at the testiculo-renal junction are the two factors responsible for maintaining continuous blood flow into the renal vein, regardless of valves. Reflux down the testicular vein was found to be secondary to a change in the pressure gradient and negative pressure at the testiculo-renal junction, as could be demonstrated by pressure measurement [12, 13].

## The Role of the Fasciomuscular Tube in Testicular Thermoregulation

The fasciomuscular tube, by its sphincteric action on the cord veins, plays an important role in the temperature-regulating mechanism of the testicle [5, 8]. On contraction the tube, especially its cremasteric component, compresses the cord veins. The resulting diminished blood volume and shrinking of the exposed cord surface area prevent heat loss. Meanwhile, the testicular temperature increases owing to the testicle being drawn near the warmer body surface. Tubal relaxation leads to an increased blood volume within the cord veins, a greater exposed cord surface area, and a lowered testicular position, a mechanism which favors heat loss. The active contractile function of the cremasteric muscle is so marked in some animals that the testicle can be retracted within the abdomen and extruded into the scrotum voluntarily. This occurs in most rodents, many insectivores, and bats [22].

## The Role of Dartos in Testicular Thermoregulation

The dartos muscle bundles, being arranged in a crisscross "plywood" pattern, constitute potential "sphincters" around the blood vessels between their decussations [6]. When the muscle contracts, the intervening vessels are constricted, resulting in diminished scrotal blood flow, a mechanism which minimizes heat loss (Fig. 9). Meanwhile, the testicle is elevated close to the warm body surface. With dartos relaxation the potential sphincters relax, leading to increased blood circulation and heat radiation (Fig. 9). The testicle is drawn away from the body surface by the scrotal ligament.

Another mechanism is provided by the dartos meshwork arrangement. Mesh obliteration by dartos contraction helps to preserve the intrascrotal temperature, while the opening of meshes encourages heat loss.

## Cremasterico-Darto-Venous Complex

The testicle possesses a "thermoregulatory apparatus" which maintains a constant scrotal-rectal temperature difference. It consists of the cremasterico-darto-venous complex which includes the cremasteric muscle and the dartos, as well as the cord and scrotal vessels [8]. The complex is structurally adapted to serve as a thermostat for the testicle. Its components work collectively and spontaneously in reaction to changes in environmental temperature. The cremasterico-dartos component through its sphincteric action, regulates the amount and rate of blood flow within the cord and scrotal vessels. In addition, it changes the exposed surface areas of both the scrotal skin and spermatic cord. Thus, on exposure to heat, cremasterico-dartos relaxation increases the cord and scrotal blood flow and exposed surface area, factors which encourage heat radiation. Conversely, cremasterico-dartos contraction preserves the testicular temperature by diminishing the cord and scrotal blood flow and surface area.

The aeration induced by the dartos meshes or "windows" [6], and the heat radiated from the warm abdominal wall are accessory factors to the cremasterico-darto-venous complex. Closure of the dartos windows on dartos contraction transforms the scrotum into a thermoisolated cavity; the testicles being high in the scrotum gain warmth from the abdominal wall. Opening of the dartos windows by dartos relaxation, while the testicles are in a lowered position, helps testicular aeration and heat radiation.

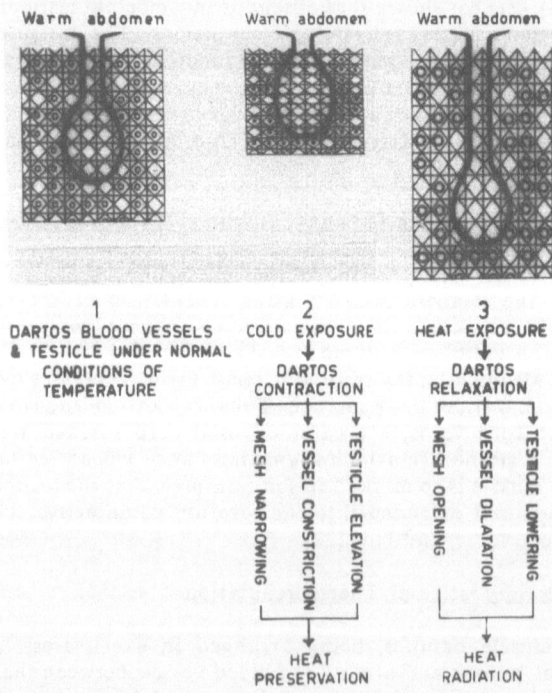

Fig. 9. Mechanism of thermoregulatory function of dartos. (From Shafik [6]).

## Weight-Bearing Mechanism of the Testicle and Varicocele

Under normal temperature conditions, the testicular weight is carried by the fasciomuscular tube, especially its cremasteric component, and not by the dartos or the spermatic cord [6]. With variations in scrotal sac temperature, the dartos goes into action. Thus, on exposure to cold, both the dartos and the cremasteric muscle contract, carrying the testicle in close proximity to the warmer abdominal surface. Under the influence of heat, both muscles relax, the testicle being carried by the cremasteric muscle only, the dartos functioning to pull the testicle away from the abdomen by the scrotal ligament [10]. With excessive scrotal elongation, as in varicocele, the testicular weight is taken off the cremasteric muscle by the spermatic cord, which thus becomes stretched. The scrotal pain in varicocele is attributable to overstretching of both the cremasteric muscle and the spermatic cord because of excessive scrotal elongation [6].

## PATHOGENESIS OF VARICOCELE

### The Role of the Fasciomuscular Tube in Varicocelogenesis

A study of the fasciomuscular tube in primary varicocele showed that it was flabby and capacious, with degeneration and atrophy of the cremasteric muscle and the elastic fibers [5, 19]. Due to tubal subluxation, the sphincteric action of the tube is lost, efficiency of the pumping mechanism is reduced, and the cord veins are supported less and cannot withstand increased venous pressure during muscular contraction. In addition, the squeezing effect of the elastic fibers is lost, owing to their atrophy. As a result, the tubovalvular antireflux mechanism is disturbed: the veins become engorged with blood and dilated, which in turn renders the valves incompetent. This throws an additional load onto the already weakened tube and pumping mechanism. The chain of events which lead to a varicocele is thus initiated.

The tube, in addition to its function as a venous pump, suspends the testicle in its normal position in the scrotum. Tubal subluxation leads to a lowered testicular position which results in further blood stagnation and varicosity.

In light of this new concept, a varicocele is considered to result from dysfunction of the fasciomuscular venous pump, attributable to a subluxated tube rather than to a disease of the veins themselves [5, 19]. An operation [3] in which the tube is plicated by multiple purse-string stitches has been successfully practiced for the correction of the atrophic subluxated tube and the cure of a varicocele. In severely atrophic tubes, as in advanced and recurrent varicoceles, a fascia lata graft of the tube produced satisfactory results [7].

### Experimental Varicocele Model

The fasciomuscular tube subluxation in the genesis of varicocele could be developed experimentally. Varicocele was induced in dogs by detubation of the spermatic cord [19, 20]. The animals with induced varicocele had higher testicular temperature than the controls. They showed microscopic variceal and testicular degenerative changes with similar changes in the contralateral testicle. Radioimmunoassay showed a significant decrease of serum testosterone and an increase of prolactin postoperatively. Follicle-stimulating and luteinizing hormones showed no significant change from the preoperative level.

### Venous Tension in Varicocele

A study on venous tension patterns in the cord veins [12, 13] has shown that the average normal venous tension with the individual at rest is 58.7 mm Hg on the right side and 59.9 mm Hg on the left side. In varicocele patients, venous tension on the right, nonvaricose side is slightly above normal with an average of 59.6 mm Hg whereas on the left, varicose side, it is considerably higher, the average being 79.6 mm Hg. During Valsalva's maneuver, the venous tension on the left, varicose side is elevated to an average of 83.8 mm Hg.

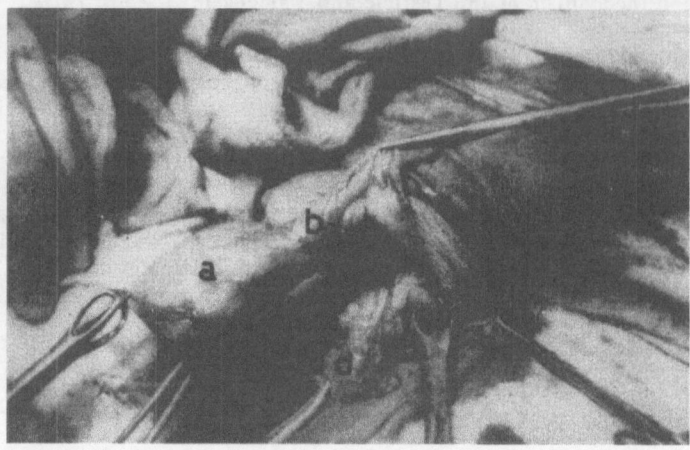

Fig. 10. Operative view showing normal fat pattern in infertile subject. (a) testicle; (b) spermatic cord; (c) fasciomuscular tube, opened; (d) posterior extratunicary pad of fat. (From Shafik and Olfat [17]).

These figures demonstrate conclusively the presence of venous hypertension in varicose veins, which is an effect of venous backflow as evident from the marked increase in venous tension after Valsalva's maneuver. These data further show that the high venous tension is not transmitted across the communicating veins from the varicose to the contralateral side, as indicated by the nearly normal tension values in the latter.

Venous reflux and hypertension eventually lead to congestion, dilatation, and varicosities of the cord veins. Cord congestion results in impaired venous drainage of the testicle with a consequent testicular congestion and possibly accumulation of toxic testicular metabolites. The latter could be responsible for inhibiting the spermatogenic function of the testicle.

### Local Circulation Theory in Bilateral Effect of Varicocele

The cause of bilateral effect of varicocele leading to infertility is not known. Our studies exclude the effect being caused by contralateral venous reflux and hypertension,

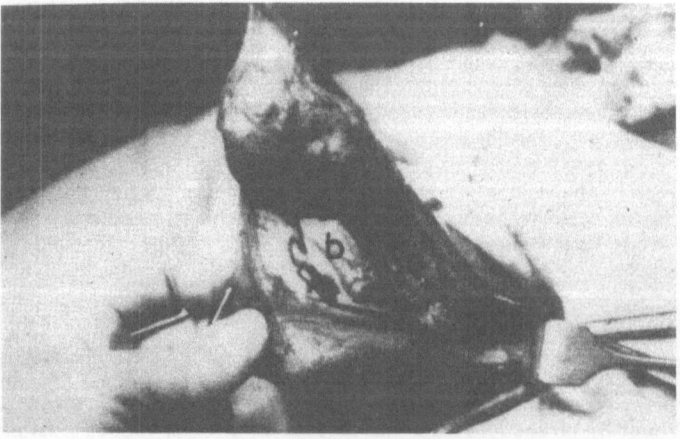

Fig. 11. Operative view of lobular lipomatosis in infertile subject showing the posterior extratunicary pad of fat containing tortuous veins. (a) spermatic cord; (b) extratunicary fat. (From Shafik and Olfat [17]).

162

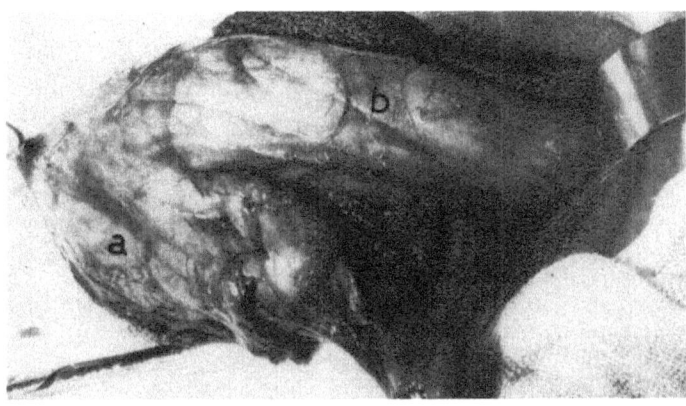

Fig. 12. Operative view showing diffuse lipomatosis of the spermatic cord in infertile subject. The fat extends down to cover the testicle. (a) fat pads covering the testicle; (b) spermatic cord. (From Shafik and Olfat [17]).

since venous tension on the non-varicosed side was found either normal or negligibly elevated [13]. However, it could be the difference in venous tension between the two sides that allows the blood carrying toxic metabolites to flow across the communicating veins between the 2 testicular veins from the varicosed high tension side to the other lower tension side [21]. Furthermore, it seems that these metabolites should reach a certain concentration to induce a contralateral inhibitory effect [13].

## Scrotal Fat and Varicocele

The anatomy of the scrotal fat was studied histologically in 28 normal cadavers and in 44 infertile subjects [17, 18]. Two fat patterns were described, normal and infertile. According to its relation to the fasciomuscular tunics of the spermatic cord, fat can be extra-, inter-, and intra-tunicary.

In the normal fat pattern, a small extratunicary triangular pad of fat is consistently encountered posterior to the cord (Fig. 10). It is continuous over the pubic ramus with the suprapubic fat. Intratunicary fat occurs as small granules between the cord veins.

Of the 44 infertile subjects, 38 showed excess and abnormally distributed fat, a condition I describe as "scrotal lipomatosis" of which two types are recognized: extratunicary and intratunicary [17]. In the former a thick pad of fat containing multiple small tortuous veins exists posterior to the spermatic cord (Fig. 11). The latter presents two patterns, diffuse and lobular. The diffuse pattern occurs both in obese subjects and in those of normal build. The fat takes the form of a diffuse sheet which covers and is adherent to the cord veins (Fig. 12). The lobular pattern (Fig. 13) occurs exclusively in the obese. The fat lobules are easily separable from the cord veins. Each lobule has a pedicle which can be traced along the inguinal canal to the extraperitoneal fat. The pedicles gather in the inguinal canal to form a "lipomatous cord" which suspends the testicle high in the abdomen.

At operation, the cremasteric veins were found to be dilated and tortuous in all of the 38 cases of scrotal lipomatosis [17]; pampiniform varicosity was also present in 20 cases (Fig. 14). Cord varices were not palpable clinically, being masked by lipomatosis. However, in seven patients with diffuse lipomatosis there was scrotal redundancy which could indicate cord varicosity.

Cord varices in scrotal lipomatosis seem to be due to venous stasis, resulting from the intratunicary fat compressing the veins and interfering with the venous pumping mechanism [17]. Scrotal redundancy in diffuse lipomatosis is due to cord varicosity.

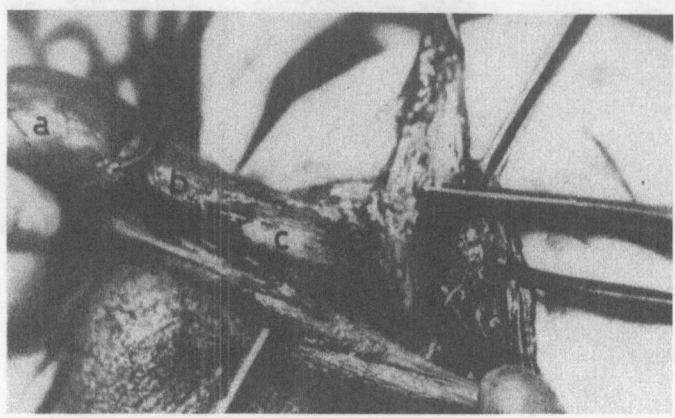

Fig. 13. Operative view showing lobular lipomatosis in infertile subject. (a) testicle; (b) spermatic cord; (c) fat lobule; (d) posterior extratunicary pad of fat. (From Shafik and Olfat [17]).

However, it is not evident in lobular lipomatosis, despite cord varicosity, as the testicle is supported by the "lipomatous cord".

## PATHOLOGY OF VARICOCELE

A study of the histopathological changes of the cord veins in 28 patients with left-sided varicocele was performed [14]. Twenty-five patients were infertile and three had fathered children.

The cremasteric veins were varicose in all cases. They were best identifiable over the anterior aspect of the tunica vaginalis, being the only veins in this area. The veins were small and tortuous. They lay under the glistening external spermatic fascial tube, and varicosity became less manifest as the veins were followed up the cord. The pampiniform plexus, which was revealed after the internal spermatic fascial tube had been opened, occupied the anterior cord compartment. The varicosity was maximal at

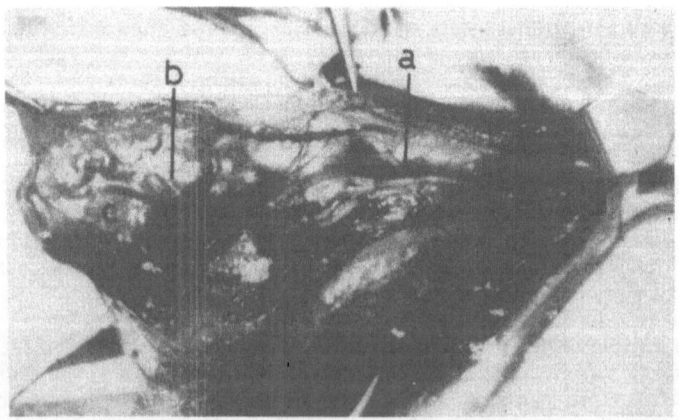

Fig. 14. Operative view showing cremasteric and pampiniform varicosity in infertile subject with lobular lipomatosis. Extratunicary and intratunicary fat were removed. (a) pampiniform varices; (b) cremasteric varices; (c) testicle. (From Shafik and Olfat [17]).

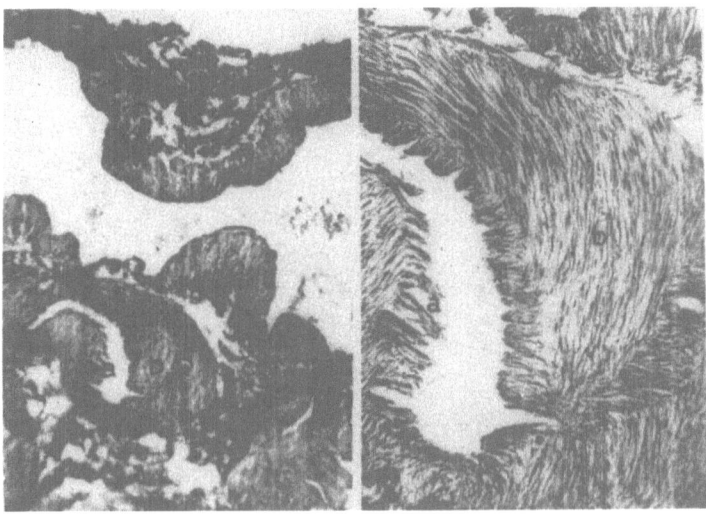

Fig. 15. Photomicrograph of varicose pampiniform veins. Intima shows papillary projections. The media is greatly thickened and its muscle bundles are hypertrophied. (a) intima; (b) media. Hematoxylin and eosin; left x 50, right x 185. (From Shafik and Olfat [14]).

the back of the testicle, and diminished gradually as it extended upwards. Pampiniform varicosity occurred in 25 patients.

At operation pampiniform varices were distinguished from cremasteric ones in that (a) they covered the testicular back and not the front as occurs in cremasteric varices, (b) they were under the cover of the internal fascial tube, and (c) they were larger.

The vasal plexus was not varicosed in any patient, and in the inguinal canal, no varicosities whether cremasteric, pampiniform, or vasal could be detected.

Microscopic study of the pampiniform veins showed the intima thrown into folds in some specimens and with papillary formations in others (Fig. 15). The tunica media was greatly thickened due to muscular hypertrophy; in some specimens it was hyalinized (Fig. 16) and in others it was fragmented in its outer aspect. The adventitia was slightly thickened. In hugely dilated and tortuous veins septa could be demonstrated transecting, partially or completely, the venous lumen (Fig. 17). They consisted of hypertrophied muscle bundles continuous with those of media, and were lined with intima endothelium.

The cremasteric veins had regular intima, while the media was hypertrophied, with hyalinized areas. The vasal veins were microscopically normal.

Pathological Staging of Varicocele

Whatever the primary etiological factor of varicocele may be, venous hypertension in the cord veins is a constant feature and is responsible for the different pathological changes which occur in both the cord veins and the testicle (Fig. 18) [13]. In varicocele development, the increased venous pressure leads to hypertrophy of the media. In the early stage, the veins are thickened but not dilated. Venous stasis does not occur because the venous pumping mechanism efficiency is augmented due to the medial hypertrophy.

With further increase of venous tension there is increased medial hypertrophy and venous pumping power until the muscle bundles overstretch and fail. During this stage the pumping mechanism efficiency is reduced and blood stagnates. This results in more venous tension which burdens the already failing pumping mechanism. A vicious circle which leads to cord vein congestion is thus created. With prolonged venous hypertension,

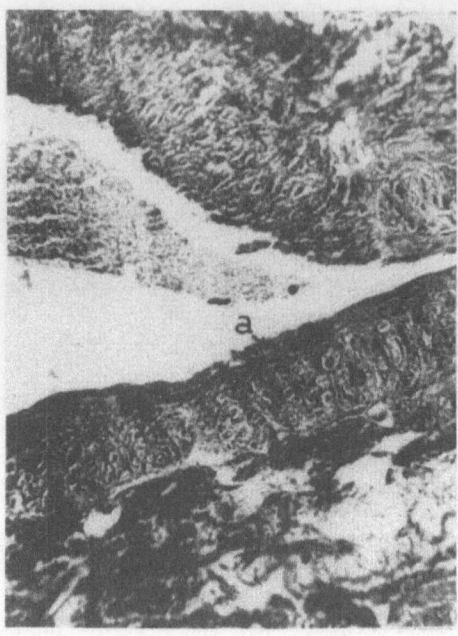

Fig. 16. Photomicrograph of a varicosed pampiniform vein showing hyalinization of the media. (a) intima (b) media. Hematoxylin and eosin, x 62. (From Shafik and Olfat [14]).

degenerative changes as evidenced by hyalinization and fragmentation affect the venous wall. Later on, atrophy and fibrosis lead to tortuosity and sacculation of the veins.

The development of septa in the hugely dilated veins seems to be a mechanism provided by nature to fractionate the large, heavy blood column within these veins in an attempt to augment the efficiency of the venous pumping mechanism and to minimize

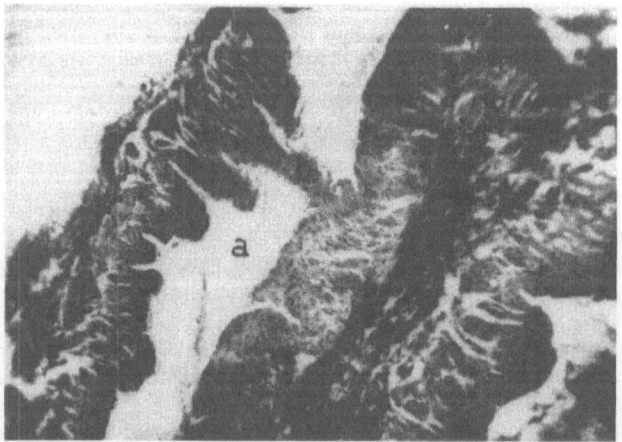

Fig. 17. Photomicrograph of a varicosed pampiniform vein showing a complete septum transecting the venous lumen. (a) venous lumen; (b) complete septum. Hematoxylin and eosin, x 62. (From Shafik and Olfat [14]).

166

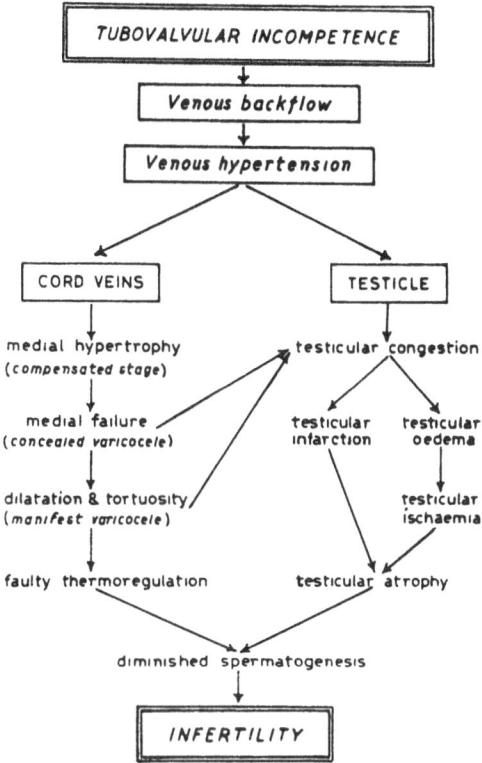

Fig. 18. Mechanism of varicocele formation.

venous stasis [14]. The mechanism of septa formation is mysterious. It could be that kinking of the veins due to tortuosity results in the formation of projections within the venous lumen. The blood backflow and venous hypertension exert their effects on these projections leading to their hypertrophy and incomplete septa formation. These septa may fuse with denuded intimal areas in the venous wall, with a resulting complete septa formation. The septa seem to protect the testicle as much as possible from the detrimental effect of the venous backflow and hypertension which occur in advanced varicocele.

Thus, three pathological stages can be identified in varicocele: hypertrophy, failure, and dilatation [14].

Hypertrophy (Compensated Stage): The earliest stage of varicocele is hypertrophy. The veins are thickened. There is no blood stasis as the venous return is kept normal by the compensatory medial hypertrophy. No testicular congestion or spermatogenic depression occurs at this stage. Clinically, the patient is symptomless and cord veins are not dilated.

Failure (Concealed Varicocele): The continuous rise of venous tension results in medial failure and venous stasis. At this stage, there is cord and testicular congestion, although the cord veins are clinically not varicose. Palpable thickening of the veins may be detected. I call this stage "concealed varicocele" [15]. It is of special clinical interest since there is testicular congestion and possible spermatogenic depression although varices are clinically undetectable. "Concealed varicocele" could thus be responsible for idiopathic infertility in some patients in whom no varicocele can be detected by clinical examination.

Dilatation (Manifest Varicocele): With prolonged and sustained venous hypertension, degeneration and atrophy of cord veins result in dilatation and varicosity.

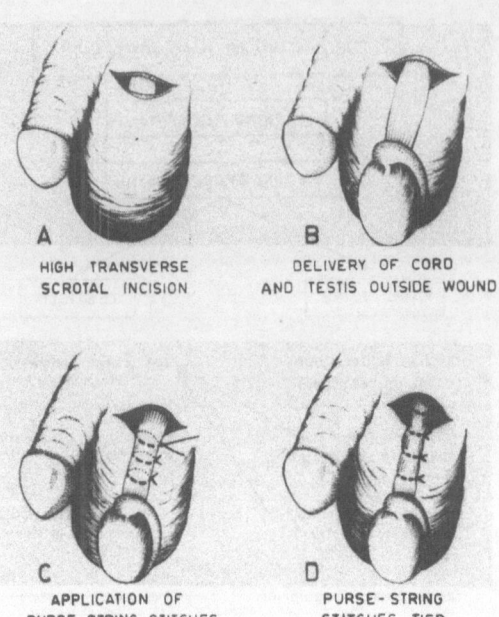

A — HIGH TRANSVERSE SCROTAL INCISION

B — DELIVERY OF CORD AND TESTIS OUTSIDE WOUND

C — APPLICATION OF PURSE-STRING STITCHES

D — PURSE-STRING STITCHES TIED

Fig. 19. Plication operation for varicocele. The spermatic cord is plicated with multiple purse-string stitches. (From Shafik [3]).

This is the stage of "manifest varicocele" in which testicular congestion and spermatogenic depression are maximal. Clinically, the cord veins present the classical varicocele picture.

### Anatomical Staging of Varicocele

With venous hypertension and medial failure, congestion and dilatation manifest primarily in the cremasteric plexus, being the least supported. When this plexus is fully engorged, the pampiniform veins are the next to congest and dilate. Vasal plexus varicosity occurs only late, as the plexus is well supported.

Accordingly, a varicocele has three anatomical stages: cremasteric, pampiniform, and vasal [14]. They are not separate types, but different stages in one process of venous congestion.

Cremasteric Varicosity (Stage I Varicocele): In the earliest stage of varicocele, the cremasteric is the first plexus to react to venous hypertension in the cord veins because (a) it is less supported, owing to its location outside the rigid internal spermatic tube; (b) its veins are poorly muscularized [4]; and (c) the communicating veins between the cremasteric and other venous plexuses, being numerous, transmit the high venous tension from inside to outside the internal spermatic tube [4].

Pampiniform Varicosity (Stage II Varicocele): With maximal cremasteric plexus dilatation and continuously increasing venous tension, the pampiniform plexus starts to congest and dilate. The pampiniform veins can absorb high venous pressure for long periods because (a) they are numerous with ample collaterals; (b) they are well muscularized and thus effect appreciable resistance to venous tension [4]; and (c) they are adequately supported, being enclosed within the internal spermatic tube.

Vasal Varicosity (Stage III Varicocele): Vasal varicosity occurs late in varicocele. Being well supported by the ductus deferens and its location inside the narrow vasal compartment, the vasal plexus requires a rather high venous tension to varicose. For this reason vasal varicosity was not encountered in a series of 28 varicocele patients [14]. It seems to occur in secondary varicocele owing to testicular vein obstruction by a tumor; this would create venous tension high enough to overcome the extra support provided to the vasal veins.

### Dartos Muscle in Varicocele

The surgical anatomy of the dartos was studied in both normal and varicocele subjects [6]. Normally, the muscle bundles are arranged separately and not in fasciculi. They decussate in a crisscross pattern, and the blood vessels occupy the spaces between the decussating bundles. Each scrotal compartment has its own dartos, and both muscles share in the scrotal septum formation.

In varicocele the scrotal skin is thinner on the affected side than on the control side, and the dartos is attenuated and acquires a more superficial position. The spaces between the decussating bundles are wide and contain dilated and thin-walled vessels.

## TREATMENT OF VARICOCELE

### Plication Operation

Previous studies [4-6, 19] have shown that the fasciomuscular tube of the spermatic cord plays a significant role in the venous pumping mechanism of the cord and that it is consistently subluxated in varicocele. Accordingly, the plication operation [3] has been devised to correct the subluxated tube and the disordered pumping mechanism. The technique has by now been performed in 380 varicocele patients.

Technique: The spermatic cord within its tube and the testicle are brought out through a transverse scrotal incision. The tunica vaginalis is everted. The fasciomuscular tube is plicated by multiple purse-string stitches, 0.5 inch apart, starting close to the testicle and continuing up to one finger"s breadth below the superficial inguinal ring (Fig. 19). The stitches, 3/0 silk placed by an atraumatic needle, should include the whole tubal thickness, and in the lower part of the cord the everted tunica as well. They are cautiously tightened to afford support but overcorrection is avoided.

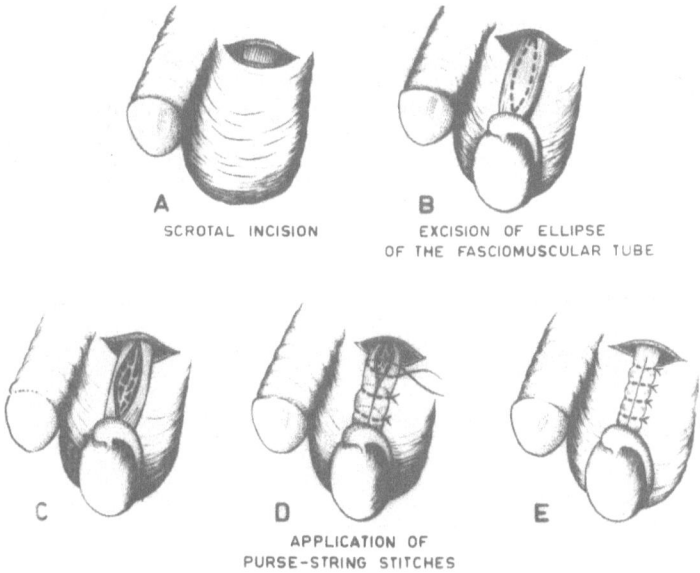

Fig. 20. Plication operation. Excision of longitudinal strip of fasciomuscular tube prior to application of purse-string stitches. (From Shafik [3]).

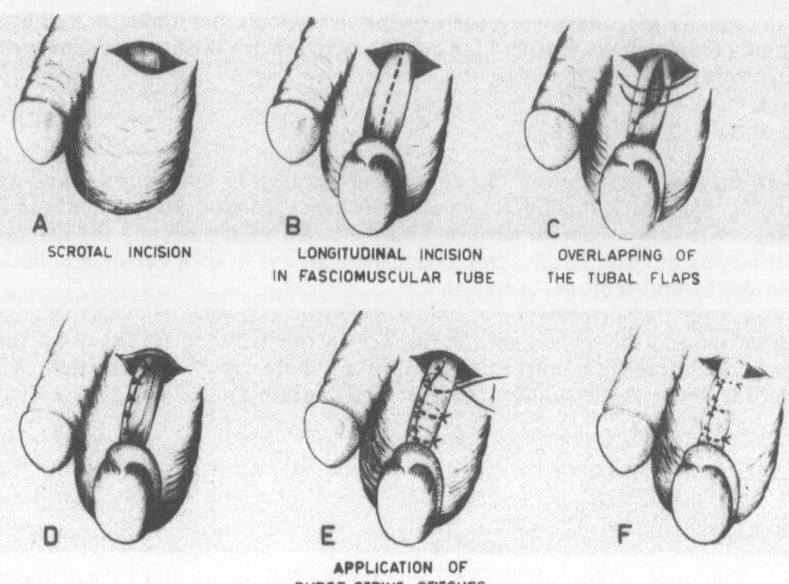

A. SCROTAL INCISION
B. LONGITUDINAL INCISION IN FASCIOMUSCULAR TUBE
C. OVERLAPPING OF THE TUBAL FLAPS
D.
E. APPLICATION OF PURSE-STRING STITCHES
F.

Fig. 21. Plication operation. Incision and overlapping of the fasciomuscular tube prior to application of purse-string stitches. (From Shafik [3]).

If the tube is too capacious, as occurred in 83 cases, a longitudinal strip is excised prior to the application of the purse-string stitches (Fig. 20). In another 62 cases the capacious tube was incised longitudinally and its two flaps were overlapped before the purse-string stitches were applied (Fig. 21). The testicle is then replaced into the scrotum, and the wound closed with drainage. The operated scrotal compartment is suspended to the abdominal wall by a silk stitch for 48 hours.

Results: The convalescent period was uneventful. Intrascrotal hematoma did not occur in any patient. Of the 380 patients operated upon, 348 were followed up for periods varying between 1.5 and 5 years. No testicular atrophy was encountered. Recurrence of varicocele occurred in 4 patients (1.1%).

Pre- and postoperative measurement of the venous tension in the left and right cord veins was performed in 38 patients of the series [12]. All patients showed preoperative venous hypertension and reflux in the left cord veins (Table 1). The readings with patients at rest varied from 70 to 92 mm Hg with an average of 81.8 mm Hg, and during

Table 1. Preoperative venous tension (mm Hg) in 38 infertile varicocele patients

|  | Left side | | Right side | |
|  | At rest | During Valsalva's maneuver | At rest | During Valsalva's maneuver |
| --- | --- | --- | --- | --- |
| Mean | 81.8 | 85.8 | 60.5 | 61.7 |
| SD | ± 5.1 | ± 5.2 | ± 5.3 | ± 5.8 |
| Range | 70-92 | 76-96 | 52-74 | 54-80 |

SD = Standard deviation
$p \leq 0.05$

Table 2. Postoperative venous tension (mm Hg) in 38 patients

| | Left side | | Right side | |
| | At rest | During Valsalva's maneuver | At rest | During Valsalva's maneuver |
|---|---|---|---|---|
| Mean | 60.8 | 62.5 | 59.5 | 60.4 |
| SD | ± 1.8 | ± 1.82 | ± 5.1 | ± 5.5 |
| Range | 55-64 | 58-66 | 52-74 | 53-78 |

SD = Standard deviation
p ≤ 0.05

Valsalva's maneuver from 76 to 96 mm Hg with an average of 85.8 mm Hg. Venous tension in the right cord veins was within normal range in all patients (Table 1) with the exception of four who showed venous hypertension despite absence of both cord varicosities and reflux.

Postoperatively, venous tension on the left side normalized in all patients, with disappearance of both varicosities and reflux. With the patient at rest, venous tension varied between 55 and 64 mm Hg with an average of 60.8 mm Hg, while it varied between 58 and 66 mm Hg with an average of 62.5 mm Hg during Valsalva's maneuver (Table 2). In the right cord veins, postoperative venous tension remained at normal levels (Table 2). The four patients with bilateral venous hypertension showed left venous tension normalization but no change in the right venous tension. Subsequent right-sided plication operation corrected venous hypertension on the right side, too.

Merits of the Operation: Plication aims at the correction of tubal subluxation in order to reestablish an efficient fasciomuscular pump [3]. The improvement which follows the operation may be attributed to (a) the better support of the cord venous plexuses, preventing their dilatation and improving the pumping mechanism efficiency; and (b) venous decongestion, which is achieved early by the support of the purse-string stitches and later by the fibrosis induced by them.

The effectiveness of the operation was demonstrated in the present series by the disappearance of both varicosities and reflux, and by normalization of venous tension. The operation does not disturb the cord contents, especially the arterial supply to the testicle and epididymis. It is worth mentioning that the different operations devised for varicocele therapy have always been directed towards the treatment of the mechanical effects of varicosities rather than towards the primary abnormality. In this respect, the plication operation may be regarded as an attempt to correct a basic factor responsible for varicocele formation.

ACKNOWLEDGEMENT

The valuable assistance of Mrs. Margot Yehia in revising the manuscript is highly appreciated.

REFERENCES

1. Kohler F. P. (1967) On the etiology of varicocele. J. Urol. 97: 741
2. Rivington W. (1873) Valves in the renal veins. J. Anat. Physiol. Norm. Pathol. Homme Anim. 7: 163
3. Shafik A. (1972) Plication operation. A new technique for the radical cure of varicocele. Br. J. Urol. 44: 152

4. Shafik A. (1973) Venous plexuses of the spermatic cord. Anatomy and role in varicocele. Urologia 40: 419

5. Shafik A. (1973) The cremasteric muscle. Role in varicocelogenesis and in thermoregulatory function of the testicle. Invest Urol. 11: 92

6. Shafik A. (1973) The dartos muscle. A study of its surgical anatomy and role in varicocelogenesis. A new concept of its thermoregulatory function. Invest Urol. 11: 98

7. Shafik A. (1973) Fascial grafting of the spermatic cord. A new operative procedure for the treatment of varicocele. Urologia 40: 530

8. Shafik A. (1974) Thermoregulatory apparatus of the testicle. A review. Urologia 41: 473

9. Shafik A. (1976) Cremasteric internus muscle. Further study. Urologia 43: 25

10. Shafik A. (1977) Anatomy and function of scrotal ligament. Urology 9: 651

11. Shafik A. (1977) The cremasteric muscle. In: Johnson A.D. and Gomes W.R. (eds) The testis. Academic Press, New York, p. 487

12. Shafik A. (1983) Venous tension patterns in cord veins. II. After varicocele correction. J. Urol. 129: 749

13. Shafik A. and Bedeir G.A.M. (1980) Venous tension patterns in cord veins. I. In normal and varicocele individuals. J. Urol. 123: 383

14. Shafik A. and Olfat S. (1977) Venous plexuses of the spermatic cord in varicocele. A histopathological study. Urologia 44: 56

15. Shafik A. and Olfat S. (1977) Concealed varicocele. A clinicopathological entity. Urologia 44: 465

16. Shafik A. and Olfat S. (1979) Aligamentous testicle. New clinicopathologic entity in genesis of male infertility and its treatment by orchiopexy. Urology 13: 54

17. Shafik A. and Olfat S. (1981) Scrotal lipomatosis. Br. J. Urol. 53: 50

18. Shafik A. and Olfat S. (1981) Lipectomy in the treatment of scrotal lipomatosis. Br. J. Urol. 53: 55

19. Shafik A., Khalil A.M. and Saleh M. (1972) The fasciomuscular tube of the spermatic cord. A study of its surgical anatomy and relation to varicocele. A new concept for the pathogenesis of varicocele. Br. J. Urol. 44: 147

20. Shafik A., Wali M.A., Abdelaziz Y.E. et al. (1989) Experimental model of varicocele. Eur. Urol. 16: 298

21. Shafik A., Moftah A., Olfat S., Mohi-el-Din M., el-Sayed A. (1990) Testicular veins. Anatomy and role in varicocelogenesis and other pathologic conditions. Urology 35: 175

22. Sisson S. and Grossmann, J.D. (1968) The anatomy of the domestic animals, 4th edition. Saunders, Philadelphia, p. 584

# EFFECT OF LOCAL HEATING ON SCROTAL TEMPERATURE

R. Hsiung[1], B. Bothorel[2], G. Dewasmes[2],
V. Candas[2] and A. Clavert[1]

[1]CECOS of Alsace, 1 Place de l'Hôpital
F67091 Strasbourg Cedex, France
[2]LPPE of CNRS, 21 Rue Becquerel
F67087 Strasbourg Cedex, France

Spermatogenesis needs special thermic conditions. The deleterious effect of hyperthermia on testicular function has been largely demonstrated (Kandeel and Swerdloff, 1988). The low testicular temperature is due to numerous regulatory systems (Setchell, 1978). In some pathologic cases, infertility is the consequence of hyperthermia (Zorgniotti and MacLeod, 1973; Mieusset et al., 1987). The aim of this study was to try to improve our understanding of human testicular regulation by a continuous measurement of scrotal skin temperature during local heating. It has been demonstrated that scrotal temperature reflects intratesticular temperature (Kurtz and Goldstein, 1986).

## MATERIAL AND METHODS

A population of 120 infertile men were submitted to a continuous measurement of scrotal skin temperature during local heating. A thermography was previously performed with a heat reactive crystal plate after 10 minutes equilibration at room temperature of 20°C - 25°C (Hsiung, 1990). The scrotum of the subject in supine position, was placed on a plate. Two skin temperature sensors were fixed with a Band-Aid on the right and the left scrotum. A third sensor was placed on the abdominal skin above the pubis. The latter was considered as the reference of normal skin behaviour during the local heating protocol. The lower part of the body was unclothed whilst the upper part of the body remained clothed normally. The basic temperature was recorded for a few minutes.

A plastic bag containing thermostatically controlled water (40°C) was used to heat the scrotum and the abdominal skin. The heating bag was kept on the skin until stabilization of temperature, which was obtained after 7-10 minutes. The highest temperatures were noted and then the plastic bag was removed. The temperatures were recorded for 10 more minutes during the spontaneous temperature decrease after removal of the heating bag. Temperatures at the end of the experiment were noted.

## RESULTS

The results are shown in Fig. 1 and Table 1. The mean abdomino-scrotal gradient between abdominal skin and right scrotal skin was 0.7°C, in thermoneutral conditions (p<0.001). During the heating phase we observed a generally slower rise for scrotal skin temperature than abdominal skin temperature. Mean abdomino-scrotal gradient was 0.5°C (p<0.001) at the end of the heating phase; the gradient was thus reduced compared

Table 1. Mean values of abdominal and scrotal skin temperature before, during and after the local heating

| | Skin Temperatures (°C) | | |
| | Before heating (n = 120) | Maximal value during heating (n = 120) | End value of cooling phase (n = 120) |
| --- | --- | --- | --- |
| Abdomen | 33.3 ± 1.0 | 38.9 ± 0.6 | 34.5 ± 0.8 |
| Right scrotum | 32.6 ± 0.9 | 38.4 ± 0.6 | 34.4 ± 0.9 |
| Left scrotum | 32.9 ± 0.9 | 38.6 ± 0.6 | 34.5 ± 0.9 |
| Abdomino-scrotal gradient | 0.7 | 0.5 | 0.1 |
| (Student test) | (p<0.001) | (p<0.001) | (NS) |

to the initial gradient observed in thermoneutral conditions. During the cooling phase, after removal of the heating bag, a slower fall of the scrotal skin temperature was observed compared to that of the abdominal skin. We defined "half time t", the time necessary to obtain half of the decrease of the change observed during the heating phase (Table 2). "Half time t" was 3 min 30 sec for right scrotal skin and only 1 min 45 sec for abdominal skin this difference is highly significant (p<0.001). Left scrotal skin behaved similarly as the right with "half time t" of 3 min 7 sec. At the end of cooling, the mean abdomino-scrotal gradient was reduced to 0.1 °C (not significant).

In ten cases (out of the 120 studied) we observed a similar increase and decrease of both abdominal and scrotal skin temperatures as shown in Fig. 2. Scrotal "half time t" was equal or lower than abdominal "half time t". All cases corresponded to hyperthermia

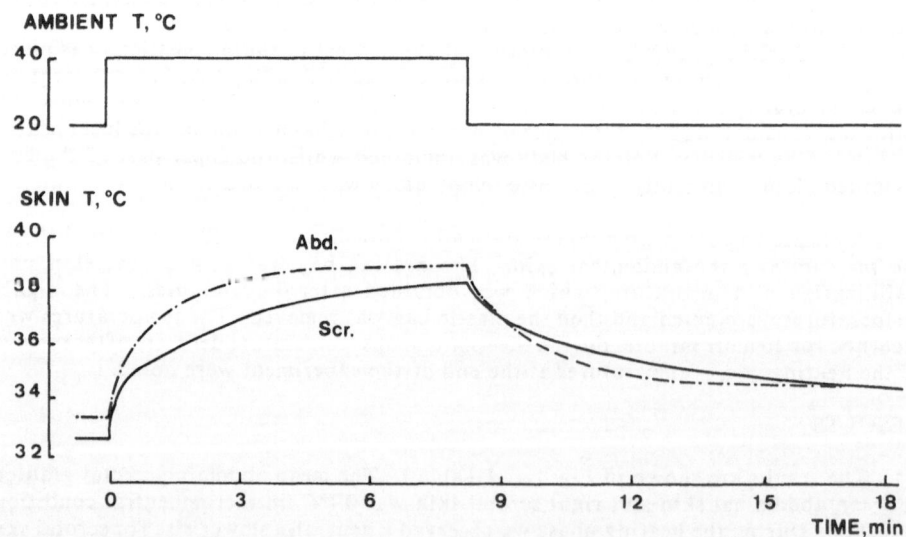

Figure 1. Mean values of the abdominal (abd.) and scrotal (scr.) temperatures during the local heating in our population of 120 infertile men.

Table 2. Mean value of "half time t" (see text) during the recovery phase

| | "half time t" (n = 120) |
|---|---|
| Abdomen | 1 min 45 sec |
| Right scrotum (Student test) | 3 min 30 sec (p<0.001) |
| Left scrotum | 3 min 7 sec |

based on thermographic criteria as described by Hsiung (1990), except in one case of severe testicular atrophy. In fact, 48 cases (out of the 120 studied) were hyperthermic and only 9 of them (in other words 20%) were "detected" by an abnormal temperature time course pattern during our heating test.

## DISCUSSION

The present experiment was performed to study the scrotal thermoregulation, especially under warm local conditions. Surprisingly the scrotum seems to be unable to struggle against heat stress as shown by the abdomino-scrotal gradient which was reduced during heating. An efficient local regulation would maintain or even increase the abdomino-scrotal gradient, but this is not the case. Lazarus and Zorgniotti (1975) also showed in other conditions (under fever) that abdomino-scrotal gradient was not maintained when core-temperature increases over 37.6 °C - 37.9 °C. Further analysis shows that proper mechanisms are involved during the heating process so that the scrotal temperature increases slower than the abdominal one during the heating and also decreases slower during the recovery phase.

It seems that a subcutaneous vasodilatation is involved:

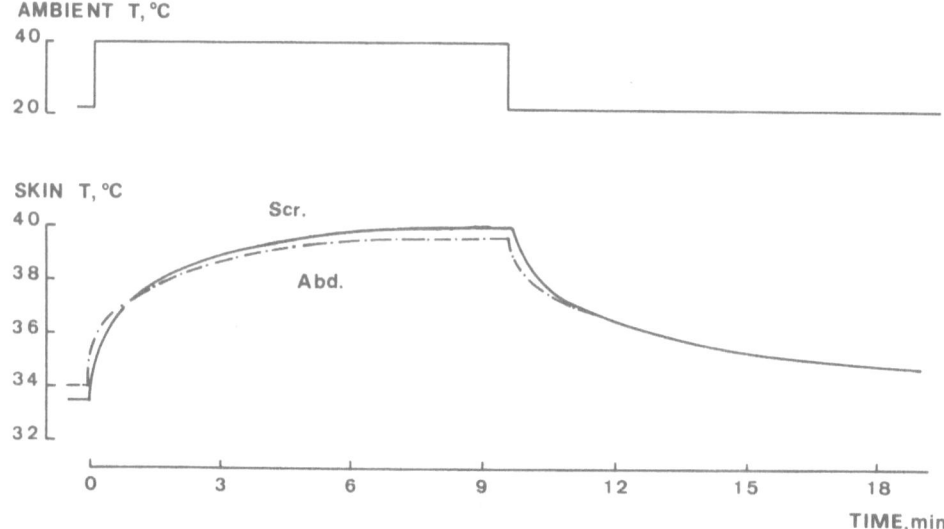

Figure 2. An example of the temperature curves observed in a pathological case of steep increase and decrease of scrotal temperature during the heating test.

·During heating, calories can be evacuated with the blood flow so that local skin temperature increases slowly.

·During the recovery phase, vasodilation still persists, maintaining a flux of calories from the testicular core to skin surface so that the local skin temperature decreases slowly.

Other mechanisms may be involved since apocrine sweat glands have been described by Waites (1963) in ram (Setchell, 1978). These mechanisms seem to be specific of the scrotal skin, and their common objective is to struggle against heat stress as demonstrated by our curves.

The efficiency of the thermoregulatory system of the testis depends on the thermal heat exchanges between arterial and venous blood in the pampiniform plexus. This phenomenon needs a pre-cooling of the venous blood by means of the scrotal skin. Any heating of the scrotal skin increases the venous blood temperature, inducing a reduction of the arteriovenous gradient and thus the efficiency of the thermoregulatory system. In cases of very high thermic conditions, this can induce a reversion of the pampiniform heat exchange resulting in a dramatic increase of the testicular temperature (Fig. 3). In some pathologic cases (20% of our observations), some alterations of these mechanisms can be responsible for scrotal hyperthermia even in basic conditions.

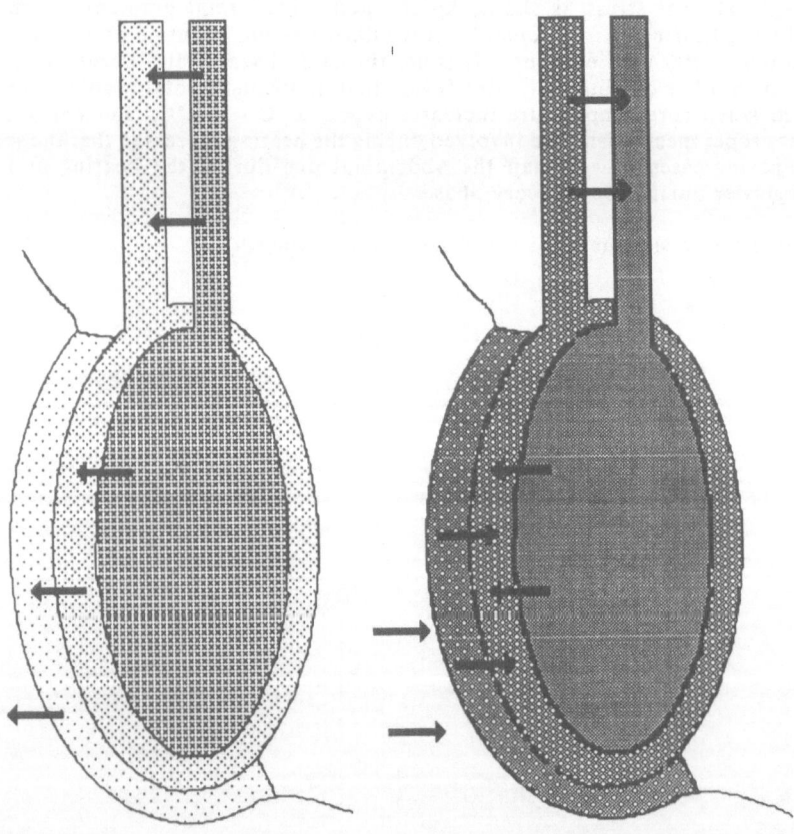

Figure 3. Normal heat exchange (left). Abnormal heat exchange under hot conditions (right).

## CONCLUSION

Due to its anatomo-physiological condition, the testis can be maintained at lower temperature than the body core. This is necessary for an efficient spermatogenesis. There are specific regulatory systems involved in the struggle against heat but their efficiency is decreased when heat stress is too strong, or under pathological conditions leading to disturbances of the spermatogenesis. Subcutaneous vasodilatation of the scrotal skin seems to be the main mechanism involved in the local short term regulation, but other mechanisms such as sweating may be involved and need further analysis.

## REFERENCES

Hsiung, R., Nieva, H., Clavert, A. 1990. Scrotal hyperthermia and varicocele. In: Temperature and other environmental factors on the testis, A.W. Zorgniotti, ed., Plenum, New York.

Kandeel, F.R. and Swerdloff, R.S. 1988. Role of temperature in regulation of spermatogenesis and the use of heating as a method for contraception, Fertil. Steril., 49: 1-23.

Kurz, K.R. and Goldstein, M. 1986. Scrotal temperature reflects intratesticular temperature and is lowered by shaving, J. Urol., 135:290-292.

Lazarus, B.A. and Zorgniotti, A.W. 1975. Thermoregulation of the human testis, Fertil. Steril., 26: 757-759.

Mieusset, R., Bujan, L., Mondinat, C., Mansat, A., Pontonnier, F., Grandjean, H. 1987. Association of scrotal hyperthermia with impaired spermatogenesis in infertile men, Fertil. Steril., 48 : 1006-1011.

Setchell, B.P. 1978. The scrotum and thermoregulation, In: The mammalian testis. B.P. Setchell, ed., Paul Elek, London.

Waites, G.M.H. and Voglmayr, J.K. 1963. The functional activity and control of the apocrine sweat glands of the scrotum of the ram, Aust. J. Agric. Res., 14: 839-851.

Zorgniotti, A.W. and MacLeod, J. 1973. Studies in temperature, human semen quality and varicocele, Fertil. Steril., 24: 854-863.

# IS TESTICULAR FUNCTION IN IMMATURE RATS INCREASED RATHER

# THAN DECREASED BY A MODERATE INCREASE IN TEMPERATURE?

Anders R.J. Bergh

Department of Pathology
University of Umeå
Umeå 90187, Sweden

## INTRODUCTION

Spermatogenesis cannot be maintained at normal body temperature, but the cellular mechanisms responsible are unknown. The most common experimental approach to study this is to examine the effects of experimental cryptorchidism or heat application in adult animals. Experimental cryptorchidism is generally obtained by returning a mature descended testis back into the abdomen. This approach has been used in a great number of studies and the general conclusion is that it results in a rapid degeneration of early spermatids and spermatocytes (van Demark and Free, 1970; Setchell, 1978). It has been suggested that these particular cell types are directly sensitive to increased temperature in vitro, possibly by effects on their cell membranes (Lee and Fritz, 1973) or on their protein synthesis (Nakamura et al., 1978). When experimental cryptorchidism is induced in newborn pigs, it eventually results in decreases in testis weight, number of gonocytes and in total Leydig cell mass (identical changes as in congenital cryptorchid pigs; van Straaten, 1978), suggesting that cell types other than developed germ cells are sensitive to increased temperature.

## EXPERIMENTAL PRIMARY CRYPTORCHIDISM IN RATS

By cutting the lower part of the gubernaculum testis some hours after birth testicular descent and scrotal development, which in normal rats are completed at about 40 days of age, are prevented (Bergh et al., 1978). By using this experimental model, called experimental primary cryptorchidism, it is possible to examine the differences that gradually occur between the descending testis and the one that remains in the abdomen. There are important principal differences between this model and the procedure wherein descended mature testes are returned to the abdomen (secondary experimental cryptorchidism; Bergh, 1989), and they are summarized in Table 1. Considering these differences, it is not surprising that the early effects of increased temperature on testis morphology and function may be different in these two experimental conditions.

As soon as it is possible to detect a temperature difference between the abdominal cavity and the developing scrotum, morphological changes are noted in the abdominal testis in primary experimentally cryptorchid rats. At this age (16 days), testis weight is slightly increased, more tubules have acquired a lumen, the Sertoli cells contain more lipid droplets and the activity of the enzyme cholesteryl ester hydrolase (CEH, an enzyme involved in mobilizing cholesterol from lipid droplets and possibly localized in Sertoli cells, see Hoffmann et al., 1988) is increased in abdominal compared to that in

*Temperature and Environmental Effects on the Testis*
Edited by A. W. Zorgniotti, Plenum Press, New York, 1991

**179**

Table 1. Principal differences between primary and secondary
experimental cryptorchidism

|  | Primary | Secondary |
|---|---|---|
| Temperature increase | gradual over several days/weeks from 0 to 5°C | immediate 5°C |
| Cell types present in the testis | immature Sertoli and Leydig cells; no developed germ cells | mature Sertoli and Leydig cells; sperm |

descending testes. Signs of germ cell degeneration or altered Leydig cell function are not observed at this age (Bergh, 1983a; Bergh et al., 1984; 1987a; Hoffmann et al., 1988). Some aspects of testis function are apparently increased rather than reduced in abdominal testes at this age! At 20 days of age, germ cells (principally pachytens) are

Table 2. Summary of morphological and functional differences between the scrotal and abdominal testis in immature unilaterally cryptorchid rats

|  | Age of Animals | | |
|---|---|---|---|
|  | 12 days | 16 days | 20 days |
| Temperature difference | 0 | 0.7°C | 1.5°C |
| Abdominal testis weight | normal | increased | normal (decreased from 25 days) |
| Germ cell degeneration (in abdominal testis) | 0 | 0 | pachytens, inside the newly formed blood-testis barrier |
| Sertoli cell function (in abdominal testis) | normal | increased (tubule lumen formation, CEH) | (a) decreased (CEH, ABP, E2) (b) increased (tubule fluid secretion) |
| Leydig cell function (in abdominal testis) | normal | normal | decreased (gonadotropin-stimulated steroid hormone secretion) |

(Data from Bergh, 1983a; Bergh et al., 1984, 1987; Hoffmann et al., 1988 and unpublished observations).

starting to degenerate, particularly in the tubule segments where lipid accumulation was noted 4 days earlier (Bergh, 1983a). At this age some aspects of Sertoli cell function are decreased (the secretion of estradiol-17ß, E2, and androgen binding protein, ABP), while others are increased (tubule fluid secretion, Bergh et al., 1984). The gonadotropin-stimulated secretion of steroid hormones from Leydig cells is also decreased at this age (Bergh et al., 1984; 1987a). The observations are summarized in Table 2.

DISCUSSION

In experimental primary cryptorchidism it appears that changes in Sertoli cells precede those in germ and Leydig cells. It could be argued that the altered Sertoli cell function could be secondary to undetected changes in germ or Leydig cells. This cannot be fully excluded, but Sertoli cell function is changed by abdominal temperature in vitro (Hagenäs et al., 1978; Steinberger, 1981). Experimental cryptorchidism induces the same functional (Hagenäs et al., 1978) and morphological (Bergh, 1981) changes in Sertoli cells in testes lacking germ cells (as a result of fetal irradiation) as in Sertoli cells in intact testes. Leydig cell function in vitro is not impaired by abdominal temperature (Rommerts et al., 1980). I have therefore suggested that a temperature-induced change in Sertoli cell function is the primary event, and that germ cell degeneration and altered Leydig cell function are secondarily caused by changed paracrine influence from these functionally altered Sertoli cells (Bergh, 1983a,b). However, it cannot be excluded that more developed germ cells, not present in immature testes, are directly sensitive to abdominal temperatures (see introduction). If so this could be of importance when trying to understand the effects of increased temperature on mature testes, but not for understanding of the early changes in congenitally cryptorchid testes. Moreover, it appears that Sertoli cell morphology and function can be altered in vivo by an increase in temperature less than 1 °C (Bergh, 1981;1983a), but the in vitro effects on developed germ cells are, as yet, only demonstrated after increases of at least 4 °C (Lee and Fritz 1973; Nakamura et al., 1978).

Another surprising observation obtained in this model is that the initial effect of a slightly increased temperature may be an increased testis function or accelerated testis development. The early signs of increased function are moderate and their significance could therefore be questioned. A supranormal testis weight, however, is observed in immature mice where cryptorchidism was induced by fetal estradiol administration (Jean, 1973). There are also other data suggesting that some aspects of Sertoli cell function such as FSH-stimulated tubule plasminogen activator (Bergh et al., 1987b) and inhibin secretion (Gonzales et al., 1989) in vitro are increased by experimental cryptorchidism. Thus, it cannot be excluded that the general dogma that spermatogenesis is impaired by increased temperature because some particular cell function is decreased may not be true. We should perhaps start to consider the possibility that an increase in some particular cell function later results in damaged spermatogenesis.

ACKNOWLEDGEMENTS

This work has been supported by grants from the Swedish Medical Research Council (project 9535) and the Maud and Birger Gustavsson Foundation.

REFERENCES

Bergh, A. (1981) Morphological signs of a direct effect of experimental cryptorchidism on the Sertoli cells in rats irradiated as fetuses. Biol. Reprod. 24: 145.

Bergh, A. (1983a) Early morphological changes in the abdominal testes in immature unilaterally cryptorchid rats. Int. J. Androl. 6: 73.

Bergh, A. (1983b) Paracrine regulation of Leydig cells by the seminiferous tubules. Int. J. Androl. 6: 57.

Bergh, A. (1989) Experimental models of cryptorchidism. In: The Cryptorchid Testis. Eds: Abney, T.O. and Keel, B.A. CRC Press, Boca Raton. p 15.

Bergh, A., Helander, H.F. and Wahlquist, L. (1978) Studies on factors governing testicular descent in the rat, particularly the role of the gubernaculum testis. Int. J. Androl. 1: 342.

Bergh, A., Damber, J-E., Ritzén, E.M. (1984) Early signs of Sertoli and Leydig cell dysfunction in the abdominal testis in immature unilaterally cryptorchid rats. Int. J. Androl. 7: 389.

Bergh, A., Damber, J-E., Huhtaniemi, I. (1987a) Intratesticular steroids and gonadotropin receptor concentrations in the testes of immature unilaterally cryptorchid rats. Int. J. Androl. 10: 803.

Bergh, A., Damber, J-E., Jacobsson, B.H., Nilsson, T.K. (1987b) Production of lactate and tissue plasminogen activator in vitro by seminiferous tubules obtained from unilaterally cryptorchid rats. Arch. Androl. 19: 177.

Gonzales, G.F., Risbridger, G.P., de Kretser, D.M. (1989) In vivo and in vitro production of inhibin by cryptorchid testes from adult rats. Endocrinology 124: 1661.

Hagenäs, L., Ritzén, E.M., Svensson, J., Hansson, V. (1978) Temperature dependence of Sertoli cell function. Int. J. Androl. Suppl 2: 449.

Hoffmann, A-M., Bergh, A., Olivecrona, T. (1988) Changes of testicular cholesteryl ester hydrolase activity in experimental cryptorchid rats. J. Reprod. Fert. 86: 11.

Jean, C. (1973) Croissance et structure des testicules cryptorchides chez les souris nees de meres traitees a l'oestradiol pendant la gestation. Ann. Endocr. (Paris) 34: 669.

Lee, L.P. and Fritz, I.B. (1973) Studies on spermatogenesis in the rat. V. Increased thermal lability of lysosomes from testicular germ cells and its possible relationship to impairment in spermatogenesis in cryptorchidism. J. Biol. Chem. 247: 7956.

Nakamura, M., Romrell, L.J. and Hall, P.F. (1978) Influence of temperature and glucose on protein biosynthesis by immature (round) spermatids. J. Cell. Biol. 79:1.

Rommerts, F.F., de Jong, F.H., Grootegoed, J.A. and van der Molen, H.J. (1980) Metabolic changes in testicular cells from rats after long-term exposure to 37°C in vivo or in vitro. J. Endocr. 85: 471.

Setchell, B.P. (1978) The Mammalian Testis. Paul Elek, London, p 360.

Steinberger, A. (1981) Regulation of inhibin secretion in the testis. In: Intragonadal Regulation of Reproduction. Eds: Franchimont, P. and Channing, C.P. New York, Academic Press, p 283.

van Demark, N.L. and Free, M.J. (1970) Temperature effects. In: The Testis, vol III. Eds: Johnson, A.D., Gomes, W.R. and van Demark, N.L. Academic Press, New York, p 233.

van Straaten, H.W.M. (1978) Lack of a primary defect in maldescended testis of the neonatal pig. Biol. Reprod. 19: 994.

# THE EFFECT OF INTERMITTENT SCROTAL HYPERTHERMIA

# ON THE SPRAGUE-DAWLEY RAT TESTICLE

Kevin R. Loughlin, Kelledy Manson, Russell Foreman,
Beth Schwartz and Phyllis Heuttner

Division of Urology
Brigham and Women's Hospital
Harvard Medical School
Boston, Massachusetts

ABSTRACT

Twenty-four mature male Sprague-Dawley rats were divided into 2 groups of 12. One group was exposed to intermittent hot baths for 1 month, while the other group was exposed to room temperature baths. The group that was exposed to the elevated temperature demonstrated histological testicular changes that included decreased tubular diameter, basement membrane thinning, and decreased spermatogenesis. Fertility rates were markedly reduced in the heat exposed group, but complete recovery of fertility occurred by 10 weeks after completion of the heat treatments.

INTRODUCTION

The deleterious effects of chronic heat on the rodent and canine testis have been well established using an experimental varicocele model[1,2]. This study was undertaken to examine what effects brief, intermittent heat exposure had on the rodent testicle.

MATERIALS AND METHODS

Twenty-four sexually mature male Sprague-Dawley rats (300-350 gms weight) were tested for fertility 3 weeks prior to the start of the experiment by cohabiting each male with 2 female rats in estrus for 7 days. The females were then sacrificed after 7 additional days and checked for the presence of fetuses.

Three days before the baths began, 3 animals from the same shipment of rats used in the experiment were selected at random and were perfused and sacrificed to establish normal histology. All testicular specimens were prepared in an identical way throughout the experiment following the perfusion methods of Hoffer[3].

After all 24 rats had been verified as being fertile, they were randomly divided into 2 groups of 12. Twelve rats were assigned to the warm bath group: range 39.9 - 41.1°C. Another twelve were assigned to the room temperature bath group: range 20.2 - 23.4°C. The bath protocol was as follows:

Week 1. Each animal in both groups was placed in a bathing sling and bathed in their respective temperature baths for 30 minutes on Monday, Wednesday, and Friday. Animals were anesthetized before each bath with nembutal (.025-.05 cc/kg).

*Temperature and Environmental Effects on the Testis*
Edited by A. W. Zorgniotti, Plenum Press, New York, 1991

Table 1. Results of fertility testing

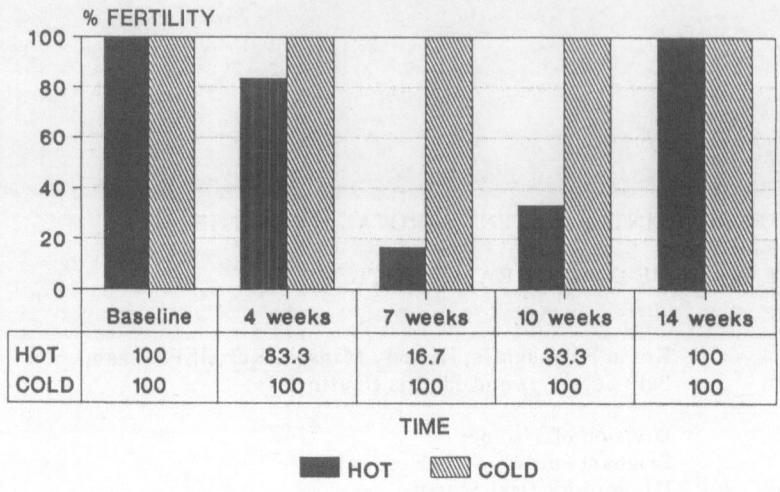

| | Baseline | 4 weeks | 7 weeks | 10 weeks | 14 weeks |
|---|---|---|---|---|---|
| HOT | 100 | 83.3 | 16.7 | 33.3 | 100 |
| COLD | 100 | 100 | 100 | 100 | 100 |

TIME

■ HOT ▨ COLD

<u>Week 2</u>. Each animal was bathed as per week 1. Testicular perfusions in preparation for histology were performed on 3 animals in each group following the second week of baths.

<u>Week 3</u>. The remaining animals in each group were bathed as per week 1.

<u>Week 4</u>. Each animal was bathed as per week 1. Three animals in each group were again perfused and sacrificed for histology.

Fertility testing was repeated at the end of the 4, 7, 10, and 14 weeks. All histological specimens were read in a blinded fashion.

RESULTS

The fertility rate in the room temperature bath group remained at 100% throughout the experiment. The histology in this group was unchanged as compared to baseline. The fertility rate in the hot bath group was 83% at 4 weeks and dropped to 16% at 7 weeks. By the end of 14 weeks the fertility rate in the hot bath group had returned to 100%. The fertility rates are displayed in Table 1.

Histological changes were observed in both groups, but were more pronounced in the hyperthermia group. The following histologic changes were observed.

1. Tubular diameter range steadily decreased with progressive intermittent heat exposure.
2. Intracellular vacuolization occurred.
3. Basement membrane thinning was observed.
4. Progressive loss of spermatogenesis was seen.
5. No significant changes appeared in the Sertoli and Leydig cells.

Serum testosterone levels remained normal throughout the experiment in both the cold and hot bath groups. Serum gonadotropin levels were not measured.

DISCUSSION

Zorgniotti and MacLeod reported the relationship between varicoceles and increased intrascrotal temperature in humans in 1973[4]. Experimental work in the canine and rodent model by Saypol and associates[1] further delineated the relationship between unilateral varicocele formation and persistent bilateral intrascrotal temperature elevation.

Our study was conceived to determine if very brief exposure to heat, such as that men might experience in saunas or hottubs, could result in transient or permanent fertility impairment. The study was designed to examine both functional and histological changes.

Although the rats began to demonstrate impaired fertility, at the end of their exposure to hot baths, the maximum fertility impairment occurred 3 to 6 weeks after the cessation of the hot baths. This observation is consistent with the length of spermatogenesis (45-52 days) and sperm turnover in the Sprague-Dawley rat. Histologic changes appeared to parallel the changes in functional fertility. The most striking changes observed were decrease in the tubular diameters, basement membrane thinning, and progressive loss of spermatogenesis. Sertoli and Leydig cells appeared essentially unchanged. All rats that were exposed to the hot baths regained fertility by 10 weeks following completion of the baths.

Our study supports the observation that exposure to heat is detrimental to testicular function and fertility. What is striking is that even extremely brief heat exposure appeared to have profound effects on both testicular histology and reproductive function. The deleterious effects of intermittent heat in our study appeared to be transient and reversible.

REFERENCES

1.   Saypol, D.C., Howards, S.S., Turner, T.T., and Miller, E.D. Jr.: Influence of surgically induced varicocele on testicular blood flow, temperature and histology in adult rats and dogs. J. Clin. Invest. 68:39 (1982).

2.   Saypol, D.C.: Varicocele. J. Androl. 2:61 (1981).

3.   Hoffer, A.P., Agarwal, A., Meltzer, P., Nagvi, R., and Matlin, S.A.: Antifertility, spermicidal and ultrastructural effects of gossypol and derivatives administered orally by intratesticular injections. Contraception 37:301 (1988).

4.   Zorgniotti, A.W. and MacLeod, J.: Studies in temperature, human semen quality and varicocele. Fertil. Steril. 24:854 (1973).

# HUMAN SCROTAL TEMPERATURE DURING HEAT EXPOSURE

## ASSOCIATED WITH PASSIVE LEG HEATING

G. Dewasmes[1], B. Bothorel[1], R. Hsiung[2], A. Clavert[2]
and V. Candas[1]

[1]LPPE of CNRS, 21 rue Becquerel
F67087 Strasbourg Cedex, France
[2]CECOS of Alsace, 1 Place de l'Hôpital
F67091 Strasbourg Cedex, France

## INTRODUCTION

In contrast to the numerous studies carried out for investigating the temperature regulation of the animal testis (especially in the ram), there are only a few reports dealing with human testis thermoregulation (Brindley, 1982; Kandeel and Swerdloff, 1988 for a review). In man exposed to external heat load, the temperature of the testis results from the thermal equilibrium between the local heat production, the heat exchanges with the surroundings and the heat transfer occurring via the local vascular system. According to the investigation by Buettner (1969), there are no active sweat glands in the human scrotal skin, although the water loss through skin diffusion can be larger than in other areas.

Considering the relatively reduced means of heat dissipation of the human scrotum, it was interesting to examine the scrotal thermal response to the environment when the latter becomes much warmer than in normal conditions. Thus, we carried out this preliminary experimental series in which we measured core, mean skin, leg and scrotal temperature when:

·the thermal load was either slowly or abruptly imposed to normal subjects.
·the legs were exposed to additional passive heating due to absence of local evaporation.

## METHODS

Five healthy young male subjects volunteered for these experiments after being informed of the nature of the experiments. At 8:00 a.m. the subject entered the climatic chamber, lay on the rubber band hammock in a thermoneutral environment: air ($T_a$) and wall ($T_w$) temperatures at 28°C, dew point temperature ($T_{dp}$) at 10°C and air velocity ($V_a$) at 0.3 m/sec.

A rectal probe inserted at a depth of 11 cm beyond the anal orifice allowed the core temperature measurement. Thirteen local skin temperature sensors stuck on the skin led to mean skin temperature determination. Right and left leg temperatures were calculated by averaging the thigh, calf and foot temperatures of each leg. A temperature sensor was also attached to the right scrotal area.

After setting up the semi-nude subject (wearing only briefs), each leg was entered into its corresponding leg chamber, which was not ventilated. Each leg was thus enclosed in its air-tight chamber to avoid occurrence of lower limb evaporation.

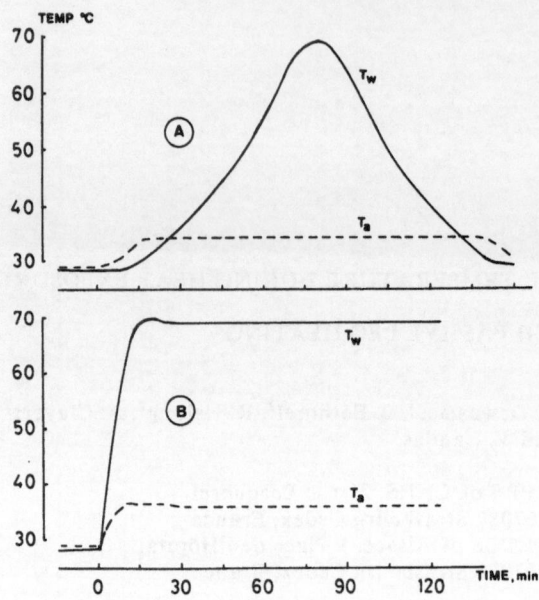

Figure 1. Temperatures of the walls ($T_w$) and of the air ($T_a$) as a function of time, under two experimental conditions (A and B).

After at least 30 min at thermoneutrality, each subject was exposed to external thermal load in the chamber when $T_a$ was set to either 34°C (condition A) or 36°C (condition B) (Fig. 1) and when the four walls surrounding the subject were elevated to 69°C (right and left walls, floor and ceiling). Depending on the experimental day, wall temperatures were either raised slowly and then decreased slowly (condition A) or abruptly increased and maintained at a high level (condition B), as shown in Fig. 1. The maximum heat stress corresponds to an operative temperature of 47°C at constant levels of humidity and air movement.

RESULTS

Thermoneutral Conditions

The average temperatures obtained for the five subjects exposed to two experimental days in the thermoneutral environment were:

abdominal temperature: 34.7°C (SD = 0.35°C)
scrotal temperature: 33.3°C (SD = 0.54°C)
rectal temperature: 36.6°C (SD = 0.16°C)

Consequently, the temperature difference between the body core and the scrotum was 3.3°C, while the difference between abdomen and the scrotum was 1.4°C.

Thermal Transients: Condition A

The results of Figure 2 show that the leg temperatures initially at 33.3°C reached 36.6°C at the end of the experimental session. Abdominal temperature rose from 34.8°C to 36.8°C and then returned to 35.3°C at the end of the radiant heat cycle. The scrotum temperature (33.3°C) was raised to 36.0°C at its maximum, which was reached 10 min later than the maximum of radiant heat. The decrease of scrotal temperature (Fig. 2) to 34.4°C occurred in parallel to the decrease in abdominal temperature.

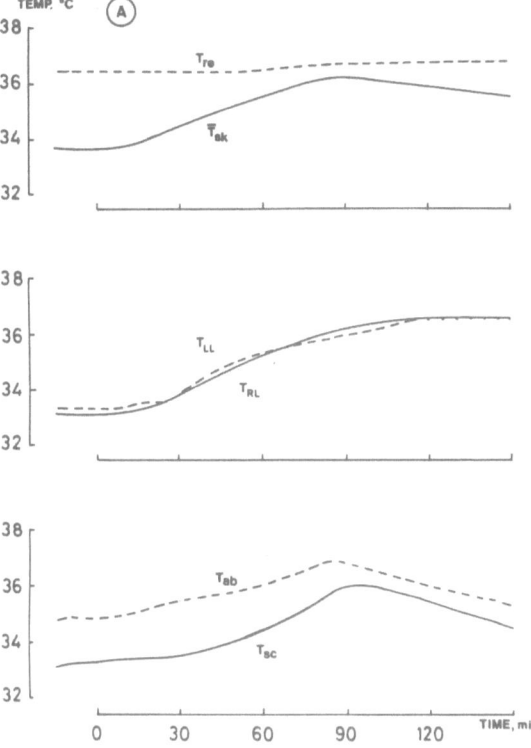

Figure 2. Body temperatures (T) as a function of time under experimental condition A.
Subscripts, re: rectal          sk: mean skin
          LL: left leg          RL: right leg
          ab: abdominal         sc: scrotal


## Steady Thermal Strain:  Condition B

Under condition B, air temperature and wall temperatures were maintained constant throughout the session at 36°C and 69°C, respectively. As a consequence, the leg temperatures inside the non-ventilated leg chambers increased more steeply than under condition A and finally reached higher levels (Fig. 3).

The abdominal temperature increased from 34.6 to 37.2°C within a few minutes and then was reset to 36.8°C due to the evaporation of sweat. The scrotal temperature was elevated dramatically from 33.2°C to 36.8°C within the first 40 min, becoming higher than both abdominal and body core temperatures. At the end of the session, the scrotal surface temperature was 37.7°C, while the core temperature was 37.2°C. Thus, during 80 min the thermal gradient between the scrotum and core was reversed, implying that the testis was continuously rewarming and was unable via the vascular system to maintain a lower temperature.


## DISCUSSION

Very few studies have examined the changes in scrotal temperature in response to changes in environmental conditions. Rock and Robinson (1965) showed that scrotal-rectal temperature difference remained fairly constant in subjects immersed in water until the bath temperature reached 39°C, after which the scrotal temperature rose steeply and rapidly reached a higher level than that of the core temperature. In such conditions, however, the high thermal conductivity of water resulted in severe heat stress.

189

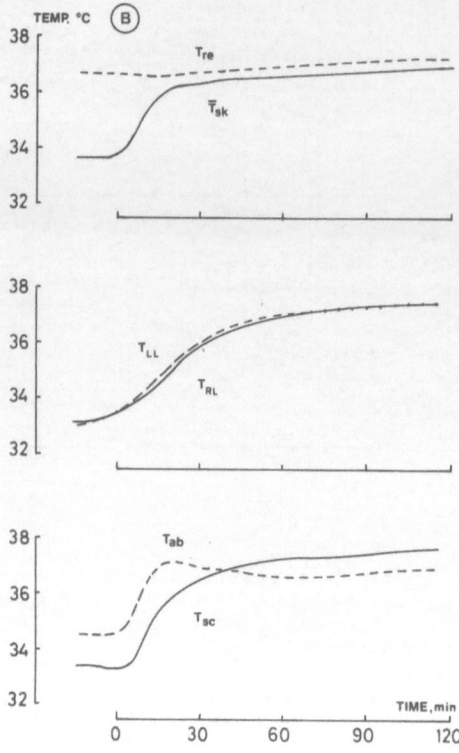

Figure 3. Body temperatures (T) as a function of time under condition B. See Figure 2 for subscript meanings.

In our conditions of moderate transient heat stress, the scrotal temperature followed the external heat load, similar to that which occurs on other skin areas and always remains below the level of internal temperature. In condition A, the fact that the difference between the abdomen and the scrotum temperatures became smaller than the initial is mainly due to the inertia of the scrotal region: the maximal temperature being observed after the end of the heat increase. This can be explained by the presence of the briefs covering the scrotal skin, delaying the heat transfer; this may also be partly due to the persisting vasodilation into the scrotum, retarding the cooling of that region. Absence of evaporation could also be a reason for the observed phenomenon. Continuous increase in the level of lower limb temperatures may also have influenced the scrotal area to some extent.

When thermal stress becomes more important, the scrotal area is unable to maintain its normal local temperature. Absence of temperature adjustments after the initial rapid rise (as observed on the abdomen) clearly illustrates the impossibility for the scrotum to maintain efficient cooling. Dramatically, the scrotal temperature level becomes higher than the core, which results in continuous heat gain into the testis, and certainly induces the well-known disturbing effects on spermatogenesis.

From the present study, it can be concluded that apparently normal subjects exposed to transient or moderate thermal stress are able to maintain a relatively low scrotal temperature, via the vascular system and the dry heat exchanges resulting from vasomotor adjustments. Nevertheless, the scrotal area and, consequently, the testis itself will not be able to struggle against prolonged heat stress. If it is assumed that a level of 35 °C in intrascrotal temperature produces dramatic disturbances in spermatogenesis (Lazarus and Zorgniotti, 1975), any exposure to severe external heat stress lasting more than half an hour may result in damaging consequences.

# REFERENCES

Brindley, G.S. 1982. Deep scrotal temperature and the effect on it of clothing, air temperature, activity, posture and paraplegia. Br. J. Urol., 54: 49-55.

Buettner, K.J.K. 1969. Scrotum, sole and areola-skins sharply differing in sweating and in transfer of alcohols and water. Fed. Proc., 28: 528.

Kandeel, F.R. and Swerdloff, R.S. 1988. Role of temperature in regulation of spermatogenesis and the use of heating as a method for contraception. Fertil. Steril., 49: 1-23.

Lazarus, B.A. and Zorgniotti, A.W. 1975. Thermoregulation of the human testis. Fertil. Steril., 26: 757-759.

Rock, J. and Robinson, D. 1965. Effect of induced intrascrotal hyperthermia on testicular function in man. Am. J. Obstet. Gynecol., 93: 793-801.

# EFFECT OF MEDULLARY LESIONS ON SCROTAL THERMOREGULATION:

# A PRELIMINARY STUDY

S. Belhamou[1], A. Chapuis[2], R. Hsiung[3]
and A. Clavert[3]

[1]Service d'urologie Centre Hospitalier General
F68057 Mulhouse Cedex, France
[2]Centre de Réadaptation Fonctionnelle
F68093, Mulhouse Cedex, France
[3]CECOS of Alsace, 1 Place de l'Hôpital
F67091, Strasbourg Cedex

Hyperthermia is an important factor of infertility as it has been demonstrated by Zorgniotti (1973) and Mieusset (1987). The low testicular temperature is regulated by numerous regulatory systems (Setchell, 1978), which are neurologically mediated. Iggo (1969) described cutaneous thermoreceptors in the scrotal skin. Kumazawa and Mizumura (1983) demonstrated a thermal dependency of chemical response of the superior spermatic nerve *in vitro*. In rats, scrotal temperature induces neurological responses in thalamus, hypothalamus (Kanosue et al., 1985) and caudate-putamen (Taylor et al., 1987). The aim of this study was to evaluate the influence of spinal injury on scrotal temperature in men.

## MATERIAL AND METHODS

Ten cases of patients with traumatic medullar lesions were submitted to thermic scrotal exploration: 3 were tetraplegic; 7 were paraplegic. The patient was placed in recumbent position. His scrotum was placed on an isolating plate to perform the thermography with a heat reactive crystal plate (Hsiung, 1990). We allowed 7-10 min equilibration at room temperature. Then we performed the test of local heating with a thermostatic heating bag at 40°C. This exploration was performed in summer from 3-17 August 1988, in the same ambient temperature of 28°C. The interpretation of both thermography and temperature curves during local heating were made in double blind.

## RESULTS

The results are shown on Fig. 1. We were surprised to find that all cases of tetraplegic subjects had a normal thermography and a normal thermic curve during the local heating protocol. On the other hand, all cases with lesional syndrome between T10 and L1 (cases Nº 6,7,9) were hyperthermic and case Nº 9 had, in addition, an abnormal temperature time course pattern during the local heating.

·Case Nº8 had incomplete paraplegia in T6, complete in T10 and was hyperthermic. We believe that this case can be considered to belong to the above group,
·Case Nº3 had incomplete paraplegia in L1, complete in L3 and was hyperthermic. We believe also that this case belonged to the above group,
·Case Nº10 had paraplegia in T9 and discreet hyperthermia,
·Case Nº1 had incomplete paraplegia in T11, complete in L2 and no hyperthermia.

The mean scrotal temperature of the group of 3 patients with spinal injury between T10 and L1 and case N°8 and N°3, was 33.4°C (10 measures). The mean scrotal temperature of the 5 other patients was 31.6°C (10 measures). The difference between the two groups was 1.8°C. The number of observations is too small for statistical analysis.

DISCUSSION

Brindley (1982) measured deep scrotal temperature by invagination thermometry with the method of Zorgniotti (1973). In a group of 23 men with spinal injury, he found a significant higher temperature than in the control group but he did not find any correlation with the level of lesion, completeness or flaccidity. These measures were performed on subjects sitting still, clothed normally with no instruction given as how to sit. On the other hand, the author measured the effect of the sitting position on his scrotum: when thighs were apart (70°) the scrotal temperature was about 1.6°C lower than when the thighs were together. He thus considers that the sitting position of paraplegic men is a much more important factor of hyperthermia than the disorders of vasomotor or sudomotor reflexes.

In our experiment all patients were placed exactly in the same conditions (ambient air, recumbent position and scrotum isolated from the thigh, equilibration at room temperature during 7-10 min). We found hyperthermia in patients with a lesional syndrome between T10 and L1. Mean temperature was 1.8°C higher than in the other group. The number of patients was too small for statistical evaluation. The sympathetic nervous system is involved in the vasomotor and sudomotor reflexes. The lesional syndrome between T10 and L1 corresponds to the perineo-scrotal area. So it is not so surprising that the spinal lesion between T10 and L1 induces an alteration of the scrotal thermoregulation.

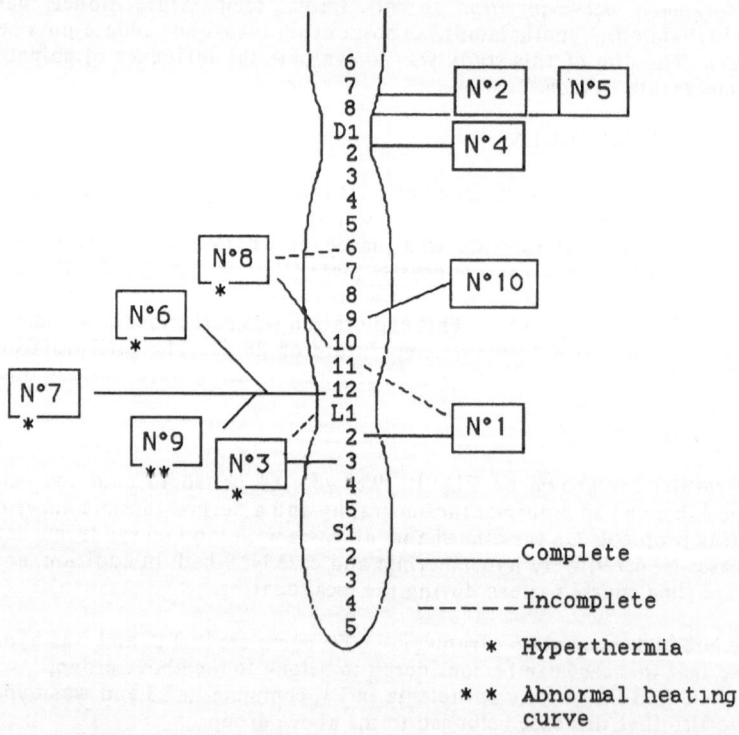

Figure 1. Correlation of the scrotal thermoregulation with the level of the medullary lesion.

# CONCLUSION

This study, performed on a short series, showed that a lesional syndrome of the sympathetic vasomotor and sudomotor centers located in the dorso-lumbar medulla between T10 and L1 may alter the scrotal thermoregulation. We observed a total hyperthermia with a mean scrotal temperature 1.8°C higher than in the other cases. In one case of lesional syndrome in D12 there was, in addition, an abnormal temperature time course pattern during our heating test. These observations need further analysis of more cases and of the sweat response in the cases of abnormal scrotal thermoregulation.

# REFERENCES

Brindley, G.S. 1982. Deep scrotal temperature and the effect on it of clothing, air temperature, activity, posture and paraplegia, Brit. J. Urol., 54:49-56.

Hsiung, R.,Bothorel, B., Dewasmes, G., Candas, V. and Clavert, A. 1990. Effect of local heating on scrotal temperature, in: Temperature and other environmental factors on the testis, A.W. Zorgniotti, ed., Plenum, New York.

Hsiung, R., Nieva, H. and Clavert, A. 1990. Scrotal hyperthermia and varicocele, in: Temperature and other environmental factors on the testis, A.W. Zorgniotti, ed., Plenum, New York.

Iggo, A. 1969. Cutaneous thermoreceptors in primates and subprimates, J. Physiol., 200:403-430.

Kanosue, K., Nakayama, T., Ishikawa, Y., Hosono, T., Kaminaga, T. and Shosaku, A. 1985. Response of thalamic and hypothalamic neurones to scrotal warming in rats: non-specific responses? Brain Res., 328:207-213.

Kumazawa, T. and Mizumura, K. 1983. Temperature dependency of the chemical response of the polymodal receptor units in vitro, Brain Res., 278:305-307.

Mieusset, R., Bujan, I., Mondinat, C., Mansat, A., Pontonnier, F. and Grandjean, H. 1987. Association of scrotal hyperthermia with impaired spermatogenesis in infertile men, Fertil. Steril., 48:1006-1011.

Taylor, D.C.M., Steele, J.E. and Gayton, R.J. 1987. An analysis of the responses of rat striatal neurones to scrotal skin temperature, Brain Res., 419:352-356.

Zorgniotti, A.W. and MacLeod, J. 1973. Studies in temperature, human semen quality and varicocele, Fertil. Steril., 24:854-863.

SECTION 5

INTRINSIC HEAT EFFECTS

# INTRINSIC TESTICULAR TEMPERATURE ELEVATION

# AND SUBFERTILE SEMEN

Adrian W. Zorgniotti

Department of Urology
New York University School of Medicine

*-germinal thermal discordance occurs in such a narrow range around the optimum that infringement of body or reproductive requirements or tolerances, may result from relatively small and otherwise inconsequential climate or bodily changes.*

*Cowles (1965)*

## INTRODUCTION

The possibility that intrinsic elevation of scrotal temperature could be related to subfertile semen has, only recently, become a matter for research interest.

## MATERIALS AND METHODS

Three hundred consecutive patients presenting with infertile marriage and 30 volunteers with normal semen (controls without proven fertility) were assigned random numbers. All patients had a history, physical examination which included the genitalia in the standing position. A diagnosis of varicocele was made only if this could be palpated upon bearing down. Other examinations included serum follicle stimulating hormone (FSH) determination where appropriate. Controls had examination of the genitalia only. All subjects had two or more semen analyses by independent laboratory specialized in fertility diagnosis. Semen was considered subfertile when decreased values were present in one or more parameters: Count, $< 20 \times 10^6$/ml; Motility, $< 40\%$ actively motile sperm; Morphology, $< 60\%$ oval forms. Azoospermic patients were excluded (Table 1).

Intrascrotal temperature readings are made by wrapping the scrotum about a calibrated laboratory mercury thermometer (Zorgniotti and MacLeod, 1973). The patient was placed supine after removal of trousers and undergarment allowing for a six minute

Table 1. Semen findings in 300 consecutive subfertile patients

| | |
|---|---|
| Count: < 20 million/ml | 161/300 (53.7%) |
| Motility: < 40% 2+ | 235/300 (78.3%) |
| Morphology: < 60% ovals | 255/300 (85.0%) |

*Temperature and Environmental Effects on the Testis*
Edited by A. W. Zorgniotti, Plenum Press, New York, 1991

**199**

Table 2. Intrascrotal temperatures: normal vs. subfertile patients

|  |  | Mean °C | Range °C | SD |
|---|---|---|---|---|
| 20 NORMAL VOLUNTEERS mean count 125 X 10⁶/ml | RIGHT | 33.4 | 32.1 - 34.5 | 0.5 |
|  | LEFT | 33.5 | 32.2 - 34.8 | 0.5 |
| 20 PATIENTS mean count 25.9 X 10⁶/ml | RIGHT | 34.7 | 33.3 - 35.7 | 0.6 |
|  | LEFT | 34.8 | 32.6 - 35.9 | 0.7 |
| 20 PATIENTS mean count 6.2 X 10⁶/ml | RIGHT | 34.9 | 33.4 - 36.1 | 0.5 |
|  | LEFT | 35.0 | 33.2 - 36.3 | 0.7 |

significance: normal vs. patients  $p = 0.01$

equilibration in an ambient 21 - 25 °C. The bulb of the thermometer was prewarmed to about 37.0 °C and then placed against the scrotum over the most prominent part of the anterior testis and the scrotal skin drawn about it. When the falling mercury column reached equilibrium this was deemed to be the intrascrotal temperature. All estimations were made by A.W.Z. or by one of three Physician's Assistants.

For comparison of means, a random sample was taken of twenty patients and twenty normals. In addition, a subgroup of twenty patients whose spermatozoa count was < 20 X 10⁶/ml was also randomly selected. Count was selected in preference to motility or morphology as it is an objective determination and least subject to interlaboratory variation. Means were compared by student's "t test" (Table 2).

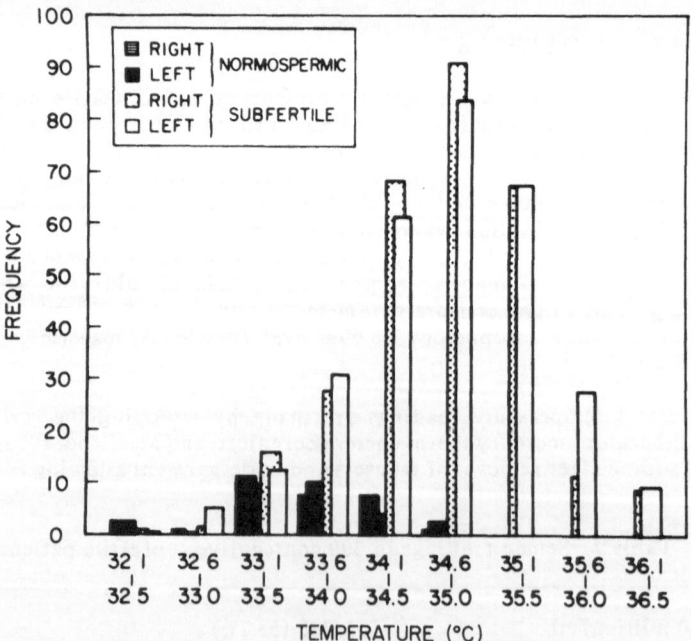

Figure 1. Distribution of intrascrotal temperatures of 30 normospermic control subjects and 300 patients with subfertile semen. Separate peaks are seen: controls 33.1 - 33.5 °C; patients 34.6 - 35.0 °C. (From J. Reprod. Fert. with permission).

Table 3. Temperature (°C) vs. FSH (mIU/ml) values

| | n | Mean FSH | Range | Mean temperature R | L |
|---|---|---|---|---|---|
| < 15.9 | 100 | 7.9 | (3.0 - 15.6) | 34.7 | 34.7 |
| > 16.0 | 22 | 27.1 | (16.0 - 65.0) | 34.8 | 34.7 |

RESULTS

Frequency distribution curves were plotted for right and left intrascrotal temperatures for 30 normals and 300 patients. The curves for patients are minimally skewed. The control curve, with a smaller cohort, is flatter. Distinct peaks are seen: patients at 34.6 - 35.0 °C; controls at 33.1 - 33.5 °C with overlap of the tails (Figure 1).

Mean intrascrotal temperatures (Table 2) of the patient groups are circa 1.5 °C higher than the control group and this difference is significant (p = 0.01). In Table 4 we see that mean temperatures of varicocele, failed varicocelectomy and idiopathic infertility patients are all higher than the controls. Count was less often affected (54% had counts less than 20 X $10^6$/ml) than motility and morphology (respectively 78% and 85% below minimum values) (Table 1). One hundred twenty two patients had FSH determinations: 100/122 had FSH < 15.9 mIU/ml, while 22/122 (18%) had FSH > 16.0 mIU/ml. Mean temperatures were higher than the control group for both FSH groups (Table 3).

DISCUSSION

The difference in means between the patient and control groups of 1.5 °C, parallels observations of intrascrotal temperature by Lazarus and Zorgniotti (1975) in men with fever unrelated to intrascrotal pathology. When rectal temperature reached 37.8 °C, there was a precipitous decrease in scrotal rectal differential of 1.5 °C. The abrupt rise

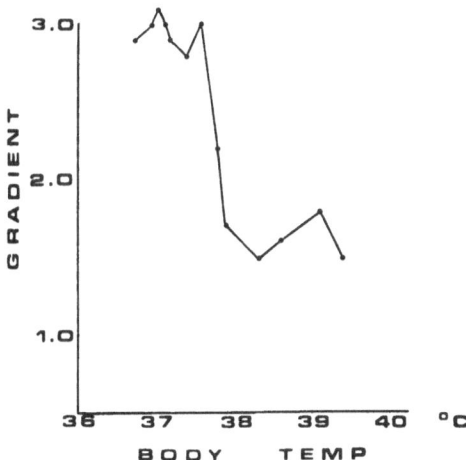

Figure 2. Mean scrotal/rectal gradients in febrile subjects. (Multiple readings taken at various rectal temperatures in each subject.) Around 37.6 - 37.9 °C there is a sharp drop in the gradient owing to failure of the countercurrent heat exchange in the pampiniform plexus. In terms of intrascrotal temperature, this represents an increase of 1.5 °C. (From Fertil. Steril. with permission).

## Table 4. Anatomical diagnoses vs. temperature (°C)

| | Number | Mean temperature (SD) | |
| --- | --- | --- | --- |
| | | Right | Left |
| All patients | 300 | 34.7 (0.8) | 34.8 (1.5) |
| Varicocele left | 76/300 (25.3%) | 34.7 (0.6) | 34.8 (0.6) |
| Varicocele right | 2/300 (0.7%) | | |
| Varicocele bilateral | 32/300 (10.7%) | 34.7 (0.5) | 34.8 (0.6) |
| Varicocelectomy failure | 86/300 (28.7%) | 34.7 (0.7) | 34.7 (0.7) |
| Idiopathic | 86/300 (28.7%) | 34.7 (0.7) | 34.8 (0.8) |
| Cryptorchidism | 10/300 (3.3%) | 34.9 (0.6) | 34.7 (0.7) |
| Other | 8/300 (2.7%) | | |

in testis temperature can be attributed to failure of the pampiniform plexus countercurrent heat exchange when the core temperature reaches a critical point. An increase of 1.5°C may represent the maximum rise in temperature possible from an intrinsic cause (Figure 2).

That there were no differences in mean temperature between the varicocele, failed varicocelectomy and nonvaricocele ("idiopathic") groups, suggests a common mechanism for elevated temperature (Table 4).

Two other conclusions are suggested by our data: (1) Spermatozoal count is less sensitive to heat than motility and morphology. (2) We could not identify a correlation between temperature and elevated FSH.

In establishing a lower limit for elevated temperature, 34.0°C represents minus one standard deviation of the patient mean. Allowing for absent testes, we found: 245/291 (84.2%) (right) and 249/298 (83.5%) (left) had intrascrotal temperature greater than 34.1°C. Thus a substantial number of patients with subfertile semen have elevated intrascrotal temperature (Table 5).

### Table 5. Values of normal and elevated temperature

| NORMAL | < 33.5°C | (32.1 - 34.6°C) |
| --- | --- | --- |
| ELEVATED | > 34.0°C | (32.6 - 36.5°C) |

REFERENCES

Cowles, R.B. 1965. Hyperthermia, aspermia, mutation rates and evolution. Quart. Rev. Biol., 40: 341.

Lazarus, B. and Zorgniotti, A.W. 1975. Thermoregulation of the human testis. Fertil. Steril., 26: 757.

Zorgniotti, A.W. and MacLeod, J. 1973. Studies in temperature, human semen quality and varicocele. Fertil. Steril., 24: 854.

# SCROTAL HYPERTHERMIA: FREQUENCY IN AN INFERTILE POPULATION

# AND ASSOCIATED ALTERATIONS IN TESTICULAR FUNCTION

Roger Mieusset, Louis Bujan, Monique Plantavid,
Arlette Mansat, Helene Grandjean and Francis Pontonnier

Centre de Sterilité Masculine, Cecos Midi-Pyrenees,
Service de Biochimie and INSERM U168
Hôpital La Grave
Toulouse, France

## POPULATIONS

Two populations were involved in this study. The infertile one was composed of all men (n = 555; 22-52 years old) presenting at our Center from December 1985 to March 1989 for conjugal infertility of at least two years duration, and whose partners had no evidence of abnormalities likely to cause infertility. The second population was made up of 78 fertile men consulting during the same period and who were either requesting a vasectomy (n = 45) or were candidate sperm donors (n = 33); these men ranged in age from 26 to 49 yr and had fathered at least one child. All these unselected men were subjected to genital examination which included research of a clinical varicocele and of testicular histories (orchitis, cryptorchidism and surgery for varicocele).

## MATERIAL AND METHODS

Measurement of width and length of each testis with a caliper allowed us to calculate testicular volumes (Macomber and Sanders, 1929). Scrotal temperature measurements were performed on both testes by means of a mercury thermometer (range 30 to 40°C; divisions of 0.05°C; available from Brooklyn Thermometer Co., Farmingdale, NY) in conditions similar to those described by Zorgniotti and MacLeod (1973), except the thermometer was prewarmed to 37°C. Measurement in duplicate 5 min later in identical conditions showed that the observed maximum deviation between two successive measurements was 0.07°C on 50 consecutive men. The mean of the observed variation, expressed as an absolute value was 0.07°C. Simultaneously, rectal and ambient temperatures were recorded by means of clinical and wall thermometers.

Blood hormone determination and sperm examination were performed later as follows. Before semen collection, serum FSH, LH and Testosterone levels were measured between 0900-1100 hr; four blood samples were taken at 20-min intervals and equal volume of serum from each sample was pooled. Kits from Hoechst Behring (standard MR68/40) and IRE (standard MRC69/104) were used respectively to measure in duplicate serum LH and FSH with RIA, and (125I testosterone) for Testosterone.

Sperm analyses included determination of semen volume and sperm count with a Mallassez cell. Sperm motility was evaluated at room temperature one hour after sperm collection; semen samples were maintained at 37°C. Spermatozoa showing a linear motility during optical examination were recorded as motile cells, and results were expressed as percentage of motile spermatozoa.

*Temperature and Environmental Effects on the Testis*
Edited by A. W. Zorgniotti, Plenum Press, New York, 1991

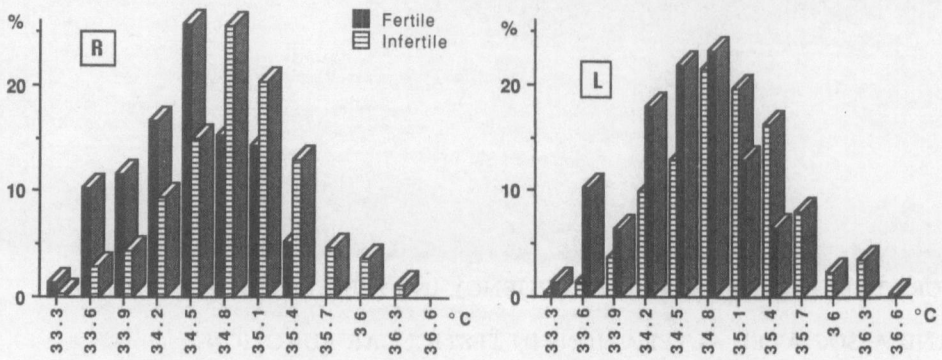

Fig. 1. Distribution of fertile and infertile men according to right (R) and left (L) scrotal temperatures.

Statistical analyses were made with the Mann-Whitney non-parametric test, except for the normally distributed temperature values (Student's t test). Chi-square test was used to compare prevalences. P < 0.05 was considered significant.

Among the 555 infertile men, ten were excluded for monorchidism and nine due to refusal of surgical exploration for azoospermia. The infertile population was therefore reduced to 536 men and included 47 azoospermic men. Subsequent surgical exploration, including bilateral testicular biopsy and bilateral epididymal exploration resulted in identification of secretory azoospermia in 27 men and excretory azoospermia in 20 men.

RESULTS

Scrotal Temperatures

Scrotal temperatures in infertile men tend to be higher, but an important overlap does exist between these two populations, as shown by the distribution of the fertile and infertile men according to the right and left scrotal temperatures (Fig. 1). Mean values

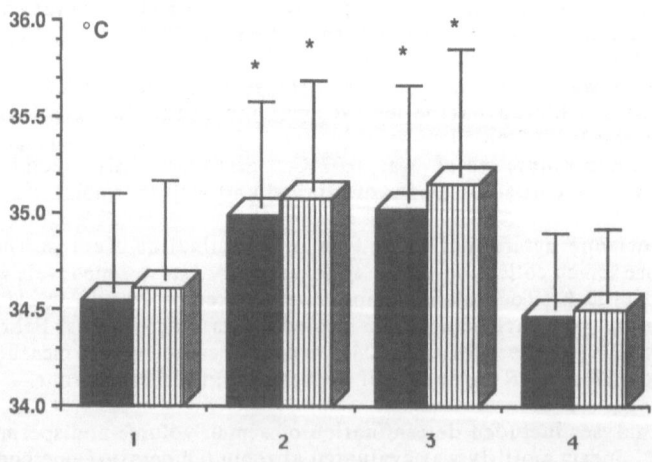

Fig. 2. Mean values (+SD) of the right and left scrotal temperatures in fertile men (1) and in infertile men without azoospermia (2), or with secretory (3) and excretory (4) azoospermia. * = p < 0.001 with groups (1) or (4).

of scrotal temperatures are significantly higher in non-azoospermic infertile than in fertile men: 0.4°C for the right and 0.5°C for the left side. Some discrepancies appear between the values in the two azoospermic groups in that the temperatures in the men with secretory azoospermia are similar to those in the non-azoospermic infertile men, while the temperature in the men with excretory azoospermia are similar to those in the fertile men (Fig. 2).

## FREQUENCY OF SCROTAL HYPERTHERMIA

The 90th percentile value for scrotal temperature in the fertile men was 35.2°C in the right and left testes. Taking this value as the upper limit for normal scrotal temperatures, we called values which were above this threshold in left, or right, or both scrotal temperatures in infertile men "hyperthermic".

According to this threshold, the frequency of scrotal hyperthermia appears to be 43% in the non-azoospermic infertile men (16% unilateral, 27% bilateral), 48% in case of secretory azoospermia (26% unilateral, 22% bilateral), and only 5% in case of excretory azoospermia. With such a definition scrotal hyperthermia is a syndrome found in 4 infertile men out of 10.

## MODIFICATIONS ASSOCIATED WITH SCROTAL HYPERTHERMIA

### General Facts

General changes associated with scrotal hyperthermia include modifications in body temperature and in testicular volumes.

Concerning the body temperature, the two infertile groups differ significantly and the hyperthermic group shows the highest value, with a mean value in rectal temperature of 0.2°C higher than in the infertile normothermic group (Fig. 3). In the hyperthermic group testicular volumes are significantly smaller than in the normothermic group or in the fertile group (Fig. 4).

### Exocrine Function of the Testis

Both total sperm count and sperm motility are significantly more depressed in the hyperthermic than in the normothermic group (Fig. 5).

### Endocrine Function of the Testis

Of the 489 non-azoospermic infertile men, serum hormone measurements were performed in 312 men. The comparison between the two infertile groups according to the scrotal temperature threshold shows that serum FSH and LH levels are significantly higher in the hyperthermic than in the normothermic group, the testosterone levels being similar in both groups (Fig. 6). It seems that serum gonadotropin levels are higher when scrotal temperature is above the normal threshold.

But spermatozoa output and serum gonadotropin levels are linked together (Franchimont et al., 1972; Handelsam et al., 1984). To determine the relationships among spermatozoa output, scrotal temperature and serum gonadotropin levels, infertile men were subdivided into two classes according to whether their spermatozoa output was less or more than 60 million per ejaculate (Table 1). Each spermatozoa output class was then subdivided again into two groups according to the scrotal temperature threshold.

The rise previously observed in gonadotropin levels and which is found again in the "less than 60 million" group, does not result from the fall in the spermatozoa output since the total sperm count does not differ between the normothermic and hyperthermic groups. These results could indicate that some relationships exist between testis temperature and Sertoli and Leydig cell functions in man. But more detailed analyses

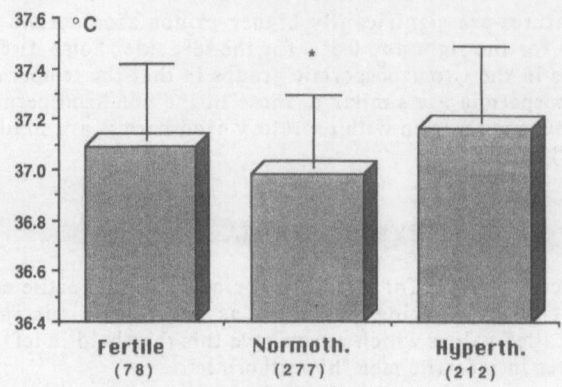

Fig. 3. Mean values (+SD) of rectal temperature in fertile men, and in infertile men with normal scrotal temperatures (Normoth.) or with uni- or bilateral scrotal hyperthermia. * = p < 0.01 with each group.

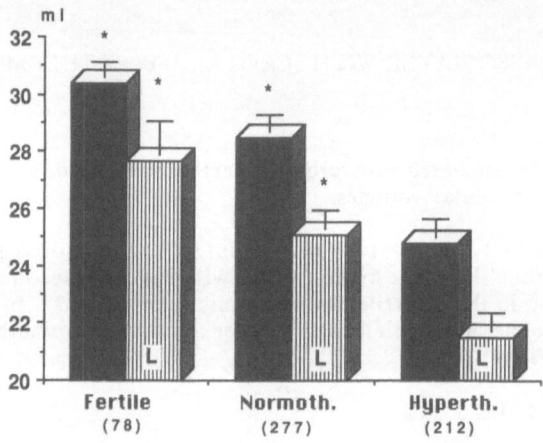

Fig. 4. Mean values (+sem) of right and left (L) testicular volumes in fertile men and in infertile men with normal scrotal temperatures (normoth) or with uni- or bilateral scrotal hyperthermia (hyperth). * = p < 0.01 with "hyperth" group.

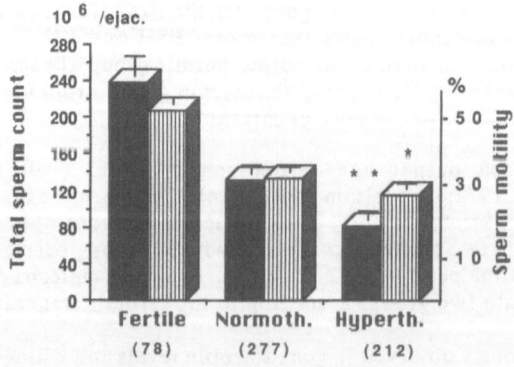

Fig. 5. Mean values (+sem) of total sperm count and sperm motility in fertile men, and in infertile men without (Normoth.) or with (Hyperth.) scrotal hyperthermia. * = p < 0.05 and ** = p < 0.001 with "Normoth." group.

206

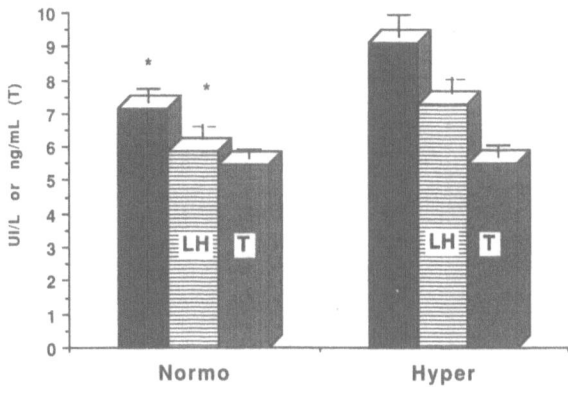

Fig. 6. Mean values (+sem) of serum FSH, LH (LH) and Testosterone (T) levels in infertile men with normal scrotal temperatures (Normo, n = 159) or with a uni- or bilateral hyperthermia (Hyper, n = 153). * = p < 0.01.

such as serum inhibin and seminal transferrin measurements are required in man to confirm such relationships already reported in animal (Demoulin et al., 1979; Steinberger, 1980; Seethalakshmi and Steinberger, 1983; Bergh and Damber, 1984; Bartlett and Sharpe, 1987).

If we consider a total sperm count of less than 60 million spermatozoa as oligozoospermia, then scrotal hyperthermia appears to be the first tool that allows us to

Table 1. Mean values (+sem) of total sperm count, serum FSH, LH and Testosterone levels, and of testicular volumes in infertile men according to both their spermatozoa output and whether their scrotal temperatures were normal (Normal) or above the threshold of 35.2°C (Hyper). * = p < 0.05 and ** = p < 0.01 with "Normal" group in each class of total sperm count

| | Total Sperm Count | | | |
| | < 60 mill./ej. | | > 60 mill/ej. | |
|---|---|---|---|---|
| Scrotal Temperature | Normal | Hyper | Normal | Hyper |
| Number | 98 | 115 | 61 | 38 |
| Total Sperm Count (mill./ej.) | 20.3 (1.7) | 18.0 (1.6) | 163 (13.5) | 153 (16.2) |
| FSH (µl/L) | 8.4 (0.6) | 10.4** (0.7) | 5.1 (0.3) | 5.3 (0.4) |
| LH (µl/L) | 6.4 (0.3) | 7.7* (0.4) | 5.0 (0.3) | 6.1* (0.5) |
| Testosterone (ng/100 ml) | 578 (17) | 546 (17) | 506 (21) | 582 (29) |
| Right Testis Volume (ml) | 25.5 (1.2) | 20.9* (0.8) | 28.1 (1.3) | 31.8 (1.7) |
| Left Testis Volume (ml) | 21.7 (1.1) | 18.3* (0.7) | 25.0 (1.1) | 25.9 (1.6) |

Table 2. Mean values (+sem) of total sperm count, sperm motility, testicular volumes and body temperature in fertile men according to whether their left scrotal temperature was less or more than 35.0°C. * = p<0.05 and ** = p<0.01

| | Left Scrotal Temperature | |
| --- | --- | --- |
| | < 35.0°C<br>n = 58 | ≥ 35.0°C<br>n = 20 |
| Total sperm count<br>($10^6$/ej) | 262<br>(15) | 168*<br>(11) |
| Sperm motility<br>(%) | 54<br>(1.7) | 46*<br>(1.4) |
| Right testis volume<br>(ml) | 31.8<br>(1.4) | 26.1*<br>(1.2) |
| Left testis volume<br>(ml) | 29.1<br>(1.5) | 23.5**<br>(1.2) |
| Rectal temperature<br>(°C) | 35.06<br>(0.05) | 35.18<br>(0.04) |

recognize the worst spermatogenesis (i.e., with the smallest testicular volumes and the highest gonadotropin levels, other things being equal). This means that the two types of oligozoospermia could exist, with different physiopathological patterns.

Some discrepant results are also reported in Table 1. Indeed, in what we would call normozoospermia, the "more than 60 million spermatozoa output class", scrotal hyperthermia differs from normothermia in nothing except an increase in serum LH levels, but we need more data to confirm this. This lack of alterations associated with the existence of a scrotal hyperthermia in case of normozoospermia is without any serious explanation today. Even if it means that modifications associated with scrotal hyperthermia could be heterogeneous, such a statement might be wrong tomorrow. However, this last fact can be put together with some relationship between scrotal temperature and spermatogenesis in fertile men.

When the left scrotal temperature of these fertile men was equal to or higher than 35°C, mean values of total sperm count, sperm motility and testicular volumes were significantly less than when scrotal temperature was under 35.0°C (Table 2). Unfortunately, we did not know what their scrotal temperature was when those fertile men fathered. But scrotal temperatures and spermatogenesis appear to be linked in either fertile or infertile men.

To conclude, scrotal hyperthermia defined as a scrotal temperature above the 90th percentile value in a fertile population, was found in 43% of an infertile population. This scrotal hyperthermia was associated with a more depressed spermatogenesis including less spermatozoa output and sperm motility, higher serum gonadotropin levels and smaller testicular volumes, confirming previous partial reports (Zorgniotti and MacLeod, 1973; Mieusset et al., 1987, 1989). However, in 25% of the cases, scrotal hyperthermia is associated with none of these alterations.

REFERENCES

Bartlett, J.M.S. and Sharpe, R.M. 1987. Effect of local heating of the rat testis on the level in interstitial fluid of a putative paracrine regulator of the Leydig cells and its relation to changes in Sertoli cells secretory function. J. Reprod. Fertil., 80: 279.

Bergh, A. and Damber, J.E. 1984. Local regulation of Leydig cells by the seminiferous tubules: Effect of short-term cryptorchidism. Intern. J. Androl., 7: 409.

Demoulin, A., Koulisher, L., Hustin, J., Hazee-Hagelstein, M.T.M., Lambotte, R. and Franchimont, P. 1979. Organ culture of mammalian testis: III Inhibin secretion. Horm. Res., 10: 177.

Franchimont, P., Millet, D., Vendrely, E., Letawe, J., Legros, J.J. and Netter, A. 1972. Relationships between spermatogenesis and serum gonadotropin levels in azoospermia and oligospermia. J. Endocrinol., 34: 1003.

Handelsam, D.J., Conway, A.J., Boylan, L.M. and Turtle, J.R. 1984. Testicular function in potential sperm donors: normal range and the effect of smoking and varicocele. Int. J. Androl., 7: 369.

Mieusset, R., Bujan, L., Mondinat, C., Mansat, A., Pontonnier, F. and Grandjean, H. 1987. Association of scrotal hyperthermia with impaired spermatogenesis in infertile men. Fertil. Steril., 48: 1006.

Mieusset, R., Bujan, L., Plantavid, M. and Grandjean, H. 1989. Increased levels of serum FSH and LH associated with intrinsic testicular hyperthermia in oligospermic infertile men. J. Clin. Endoc. Metab., 68: 419.

Seethalakshmi, L. and Steinberger, A. 1983. Effect of cryptorchidism and orchidopexy on inhibin secretion by rat Sertoli cells. J. Androl., 4: 131.

Steinberger, A. 1981. Regulation of inhibin secretion in the testis, In "Intragonadal regulation of reproduction", Franchimont, P. and Channing, C.P. (eds.), Academic Press, NY.

Zorgniotti, A.W. and MacLeod, J. 1973. Studies in temperature, human semen quality and varicocele. Fertil. Steril., 24: 854.

Zorgniotti, A.W. and Sealfon, A.I. 1988. Measurement of intrascrotal temperature in normal and subfertile men. J. Reprod. Fertil., 82: 563.

# SCROTAL HYPERTHERMIA; ETIOLOGIC FACTORS:

# FACTS AND HYPOTHESES

Roger Mieusset

Centre de Sterilité Masculine and INSERM U168
Hôpital La Grave
Toulouse, France

In animals, spermatogenesis was experimentally shown to be dependent on ambient (Cameron and Blackshaw, 1980), testicular (Keel and Abney, 1980) or scrotal (Malgrem and Larson, 1989) temperatures. In humans, occupational and living habits could influence semen quality apparently through scrotal insulation (Laven et al., 1988), as well as summer heat (Levine et al., 1988) or ambient temperature (Bornam et al., 1989). Alterations in spermatogenesis were reported to be associated with intrinsic scrotal hyperthermia (Zorgniotti and MacLeod, 1973; Mieusset et al., 1987, 1989; Zorgniotti and Sealfon, 1988). Avoidance of testicular hyperthermia, such as baths or tight clothing, was also reported to improve seminal characteristics (Lynch et al., 1986) and successful results were obtained in infertile men through chronic treatment with scrotal hypothermia (Zorgniotti et al., 1986).

From all these information and data arise the question of the etiologic factors of scrotal hyperthermia, and three points will be discussed here:

(1)  Are common andrological factors liable for the increase in scrotal temperature?
(2)  What do we know about the experimental effects of induced heat?
(3)  What hypotheses explain scrotal hyperthermia?

In an infertile population of 516 men without excretory azoospermia (Mieusset et al., 1989), the commonly studied andrological factors were either a testicular history (n = 83), including orchitis (n = 21), cryptorchidism (n = 38) and surgery of varicocele (n = 28), or a clinical varicocele (n = 191, with 147 on the left side and 29 bilateral). As these factors could be found in association, four infertile groups were distinguished, with 274 men without a history nor a varicocele, 159 men with only a varicocele, 51 men with only a testicular history and 32 with both a history and a varicocele (Table 1). In the "varicocele only" group, the mean value of the left scrotal temperature is higher than in the "no history no varicocele" group. When a testicular history is found alone, mean values of both right and left scrotal temperatures are higher than in the "no history no varicocele" group, and the left scrotal temperature is yet higher when a varicocele was added to a history.

It seems that the existence of a testicular history or a varicocele is associated with a higher scrotal temperature. But is this true for any one of these andrological factors?

For each of these factors, we looked at the percentage of testes with both this factor and a scrotal hyperthermia, the latter meaning a scrotal temperature of at least 35.2°C as previously defined (Mieusset et al., 1989). From the results reported in Table 2, three facts are evident: (1) a testis with a history or a varicocele shows a scrotal hyperthermia

Table 1. Mean values (+SD) of the scrotal temperatures in infertile men according to the presence or absence of a testicular history (orchitis, cryptorchidism, surgery for varicocele) or a clinical varicocele. * = p < 0.05 and ** = p < 0.01 with "no history no varicocele" group

| History | - | - | + | + |
|---|---|---|---|---|
| Varicocele | - | + | - | + |
| Number of Men | 274 | 159 | 51 | 32 |
| Right Scrotal Temperature (°C) | 34.93 (0.5) | 35.00 (0.4) | 35.16* (0.5) | 35.12 (0.6) |
| Left Scrotal Temperature (°C) | 34.99 (0.6) | 35.19** (0.6) | 35.18* (0.4) | 35.27* (0.5) |

twice more often than a testis without either of these two factors (55% versus 26%); (2) but that means also that when one of these factors exists, scrotal hyperthermia appears in only one case out of two; (3) and of this whole population, half of the hyperthermic testes are without any factors (152 versus 157).

To conclude, it is obvious that scrotal hyperthermia is rather a factor by itself than a specific result of the common andrological factors.

But do we have any arguments for supporting that an increase in scrotal temperature can be a factor liable to result in an alteration in human spermatogenesis? Such arguments can be found in the experiments with heat induced in man.

A short review of the literature allows us to distinguish between four different types of hyperthermia that were used (Table 3). Whatever the type, hyperthermia always induced a reversible depression in spermatozoa output. Even if all of these hyperthermic states can occur in man, the last two are closer to what was seen in the infertile populations, in the way that the rise in scrotal temperature always stays under the value

Table 2. Responsibility of the common andrological factors for scrotal hyperthermia

| | Number of men | % of testes with this factor and a scrotal hyperthermia |
|---|---|---|
| Cryptorchidism | 38 | 26/45 = 58% |
| Orchitis | 21 | 10/23 = 43% |
| Varicocele surgery | 28 | 17/31 = 55% |
| Only varicocele | 159 | 104/188 = 55% |
| | | Total = 157 |
| Neither history nor varicocele | 274 | 152/548 = 26% |

Table 3. Heat induced "hyperthermia" in man

| Type of Hyperthermia | Heated Organ | Heat Intensity | Heat Duration | Authors |
|---|---|---|---|---|
| General | Body | > 40 °C | < 1 hr | MacLeod 41 Procope 65 Brown-Woodman 84 |
| Local | Scrotum | > 38 °C | < 1 hr | Robinson 68 Watanabe 59 French 73 |
| Local | Scrotum | < 37 °C | 4-11 wk | Robinson 68 |
| Local | Testis | < 37 °C | 6-24 mo | Mieusset 85,87 |

of the physiological body temperature (Zorgniotti and MacLeod, 1973; Mieusset et al., 1988; Zorgniotti and Sealfon, 1989).

Even if this last condition (a testis or scrotal temperature under the body temperature) restricts the number of heat induced experiments that can be referred to in animals, a reversible depression in spermatozoa output was reported, as well as an increase in serum gonadotropin levels and a decrease in testicular size (Rommerts et al., 1980; Jegou et al., 1983; Seethalakshmi and Steinberger, 1983; Bergh et al., 1985; Gonzales et al., 1989).

But a gap does exist between the fact that scrotal hyperthermia appears to be a factor associated with alterations in spermatogenesis in 3 infertile men out of 10 (Mieusset et al., previous presentation), and the fact that hyperthermia induced in fertile men results in a quite similar pattern of spermatogenesis alterations. This gap gives rise to two questions which we cannot answer today.

The first question is: by which mechanism does a low intensity and long duration increase in scrotal or testis temperature induce alterations in spermatogenesis? The answers are as varied as the tools selected to perform the study. But they are all valid in as much as nobody can say today which type of cell or structure inside the testis is directly or indirectly affected by such a heat and in which sequence.

The second question is, in my opinion, the one that needs an urgent answer: how does scrotal hyperthermia occur in man? Three types of pathology can be hypothesized as involved in the occurrence of scrotal hyperthermia.

The first one is an alteration in the thermoregulatory function of the scrotum. This type of pathology was poorly supported in man until now, but several very interesting arguments will be reported in this meeting (Clavert, this book). Moreover, it seems to be obvious that most of the successes in chronic treatment with the testicular hypothermia device from Dr. Zorgniotti could be linked with a severe thermoregulatory dysfunction of the scrotum (Zorgniotti et al., 1986). A less severe alteration in this thermoregulatory dysfunction could be involved in the modifications which occur when testes are exposed to hot surroundings such as occupational and living habits (Lynch et al., 1986; Laven et al., 1988; Spira, 1989).

But a dysregulation in the thermic functions of the scrotum will be asserted only when the following questions have been answered: Do scrotal sweating and cutaneous thermoreceptors exist in man, as in animals (Waites and Voglmayr, 1963; Iggo, 1969)?

What is the thermic adjustment capacity of the scrotum to an increase in ambient as well as in body temperatures, and are these characteristics similar in both fertile and infertile men?

A scrotal hyperthermia could also be induced by a second type of pathology which is a vascular defect. Most of the authors dealing with testicular heat induced in animals failed to find any vasodilatation or modifications in the blood flow during the heating when temperatures were less than 40°C (Glover, 1966; Waites et al., 1973; Damber and Johnson, 1978). This sets the problem of the testicular vascularization as a specific system, as it has been already considered through cadmium induced injuries, for example (Setchell and Waites, 1970; Gunn and Gould, 1975). Moreover, we do not know if testicular thermoregulation appears in humans during puberty and if this occurrence is linked with the development of the testicular vascular network, as has been reported in the rat (Kormano, 1967). And since vascularization and thermoregulation are strongly tied, we can ask if the pampiniform plexus could not act like a trap for testicular heat in some situation other than fever.

The third occurrence of a scrotal hyperthermia could result from a general pathology, involving either an alteration in the vascular hemodynamics of the body or a dysregulation in the body temperature. We have no explanation for the elevated rectal temperature found in the hyperthermic infertile men (Mieusset et al., previous presentation). Moreover, such a rise is also observed with some andrological factors in which a vascular defect has been considered, such as varicocele.

In a group of infertile men without any testicular history but with a left clinical varicocele, a comparison was done between the men with a left scrotal hyperthermia and those with a normal temperature (Table 4). The spermatozoa output is less and the serum FSH level is higher in the hyperthermic group. Although only the left scrotal temperature was used to select these two groups, it was found that they had scrotal hyperthermia on both left and right sides. Moreover, rectal temperature is higher in these men. In this hyperthermic group, varicocele seems to be an indication of a vascular

Table 4. Mean values (+sem) of total sperm count and serum FSH levels, as well as scrotal and rectal temperatures (+SD) according to whether the left scrotal temperature was less or more than the threshold of hyperthermia. All these infertile men were without any testicular history but with a left varicocele

|  | Left Scrotal Temperature | |
|  | $\leq 35.2$°C<br>n = 48 | > 35.2°C<br>n = 57 |
|---|---|---|
| Total Sperm Count<br>(mill./ej.) | 72<br>(8) | 40*<br>(7) |
| FSH<br>($\mu$l/L) | 7.0<br>(0.4) | 8.7*<br>(0.5) |
| Right Scrotal Temperature<br>(°C) | 34.63<br>(0.3) | 35.40***<br>(0.4) |
| Left Scrotal Temperature<br>(°C) | 34.76<br>(0.3) | 35.56<br>(0.4) |
| Rectal Temperature<br>(°C) | 37.01<br>(0.3) | 37.18**<br>(0.3) |

* = $p < 0.05$;  ** = $p < 0.01$;  *** = $p < 0.001$

214

defect which is related to both testes; the elevation in the body temperature could be either the prime cause or the result of this vascular defect.

To conclude, the gap between the intrinsic hyperthermia in infertile men and the induced hyperthermia in normal men opens a spacious and very exciting field of physiological and biological research, the results of which could bring some new light on the testis and, therefore, adequate etiologic treatments in male infertility as well as in fertility regulation.

## REFERENCES

Bergh, A., Nikula, H., Svensson, J. Hansson, V. and Purvis, K. 1985. Altered gonadotropin, prolactin, GnRh-receptors and testicular steroid concentrations in the abdominal testes of unilateral cryptorchid rats. J. Reprod. Fertil., 74: 279.

Bornam, M., Shulenberg, G., Boomker, D., Van der Merwe, C. and Reif, S. 1989. Ambient temperature and semen quality. Fourth International Congress of Andrology, p. 23.

Cameron, R.D.A. and Blackshaw, A.W. 1980. The effect of elevated ambient temperature on spermatogenesis in the boar. J. Reprod. Fertil., 59: 173.

Damber, J.E. and Johnson, P.O. 1978. The influence of scrotal warming on testicular blood flow and endocrine function in the rat. Acta Physiol. Scand., 104: 61.

Glover, T.D. 1966. The influence of temperature on flow of blood in the testis and scrotum of rats. Proc. Roy. Soc. Med., 59: 765.

Gonzales, G.F., Risbridger, G.P. and De Kretser, D.M. 1989. In vivo and in vitro production of inhibin by cryptorchid testes from adult rats. Endocrinol., 124: 1661.

Gunn, S.A. and Gould, T.C. 1975. Vasculature of the testes and adnexa, In "Handbook of Physiology, Endocrinology V", p. 117.

Hsiung, R. 1990. Effect of medullary lesions on scrotal thermoregulation, In: "Temperature and environmental effects on the testis", A.W. Zorgniotti (ed), Plenum Press.

Iggo, A. 1969. Cutaneous receptors in primates and subprimates. J. Physiol., 200: 403.

Jegou, B., Laws, A.D. and De Kretser, D.M. 1983. The effect of cryptorchidism and subsequent orchidopexy on testicular function in adult rats. J. Reprod. Fertil., 69: 137.

Keel, B.A. and Abney, T.O. 1980. Influence of bilateral cryptorchidism in the mature rat: alterations in testicular function and serum hormone levels. Endocrinol., 107: 1226.

Kormano, M. 1967. Development of the rectum-testis temperature difference in the postnatal rat. J. Reprod. Fertil., 14: 427.

Laven, J.S.E., Haverkorn, M.J. and Bots, R.S.G.M. 1988. Influence of occupation and living habits on semen quality in men (scrotal insulation and semen quality). Eur. J. Obstet. Gynec. Reprod. Biol., 29: 137.

Levine, R.J., Bordson, B.L., Matthew, R.M., Brown, M.H., Stanley, J.M. and Starr, T.B. 1988. Deterioration of semen quality during summer in New Orleans. Fertil. Steril.

Lynch, R., Lewis-Jones, D.I., Machin, D.G. and Desmond, A.D. 1986. Improved seminal characteristics in infertile men after a conservative treatment regimen based on the avoidance of testicular hyperthermia. Fertil. Steril., 46: 476.

Malgrem, L. and Larsson, K. 1989. Experimentally induced testicular alterations in boars: histological and ultrastructural findings. J. Vet. Med. A., 36: 3.

Mieusset, R., Bujan, L., Mondinat, C., Mansat, A., Pontonnier, F. and Grandjean, H. 1987. Association of scrotal hyperthermia with impaired spermatogenesis in infertile men. Fertil. Steril., 48: 1006.

Mieusset, R., Bujan, L., Mansat, A., Plantavid, M. and Grandjean, H. 1989. Increased levels of serum FSH and LH associated with intrinsic testicular hyperthermia in oligospermic infertile men. J. Clin. Endoc. Metab., 68: 419.

Mieusset, R., Bujan, L., Mansat, A., Plantavid, M., Grandjean, H. and Pontonnier, F. 1989. Scrotal hyperthermia: frequency in an infertile population; associated alterations in testicular functions. In: "Temperature and environmental effects on the testis", A.W. Zorgniotti (ed), Plenum Press.

Rommerts, F.F.G., De Jong, F.H., Grootegoed, J.A. and Van der Mohen, H.J. 1980. Metabolic change in testicular cells from rats after long-term exposure to 37°C. J. Endocrinol., 85: 471.

Seethalakshmi, L. and Steinberger, A. 1983. Effect of cryptorchidism and orchidopexy on inhibin by rat Sertoli cells. J. Androl., 4: 131.

Spira, A. 1989. Effect of bath water heat on reproductive outcome. Presented in the conference of "Temperature and environmental effects on the testis", New York, December 8-9, 1989.

Waites, G.M.H., Setchell, B.P. and Quinlan, D. 1973. The effects of local heating of the scrotum, testes and epididymides of rats on cardiac output and regional blood flow. J. Reprod. Fertil., 34: 41.

Waites, G.M.H. and Voglmayr, J.K. 1979. The functional activity and control of the apocrine sweat glands of the scrotum of rat. Aus. J. Agr. Res., 14: 839.

Zorgniotti, A. W., Cohen, M.S. and Sealfon, A.I. 1986. Chronic scrotal hypothermia: results in 90 infertile couples. J. Urol., 135: 944.

Zorgniotti, A.W. and MacLeod, J. 1973. Studies in temperature, human semen quality and varicocele. Fertil. Steril., 24: 854.

Zorgniotti, A.W. and Sealfon, A.I. 1988. Measurement of intrascrotal temperature in normal and subfertile men. J. Reprod. Fertil., 82, 563.

# TESTICULAR HYPERTHERMIA: PHYSIOPATHOLOGY, DIAGNOSTIC

# AND THERAPEUTICAL CONCEPTS

Vincenzo Mirone[1] and Fabrizio Iacono[2]

[1]Dept. of Urology, School of Medicine
University of Catanzaro, Italy
[2]Dept. of Urology, II Medical School
University of Naples, Italy

The intrascrotal temperature has to be steadily kept lower than the corporeal temperature by 2.5°C[1], in order to let the testicle perform its normal spermatogenetic function. The testicle can keep the euthermic condition through three basic mechanisms:

· Countercurrent thermoregulation system, consisting of the testicular artery and the pampiniform plexus which wraps it, where the "cold" blood flowing down the testicle lets the "warm" blood cool off.

· Dartoic and cremasteric muscles, which adjust their contraction depending on corporeal and room temperature, by adjusting the distance between testicle and body.

· The thin thickness of the scrotal skin which allows a fast and easy thermal dispersion.

The spermatogenetic function is highly sensitive to thermal variations, even if they are seasonal[2,3]. The increase, even if temporary, in testicular temperature causes a noticeable depressive response on spermatogenesis[4]. By increasing the testicular temperature 2°C, a severe deterioration of testicular parenchyma is obtained. If such a condition lasts for more than 2 weeks, the azoospermy condition could take place.

The functional damage in cryptorchidic testicles is very similar to the one produced by experimental testicular hyperthermia[5]. The increase in the corporeal temperature, as it takes place in febrile conditions, is counterbalanced by the scrotal testicular thermoregulation system, up to a maximum of 37.7°C, which is regarded as the critical temperature point, beyond which the testicle cannot keep the thermal homeostasis anymore[1].

Another condition which can ruin the testicular thermoregulation system is the presence of a varicocele. A steady increase in testicular temperature, even of 0.6 - 1.4°C, causes a noticeable decrease in spermatogenesis and epididymis maturation[6]. Among the different etiopathogenic theories on testicular damage caused by varicocele, the thermogenetic theory is particularly successful[5,7,8,9]. As a matter of fact, in case of venous reflux, the blood, flowing down the pampiniform plexus, would interfere with the countercurrent cooling mechanism of arterial blood, leading to an increase of testicular temperature, and subsequent alteration of spermatogenesis.

Besides varicocele, other pathologies can account for an high testicular temperature: the hydrocele, whose albuginean vaginal stratum creates a thermal defense by avoiding thermal dispersion; cryptorchidism and phlogistic processes cause a too high arterial afflux.

The evaluation of scrotal temperature can be performed through a cheap and simple method, which is standardized and non-invasive: thermometry. This examination employs two kinds of thermometer: water bath thermometer and infrared rays thermometer[10]. The patient lies on his back, straddle legged, for about six minutes, at a room temperature of 20°C. The thermal range, measured by means of a water thermometer in patients affected by varicocele, ranges between 32.2 and 35°C for the right testicle and between 33.3 and 35.1°C for the left testicle. By means of the infrared rays thermometer, thermal values in patients affected by varicocele range between 30.4 and 34.6°C for the right testicle and 30.3 and 34.4°C for the left testicle[11].

We considered 2320 patients affected by infertility at our clinics. Six hundred ninety-six patients were affected by varicocele and, among them, 188 (27%) did not present remarkable seminal impairments. Another 1624 subjects were not affected by varicocele, but 406 patients (25%) presented a high testicular temperature without any evident reason and seminal parameters impairment just like the patients affected by varicocele. We defined such a condition of high scrotal temperature, without clinical nor instrumental evidence of varicocele or other infections, as essential scrotal hyperthermia[11]. The causes of the essential testicular hyperthermia are to be searched in testicular hemodynamics: endoarterites, arterial vasoconstriction and reduced cardiac input.

We considered testicular temperature in 323 patients affected by varicocele: among them 74 (23%) presented testicular normothermia and standard seminal parameters. It is likely that, as to fertile varicocele cases, a thermal dispersion system is hypertrophied, leading to subsequent normothermia and fertility. Therefore, this is a further evidence of the hyperthermal theory on testicular damage etiopathogenesis for varicocele.

The same remarks were carried out by other authors[12,13,14] who confirmed the existence of a group of sterile subjects with seminal parameters impairment, but without varicocele and high scrotal temperature. The interpretation of these data make us regard the scrotal hyperthermia as a nosological entity, which can lead to male infertility and to varicocele as one of the causes of testicular hyperthermia.

We can now bring up the concept of thermal prognostic index as regards the evaluation of surgical treatment of varicocele in case of male infertility. In a previous study we noticed that patients affected by varicocele and high testicular temperature presented a remarkable increase in seminal parameters and, therefore, in the pregnancy rate after surgical treatment, as compared to patients operated for varicocele, but without high testicular temperature. According to the thermal prognostic index, the ligature of the internal spermatic vein seems to represent the best therapeutical approach only in the case of testicular hyperthermia. In the presence of renal-spermatic reflux, on the other hand, such an approach turns out to be not so reliable a therapy against infertility.

By fully evaluating these thermal theories on the etiology of some cases of male infertility, both for varicocele and idiopathic, Zorgniotti et al.[15,16] first experienced an instrument to lower the testicular temperature through the water controlled evaporation from the scrotal surface: the THD (Testicular Hypothermia Device). This device is represented by a resilient reservoir which keeps scrotal surface steadily wet through a microinjector. It allows to adjust the downflow of the distilled water in the reservoir. THD can be worn under one's dress: for this reason patients accept to use it during the winter period, when it can be easily hidden.

Such system consists in keeping always wet the underpants, so that the steady water evaporation from scrotal surface can lower its temperature by 2°C. The THD is recommended for a period of at least 6 months during the waking hours, in presence of the following elements: seminal parameters alteration, high testicular temperature, with or without varicocele, and infertility after more than one year of non-protected attempts. The data reported in literature an improvement of seminal parameters in 73% with about 24% of pregnancies[17,18].

218

Table 1

| PAZ. | DIAGN. | SPERM CONC. | | MOTIL.% | | MOI | | RES. | PREG. |
|------|--------|------|------|------|------|------|------|------|------|
| | | PRE | POST | PRE | POST | PRE | POST | | |
| 1 | NO VARIC. | 28 | 42 | 35 | 55 | 6.37 | 17.3 | + | + |
| 2 | VARICOCELE | 20 | 37 | 40 | 50 | 4.8 | 12 | + | - |
| 3 | VARICOCELE | 15 | 30 | 45 | 50 | 4.85 | 9 | + | - |
| 4 | VARICOCELE | 25 | 38 | 40 | 55 | 5.5 | 13.5 | + | + |
| 5 | VARICOCELE | 18 | 34 | 45 | 55 | 4.9 | 9.5 | + | - |
| 6 | VARICOCELE | 27 | 40 | 40 | 50 | 5.5 | 12 | + | - |
| 7 | NO VARIC. | 30 | 45 | 33 | 53 | 6.38 | 16.4 | + | - |
| 8 | VARICOCELE | 23 | 39 | 40 | 50 | 5.2 | 13.2 | + | - |
| MEANS | | 20.4 | 38.1 | 39.7 | 52.2 | 5.4 | 11.2 | | 0.25 |

N.B. VARICOCELE MEANS LEFT SIDE IN ALL CASES.

In order to carry out a prognostic evaluation on the therapeutical possibility of THD, a mathematical index called MOI (motile oval index) was proposed. MOI is the result of the product of spermatozoa concentration by motility, divided by 100 by oval shapes, divided by 100. Patients who get the greatest benefits from hypothermal therapy with THD are patients with a MOI value higher than or equal to 4.8 million/ml. We treated a first group of 8 patients (Table 1) with MOI value higher than 4.8; these were affected by clinic varicocele and high scrotal temperature. The obtained results can be regarded as encouraging: all the cases after six months treatment experienced MOI increase, and in two cases a pregnancy took place. All the couples presented a two-year infertility story. Anyway, there are certain conditions, where THD treatment turns out to be useless: clinic varicocele without testicular hyperthermia, severe oligospermy or azoospermy (very low MOI value). If a manifest hyperthermia is not present, any hypothermal treatment turns out to be useless.

## REFERENCES

1) Lazarus, B.A. and Zorgniotti, A.W. Thermoregulation of the human testis. Fertil. Steril., 26: 757-759, 1975.

2) Mortimer, D., Templeton, A.A., Lenton, E.A. and Coleman, R.A. Annual patterns of human sperm production and semen quality. Arch. Androl., 10: 1-5, 1983.

3) Bornman, M., Schulenburg, G., Boomker, D., Van Der Merwe, C., and Reif, S. Ambient temperature and semen quality. IV International Meeting of Andrology, Florence, Book of Mini Poster, Monduzzi Press, 23,25; 1989.

4) MacLeod, J. and Hotchkiss, R.S. The effect of hyperpyrexia upon spermatozoa counts in men. Endocrinology, 28: 780, 1989.

5) Verstoppen, G.R. and Steeno, O.P. Varicocele and the pathogenesis of the associate subfertility. A review of various theories. III The theories concerning the deleterious effects of varicocele on fertility. Andrologia, 10(2): 85-102, 1978.

6) Zorgniotti, A.W. Elevated intrascrotal temperature I: A hypothesis for poor semen in infertile man. Bull N.Y. Acad. Med., 58: 535-540, 1982.

7) Davidson, H.A. Treatment of male subfertility, testicular temperature and varicocele. Practitioner, 173: 703, 1954.

8) Zorgniotti, A.W. Mechanism of alterated testis temperature and poor semen <u>In</u>: "Human Fertility Factors", Spira, A. and Jouannet, P. (eds), Paris: Inserm 103: 351-362, 1981.

9) Zorgniotti, A.W. and MacLeod, J. Studies in temperature, human semen quality and varicocele. <u>Fertil. Steril.</u>, 24: 854, 1973.

10) Zorgniotti, A.W., Toth, A. and MacLeod, J. Infrared thermometry for testicular temperature determinations. <u>Fertil. Steril.</u>, 32: 346-348, 1979.

11) Mirone, V., Iacono, F., Prezioso, D., Imbimbo, C. and Sanseverino, R. Scrotal hyperthermia and male infertility. III Corso di aggiornamento in andrologia chirurgica - Fiuggi- 1984. Libro degli atti. Acta Medica press.

12) Mieusset, R., Bujan, L., Mondinat, C., Mansat, A., Pontonnier, F. and Grandjean, H. Association of scrotal hyperthermia with impaired men. <u>Fertil.Steril.</u> 48: 1006, 1987.

13) Bujan, L., Mansat, A., Pontonnier, F. and Mieusset, R. Scrotal hyperthermia in infertile men. <u>IV Internat. Congress of Andrology</u>, Florence, 1989. Acts book, 253-257. Monduzzi press.

14) Zorgniotti, A.W. and Sealfon, A.I. Measurement of intrascrotal temperature in normal and subfertile men. <u>J. Reprod. Fertil.</u>, 82: 563-566, 1988.

15) Zorgniotti, A.W., Sealfon, A.I. and Toth, A. Further clinical experiences with testis hypothermia for infertility due to poor semen. <u>Urology</u>, 19: 636-640, 1982.

16) Zorgniotti, A.W., Cohen, M.S. and Sealfon, A.I. Chronic scrotal hypothermia: Results in 90 infertile couples. <u>J. Urol.</u>, 135: 944- 947, 1986.

17) Mirone, V., Iacono, F., Prezioso, D., Imbimbo, C. and Lotti, T. La terapia termica dell'ipertermia testicolare essenziale e da varicocele: il T.H.D. valutazione preliminare. Progressi in Andrologia chirurgica. Acta Medica, 1987.

18) Lotti, T., Mirone, V., Iacono, F. and Prezioso, D. La terapia ipotermica del varicocele. <u>V National Meeting of Andrology</u>, Bologna 1987. Acts book.

# HYPOTHESIS TO EXPLAIN SUBFERTILE SEMEN

Adrian W. Zorgniotti

Department of Urology
New York University School of Medicine

## INTRODUCTION

There is general agreement that even small increases in temperature from experimental application of external heat can cause disruption of spermatogenesis and sperm maturation with alteration of semen parameters. It is now apparent that intrinsic elevation of temperature can be identified in the presence of subfertile semen. Such increases in temperature above normal appear small: 0.5 - 1.5°C.

## INTRINSIC TEMPERATURE REGULATION

In scrotal mammals, lower than core testis temperature is maintained by two mechanisms:

(1) Countercurrent heat exchange between the internal spermatic artery and returning venous blood pre-cools arterial blood arriving at the mediastinum testis. In man this takes place in the pampiniform plexus.

(2) Heat loss by the externalized scrotum via a combination of radiation, convection, conduction and sweating.

Scrotal skin area limits heat loss into the environment. Any increased load, such as a rise in arterial blood temperature cannot be handled by the scrotum and results in increase of intrascrotal temperature.

In varicocele, elevated temperature may be explained by the presence of refluxing core temperature venous blood in the scrotum. However, this does not explain elevated temperature in men without palpable varicocele (idiopathic infertility; subclinical varicocele).

Zorgniotti (1982a) proposed that refluxing blood interferes with countercurrent heat exchange in the pampiniform plexus. This results in increase in temperature of spermatic artery blood which the scrotum is not able to dissipate and intrascrotal temperature rises (Figure 1).

It is possible that mere slowing of spermatic vein blood in the pampiniform plexus may be sufficient to interfere with countercurrent heat exchange without there being reflux present. This could be explained by compression of the left renal vein by the superior mesenteric artery, the so called "nut cracker effect" (Figure 2).

*Temperature and Environmental Effects on the Testis*
Edited by A. W. Zorgniotti, Plenum Press, New York, 1991

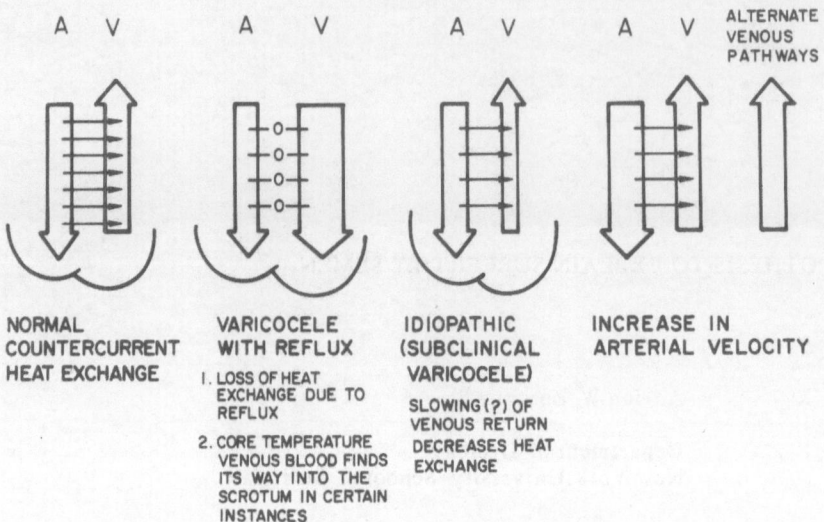

Figure 1. Countercurrent heat exchange by the pampiniform plexus. Hypothetical changes which could alter internal spermatic artery temperature. These assume that scrotal skin heat loss remains unchanged.

(a) Normal countercurrent heat exchange. Internal spermatic artery blood at core temperature loses heat by exchange with returning venous blood.

(b) Possible abolition of heat exchange in the presence of internal spermatic vein reflux with or without varicocele. With varicocele, core temperature venous blood also refluxes into the scrotum adding to the heat load.

(c) Slowing of venous return could diminish countercurrent exchange causing internal spermatic artery blood temperature to rise.

(d) Increase in arterial blood flow (unlikely) could result in elevated temperature. Returning venous blood flow could follow alternative pathways.

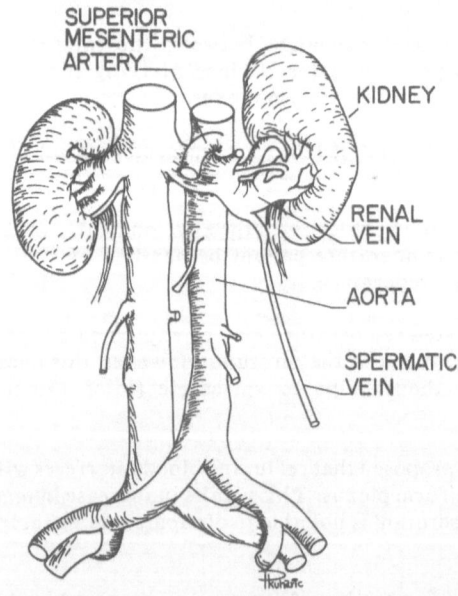

Figure 2. "Nut cracker phenomenon" whereby the superior mesenteric artery compresses the renal vein, possibly accentuated by standing. This could increase pressure in the renal vein, causing slowing of venous flow in the internal spermatic vein.

When elevation of temperature is present, it is bilateral and can be explained by the presence of a gradient between the right and left scrotum or by bilateral thermoregulatory defect.

Intrinsic thermoregulatory defect causes increases in temperature of 0.5 to 1.5 °C. This may represent the maximum elevation of temperature possible as demonstrated by Lazarus and Zorgniotti (1975) who measured scrotal/rectal temperature gradients (SRD) in men with fever. With rising body temperature, SRD remained level until rectal temperature reached 37.8 °C at which time, SRD dropped abruptly 1.5 °C and then levelled off even though rectal temperature continued to rise.

## EXTRINSIC TEMPERATURE ELEVATIONS IN MAN

(1) Clothing: Ehrenberg et al. (1957) showed that the presence of clothing elevated scrotal temperature 3.3 °C. Zorgniotti (1982b) found that clothing resulted in increases of 1.2 to 1.5 °C and that this rise was the same for normospermic and for oligospermic men, only that oligospermic men with higher starting temperatures rose proportionately higher.

(2) Climate: No testis data have been accumulated on individuals measured at different ambient temperatures.

(3) Hot baths and hot working conditions: Similarly there are no data in individuals.

## HYPOTHESIS TO EXPLAIN SUBFERTILE SEMEN (Zorgniotti and Sealfon, 1988)

(1) Intrinsic thermoregulatory defect causes elevation of testis and epididymal temperature 0.5 - 1.5 °C.

(2) These seemingly small elevations have the potential for disrupting spermatogenesis and epididymal maturation of spermatozoa.

(3) Such intrinsic elevations of temperature are chronic.

(4) Normal men with normal thermoregulation are able to compensate for the heat effects of clothing, climate, hot baths and work related heat exposure.

(5) Men with abnormal thermoregulation are not able to compensate for such extrinsic accidents and temperature rises above tolerable limits as seen in the clothing study (op cit).

(6) Chronic elevation of temperature can result in varying degrees of subfertile semen. Ultimately there is irreversible damage and azoospermia. Finding of the Sertoli Only Syndrome on testis biopsy may represent such an end point.

## REFERENCES

Ehrenberg, L. et al. 1957. Gonad temperature and spontaneous mutation-rate in man. Nature, II: 1433.

Lazarus, B. and Zorgniotti, A.W. 1975. Thermoregulation of the human testis. Fertil. Steril., 26: 757.

Zorgniotti, A.W. 1982a. Elevated intrascrotal temperature: Hypothesis for poor semen in infertile men. Bull. N.Y. Acad. Med., 58: 535.

Zorgniotti, A.W. et al. 1982b. Effect of clothing on scrotal temperature in normal men and patients with poor semen. Urology, 19: 176.

Zorgniotti, A.W. and Sealfon, A.I. 1988. Measurement of intrascrotal temperature in normal and subfertile men. J. Reprod. Fert., 82: 563.

# SECTION 6

## THERAPEUTIC USES OF HEAT TRANSFER

# CHRONIC SCROTAL HYPOTHERMIA

Adrian W. Zorgniotti[1] and Andrew I. Sealfon[2]

[1]Department of Urology
New York University School of Medicine
[2]Repro-Med Systems, Inc.
Middletown, NY 10940

## INTRODUCTION

The Testicular Hypothermia Device (THD) was conceived to test the theory that an intrinsic defect in testicular thermoregulation causes elevation of intrascrotal temperature and results in subfertile semen (Zorgniotti et al., 1980, 1986). When it was found that lowering temperature about 2.0°C by a prototype THD resulted in semen improvement and pregnancy in the wives of infertile men, the therapeutic potential was realized. The THD is a cotton scrotal covering which lowers temperature by controlled evaporation of water from its surface (Figure 1).

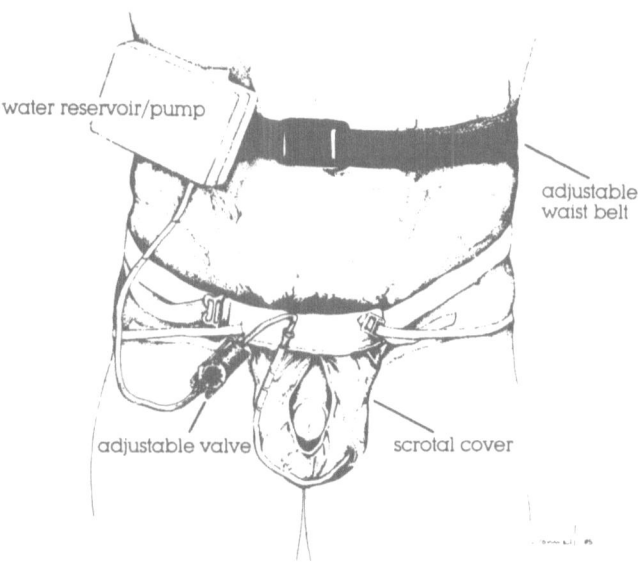

Figure 1. The Testicular Hypothermia Device (THD) lowers scrotal temperature by heat of vaporization (evaporation) about 2.0°C, bringing the testis temperature closer to normal levels. Improvement in subfertile semen and pregnancies occur depending upon the severity of the infertility at the time of initiation of treatment.

## Table 1. Criteria for the male, minimal normal values

Subfertile semen in at least one parameter:
> Less than 20 million spermatozoa/ml
> Less than 40% active forms
> Less than 60% oval forms (stained morphology)

Intrascrotal temperature above 34.1°C

## CRITERIA FOR THE THD TRIAL

Couples were required to have had non-contraceptive coitus for two years without a pregnancy. In clinical practice, this should be waived if further waiting would be harmful to the couple's chances of conception. Wives were judged able to conceive on clinical impression of a gynecologist. Husbands fell into one of three clinical entities: Varicocele, failed varicocelectomy, idiopathic infertility (sub-clinical varicocele). Excluded were patients with non-temperature caused subfertile semen (e.g., endocrine, genetic, toxic, mumps, etc.) (Table 1).

## CONDUCT OF TRIAL

Two pre-trial semen analyses were required, one not more than six months and one not more than six weeks before commencement of hypothermia. Semen analysis done at ten week intervals during the trial, included stained morphology and were performed by independent laboratories specialized in fertility diagnosis. The patient signed a consent following which he was fitted with the THD and instructed in its use.

The hypothermic effect of the device was demonstrated by infrared thermometry. After having the patient stand for six minutes with genitalia exposed, an infrared thermometer reading of the scrotal skin over the testis was made. A THD was applied and when equilibration was reached (i.e., the scrotal covering was damp), the suspensory was pulled to one side and the temperature reading repeated (Table 2).

Table 2. Mean decrease in scrotal temperature (°C) after application of the THD

n = 59

|       | DECREASE | RANGE          |
|-------|----------|----------------|
| RIGHT | 2.0      | (-0.1 to -6.2) |
| LEFT  | 2.3      | (-0.3 to -4.9) |

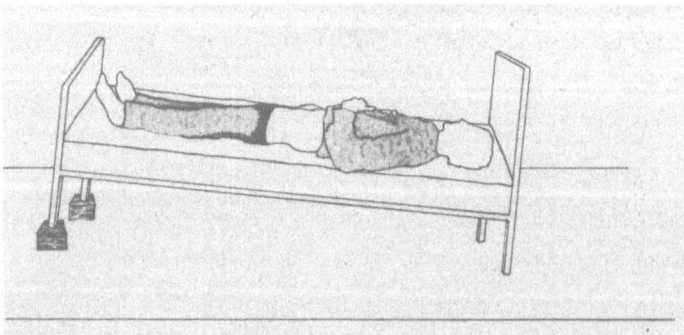

Figure 2. At night patients remove the THD and sleep without clothing covering the genitalia. They also sleep in mild Trendelenburg, which lowers intrascrotal temperature.

The patient was required to wear the THD daily at all times while up and clothed. At night, the patient removed the THD allowing for a period of dry skin after day long contact with the damp scrotal covering. He was asked to sleep without clothing over the genitalia; light covers were permitted. The patient was required to place 10 cm blocks under the legs at the foot of the bed in order to produce slight Trendelenburg (Figure 2). The THD was also removed for intercourse and bathing. In the event of a fever above 37.8°C, the patient was asked to wear the THD at all times.

RESULTS

One hundred forty men met the criteria of the protocol and were fitted with the THD. 76/140 did not wear the device as required or long enough. The most frequent reasons for dropping out were mechanical difficulties with the device which have since been remedied and discouragement over the length of time required before semen improved or pregnancy occurred. A self selected control group were 20/76 who met criteria but never wore the THD and did not have any further infertility treatment. Not conforming to the protocol were six who had non-obstructive azoospermia (having shown progressive deterioration in the semen analysis to the point of azoospermia). None showed any change in semen after 6+ months of hypothermia; the THD is not recommended for use in azoospermia.

Sixty-four wore the THD faithfully for more than 16 weeks. There was improvement in semen in 42/64 (65.6%) and pregnancy in 16/64 (25.0%). The incidence of pregnancy was further studied in terms of pretreatment semen based upon a semen index called the Motile Oval Index (MOI) (Table 3).

The MOI, an elaboration of the motile sperm count, has been used by Lee et al. (1984) in varicocelectomy studies and has proved useful to evaluate pregnancy rates with the THD. Semen with MOI of 4.8 mil/ml or higher will have one or two parameters in the normal range but at least one parameter will be subfertile. Striking increases in MOI were seen when the pretreatment MOI was above 4.8 mil/ml or having been below 4.8 mil/ml rose above it (Table 4).

Table 3. The Motile Oval Index (MOI)

MOI = Count in millions/ml X percent motile X percent ovals

e.g., using minimum normal values (Table 1),

MOI = 20 mil/ml X 40% X 60% = 4.8 mil/ml

Table 4. Couples achieving pregnancies

|  | Pregnancies | Mean MOI Increase | Range |
|---|---|---|---|
| Starting MOI < 4.8 mil/ml | 5/43 (11.6%) | 9.0 | (0.8 to 22.1) |
| Starting MOI > 4.8 mil/ml | 11/21 (52.4%) | 19.9 | (-1.0 to 57.0) |
| All Patients | 16/64 (25.0%) |  |  |
| Untreated (Met Criteria) | 1/20 (5.0%) |  |  |

Mean Years Infertile Marriage:  4.0
Mean Months to Missed Menses:  4.2
Live Births  15/16

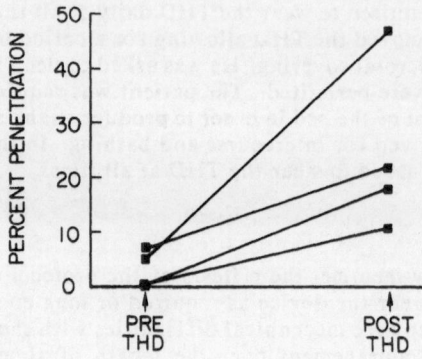

Figure 3. Hamster ovum penetration by the spermatozoa of four men who poduced a pregnancy.

## EFFECT OF THD ON HAMSTER OVUM PENETRATION

Four patients, who produced a pregnancy on the THD, had pretreatment hamster ovum penetration tests which were repeated at the time pregnancy was diagnosed. All four had subfertile starting semen: mean count 7.8 mil/ml. At the time of conception mean count had increased to 39.4 mil/ml and penetrations were also found to have increased in two instances from zero penetration (Figure 3).

## DISCUSSION

The THD removes scrotal heat by evaporation, lowering temperature about 2.0°C. The THD may have to be worn four months to a year before pregnancy occurs. The only complication can be exacerbation of tinea cruris. Removing the THD at night allows the

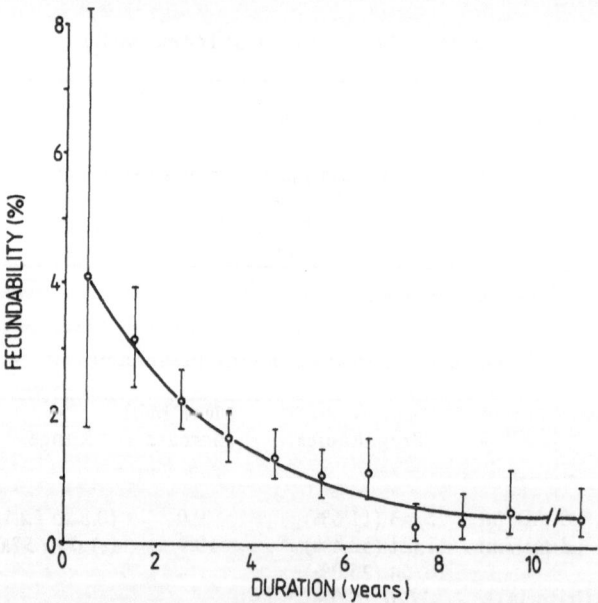

Figure 4. Decrease of fecundability (percentage conception rate/month) with respect to total "trying time" (involuntary infertility). In therapeutic trials for infertility, after three years of involuntary infertility any success must be considered the result of the trial treatment rather than a spontaneous effect. (From Urology with permission).

scrotum and adjacent skin to be dry. This can be treated by topical miconazole. Sleeping without clothing over the genitals and in slight Trendelenburg diminishes testis temperature during the sleep period.

The best result was obtained in those patients whose MOI was above 4.8 mil/ml before inception of treatment. This implies that patients with better but still subfertile semen have the best chance of producing a gestation. Changes in all semen parameters can be observed within 10 weeks of starting treatment. The mean time to pregnancy was 4.2 months; in some instances pregnancy did not occur until a year or more so that patient motivation becomes important.

Basralian et al. (1987) showed a precipitous decline in fecundability (percentage conception rate per month) over three years of involuntary infertility. Fecundability for couples "trying" less than one year is 4.17% and by 3 years has decreased to 1.65% and continues to decrease with the further passage of time (Figure 4). The authors believe that, in clinical trials with subjects who have been "trying" longer than three years, pregnancies which occur are no longer spontaneous but represent a treatment effect. In our 16 patients who produced a pregnancy, the mean involuntary infertility time was 4.0 years (range: 2 - 10 years), with 9/16 "trying" for longer than three years. The fact that pregnancies occurred suggests that hypothermia is effective (Table 5).

Failure of the pretreatment MOI of less than 4.8 mil/ml group to do as well as the pretreatment of greater than 4.8 mil/ml group, suggests that severely subfertile semen is the result of progressive and ultimately irreversible change to the germinal epithelium. The presence of an MOI of less than 4.8 mil/ml need not discourage application of the THD as improvement in semen and pregnancies do occur. Semen levels often improve to levels where In Vitro Fertilization becomes feasible and we have seen in vivo fertilization while awaiting the next IVF appointment.

As in varicocelectomy, the nature of the THD makes controlled studies with randomization, placebo or withholding of treatment virtually impossible. The THD provides a non-invasive method for treatment of subfertile semen associated with elevated testicular temperature. Patients with better, but still subfertile semen, can be expected to do better than those with extremely low counts and very poor motility. Where prior treatment has failed (e.g., varicocelectomy, hormonal therapy and manipulation of gametes, etc.), the THD is an important resource for the "hard core" infertile couple and should not be overlooked. The THD received approval of the United States Food and Drug Administration in December 1984 and must be included as a treatment alternative in informed consents.

The THD is available from Repro-Med Systems, Inc., P.O. Box 191, Middletown, N.Y. 10940.

Table 5. "Trying" time in 16 couples producing a pregnancy

| Years Trying | Calculated Fecundability*<br>% | n |
|---|---|---|
| 2 | 3.14 | 6 |
| 3 | 2.22 | 1 |
| 4 | 1.65 | 5 |
| 5 | 1.37 | 1 |
| 6 | 1.15 | 1 |
| 8 | 0.38 | 1 |
| 10 | 0.45 | 1 |
| | | 16 |

*Basralian et al. (op cit)

231

# REFERENCES

Basralian, K.R. et al. 1987. Duration of involuntary infertility and subsequent pregnancy. Urology, 29: 635.

Lee, H.W. et al. 1984. Effect of varicocelectomy on spermatogenesis. In: "Varicocele and male fertility II", M. Glezerman (ed.), Springer Verlag, Berlin.

Zorgniotti, A.W. et al. 1980. Chronic scrotal hypothermia as a treatment for poor semen quality. Lancet, I: 904.

Zorgniotti, A.W. et al. 1986. Chronic scrotal hypothermia: results in 90 infertile couples. J. Urol., 135: 944.

# HEAT INDUCED INHIBITION OF SPERMATOGENESIS IN MAN

Roger Mieusset, Louis Bujan, Arlette Mansat,
Helene Grandjean and Francis Pontonnier

Centre de Sterilité Masculine
Cecos Midi-Pyrenees and INSERM U168
Hôpital La Grave
Toulouse, France

A reversible depression in spermatozoa output was reported by several authors who have induced an increase in body or scrotal temperature in man (Table 1). In most of the experiments, this rise was above the physiological temperature of the body. But there was a possibility, as strongly suggested by Robinson and Rock (1967), that a sustained yet relatively slight increase in the testis temperature might affect fertility in a way adequate to be used as a method of male fertility control.

From these experiments and with a group of men demanding male contraception, we invented a new method of testicular heating which uses the body as a heat source. The results presented here are general and concern work that was started in 1982; some results were in part reported (Mieusset et al., 1985, 1987a, 1987b). With this method testes were pushed up into the inguinal canal and kept there. In such a situation testes stay in a hotter surrounding, since the inguinal canal temperature is 1 to 1.5°C higher than the intrascrotal one (Kitayama, 1965). When testes are in the inguinal canal, their inferior pole is located at the superior part of the root of the penis. Testes were maintained daily during waking hours in such a situation by means of two techniques.

With the first technique (Fig. 1), support was ensured by briefs provided with an orifice through which the penis and scrotum were pushed outside. With this technique, testes were relatively free to travel due to the elasticity of the fabric, and tended to descend slightly, bringing their lower pole closer to the scrotal cavity. Because such movement was possible, we thought the rise in testicular temperature was liable to fluctuate.

In the second technique, a ring of soft material was either added to encircle the orifice of the briefs, or was worn alone. Testes were therefore exposed to a more constant increase in surrounding temperature.

Twenty-one unpaid volunteers (aged 27 to 35 years) were involved in this experiment: 13 with Technique 1 and 8 with Technique 2. The timing included a 4 to 6 month-baseline study, a 6 to 24 month-heating period and an 18 month-survey after heating. Every man was subjected to semen analyses with sperm count, motility and morphology, as well as clinical examinations.

## RESULTS

The data of this study confirm the well known depression in spermatozoa output induced by an increase in testicular temperature (Fig. 2). But the results reported here

*Temperature and Environmental Effects on the Testis*
Edited by A. W. Zorgniotti, Plenum Press, New York, 1991

Table 1. Induced "hyperthermia" in man

| Authors | Number of Men | Heated Organ | Temperature | Exposure Time |
|---|---|---|---|---|
| MacLeod and Hotchkiss (1941) | 6 | Body | 41°C | 45 min 1 day |
| Watanabe (1959) | 18 | Scrotum | 44-46°C | 30 min/day 1-12 days |
| Procope (1965) | 12 | Body | +1°C in rectal temp. | 15 min/day 8 da/2 wk |
| Robinson and Rock (1967) | 10 | Scrotum | +0.8°C in scrotal temp. | waking hrs 6-10 weeks |
| Robinson et al. (1968) | 18 | Scrotum | 42.5°C | 30 min/day 14-28 days |
| French et al. (1973) | 5 | Scrotum | +2°C | 30 min/day 5 days |
| Brown-Woodman et al. (1983) | 5 | Body | +0.7°C in rectal temp. | 20 min 1 day |

were obtained with a slight and sustained increase in testis temperature. The most important inhibitory effect is given by Technique 2, where it is greater than 97% after 3 months of heating. With this technique, the depression in the spermatozoa output appears sooner, is deeper and more constant than with Technique 1. For both techniques, recovery occurs within 6 to 8 months after the heating is stopped.

Morè important is the fact that heating the testis induced a depression not only in the amount but also in the quality of the spermatozoa output. This alteration in quality was observed in sperm motility as well as in sperm morphology.

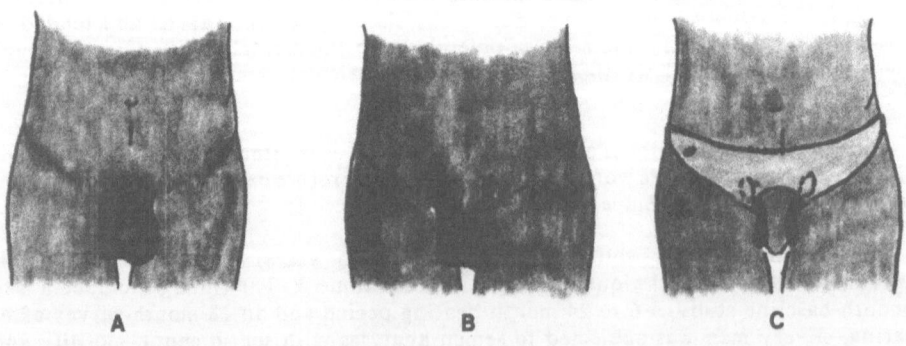

Figure 1. Draft of the first technique ($T_1$) used.
    A. Testes in scrotal position.
    B. Testes in low inguinal location, with an empty scrotum.
    C. Testes maintained in the same position than in B by means of special briefs; the empty scrotum is outside of the briefs.

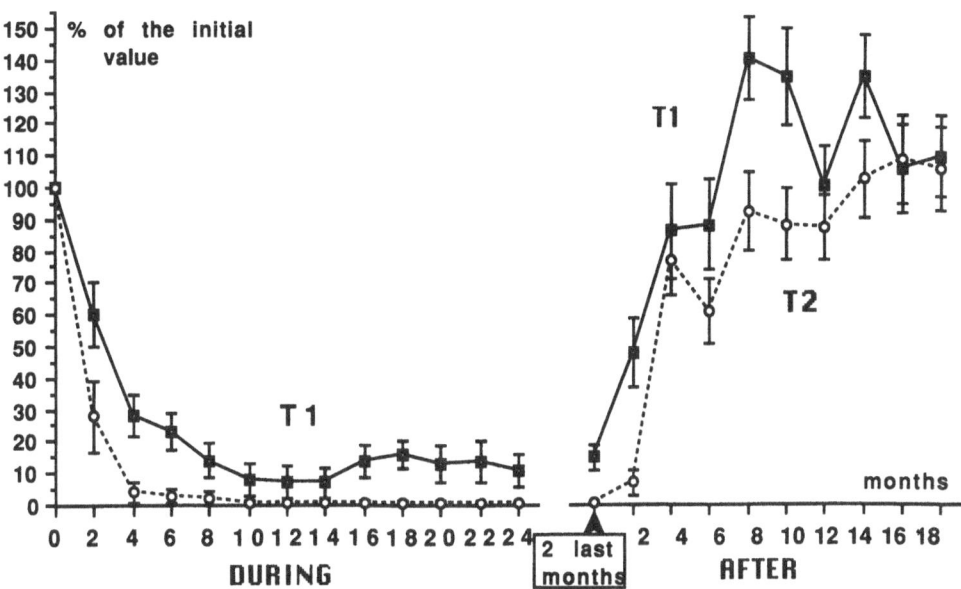

Figure 2. Spermatozoa output (mean + sem) according to Techniques 1 (T$_1$) and 2 (T$_2$) during and after heating the testes.

Sperm motility is reduced to 50% of its initial value within 6 months of heating with Technique 1, and to 80% after only three months with Technique 2. In this latter technique, the heat inhibitory effect on sperm motility, as the one on spermatozoa output, appears sooner, is deeper and more constant than with Technique 1. Recovery of the sperm motility happens within 6 to 8 months after the heating is stopped, regardless of the technique (Fig. 3).

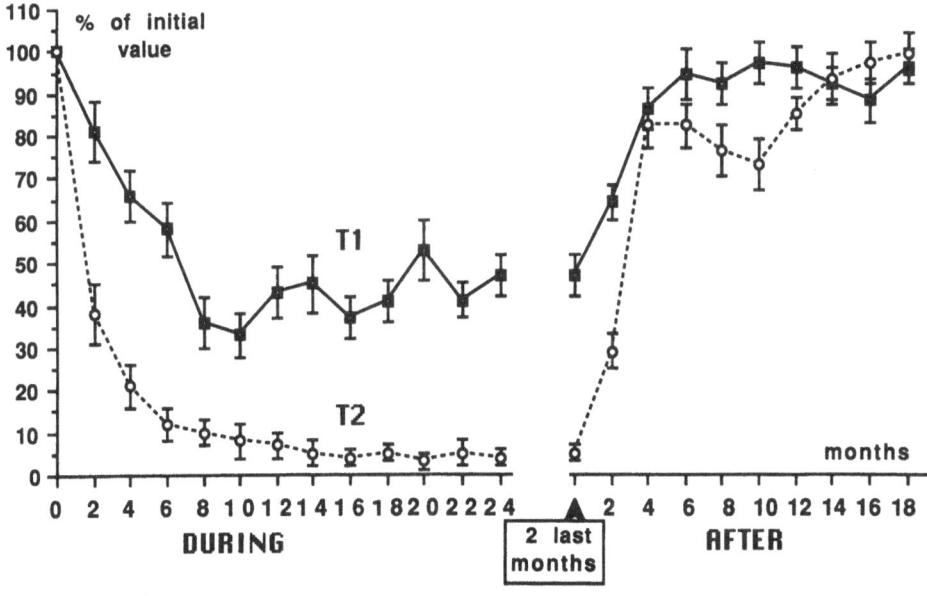

Figure 3. Sperm motility (mean + sem) according to Techniques 1 (T$_1$) and 2 (T$_2$) during and after heating the testes.

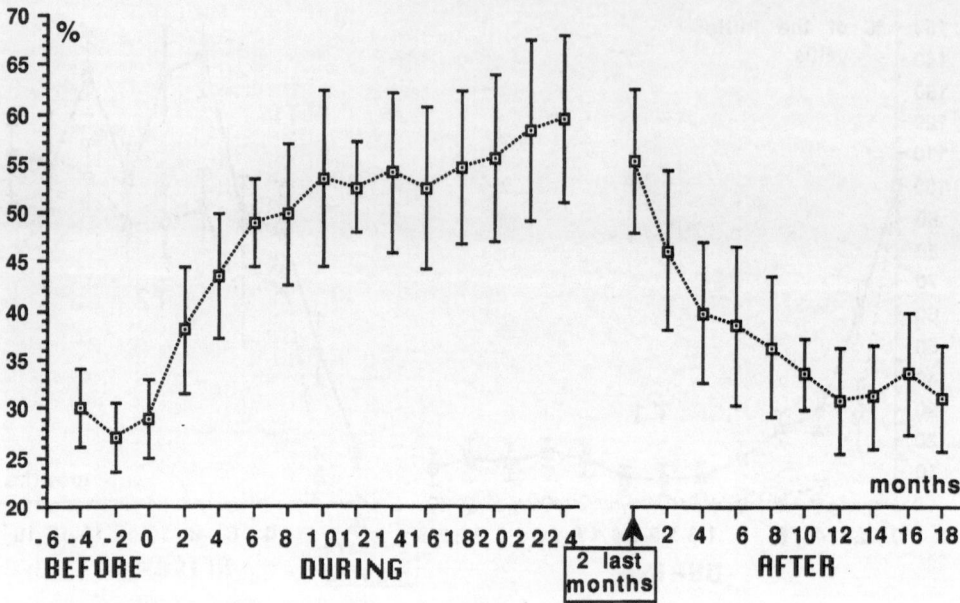

Figure 4. Percentage (mean + sem) of abnormally shaped spermatozoa before, during and after heating the testes (for both techniques).

Modification in sperm morphology during heating is characterized by an increase in the percentage of abnormally shaped spermatozoa, the mean value of which rose from 30 to 60% within 6 to 8 months of heating (Fig. 4). As reported elsewhere, these morphological alterations were particularly observed with light microscopy in the head of the spermatozoa and in the principal piece of the tail (Mieusset et al., 1987b). A study using electron microscopy revealed structural defects such as modifications in acrosomic and nuclear membranes in spermatid as well as in spermatozoa (Mansat et al., 1985; Mieusset, 1989). Recovery of the initial value is slower than for sperm count or motility, requiring 12 months after heating is stopped.

CONCLUSION

Firstly, a daily and maintained rise of 1 to 2°C in the surroundings of the testes during waking hours induces within 3 months an increased percentage of abnormally shaped spermatozoa and a severe depression in sperm count and sperm motility with respective mean values under $3 \times 10^6$/ml and 15%.

Secondly, after the heating was stopped all parameters recover within one year regardless of both the exposure time, which was between 6 and 24 months, and the values in spermatozoa output obtained during the heating period.

What were the undesirable effects of such heating?

During heating no pain and no depression in libido were reported by any of the 21 volunteers. Two years after the end of the heating, 7 of the volunteers wanted to father a child. Pregnancies occurred in partners within one to ten months. There were no miscarriages and no malformations of any kind in the 7 babies now 1 to 5 years old.

Because of all this information and data, we think that such a method of heating the testis can be used as a male contraceptive. An evaluation of its contraceptive efficacy is currently in progress. As part of this evaluation, we have defined the threshold from which this method can be used as a motile sperm count of less than 1 million per ml in two successive monthly sperm analyses. The real contraceptive efficacy of this method will be reported when the assessment still in progress will be achieved.

# REFERENCES

Brown-Woodman, P.D.C., Post, E.J., Gass, G.C. and White, I.G. 1984. The effect of a single sauna exposure on spermatozoa. Arch. Androl., 12: 9.

French, D.J., Leeb, C.S., Fahrion, S.I., Law, O.T. and Jecht, E.W. 1973. Self-induced scrotal hyperthermia in man followed by decrease in sperm count. A preliminary report. Andrologie, 4: 311.

Kitayama, T. 1965. Study on testicular temperature in man. Acta Urol. Japon., 11: 435.

MacLeod, J. and Hotchkiss, R.S. 1941. The effect of hyperpyrexia upon spermatozoa counts in men. Endocrinol., 28: 780.

Mansat, A., Mieusset, R., Bujan, L. and Pontonnier, F. 1986. Alteration morphologique des spermatozoides sous l'influence d'un facteur exogene. Bull. Ass. Anat., 211 bis: 59.

Mieusset, R. 1989. Regulation thermique de la fonction testiculaire. Rech. Gynecol., 3: 163.

Mieusset, R., Bujan, L., Mansat, A., Pontonnier, F. and Grandjean, H. 1987a. Hyperthermia and human spermatogenesis: enhancement of the inhibitory effect obtained by "artificial cryptorchidism". Int. J. Androl., 10: 471.

Mieusset, R., Bujan, L., Mansat, A., Pontonnier, F. and Grandjean, H. 1987b. Effects of artificial cryptorchidism on sperm morphology. Fertil. Steril., 47: 150.

Mieusset, R., Grandjean, H., Mansat, A. and Pontonnier, F. 1985. Inhibiting effect of artificial cryptorchidism on spermatogenesis. Fertil. Steril., 43: 589.

Procope, B.J. 1965. Effect of repeated increase of body temperature on human sperm cells. Intern. J. Fertil., 10: 333.

Robinson, D. and Rock, J. 1967. Intrascrotal hyperthermia induced by scrotal insulation: effect on spermatogenesis. Obst. Gynec., 29: 217.

Robinson, D., Rock, J. and Menkin, M.F. 1968. Control of human spermatogenesis by induced changes of intrascrotal temperature. J. Am. Med. Ass., 204: 80.

Rock, J. and Robinson, D. 1965. Effect of induced intrascrotal hyperthermia on testicular function in man. Am. J. Obst. Gynec., 93: 793.

Watanabe, A. 1959. The effect of heat on human spermatogenesis. Kyushu J. Med. Sci., 10: 101.

REFERENCES

SECTION 7

VARICOCELE: DIAGNOSIS, PATHOGENESIS AND TREATMENT

# SCROTAL HYPERTHERMIA AND VARICOCELE

R. Hsiung, H. Nieva and A. Clavert

CECOS of Alsace
1 Place de l'Hôpital
F 67091 Strasbourg Cedex, France

In clinical practice, thermography is generally performed when a varicocele is suspected (Lewis and Harrison, 1979; 1980). The concept of scrotal hyperthermia was associated with varicocele (Zorgniotti and MacLeod, 1973; Comhaire et al., 1976; Monteyne et al., 1978). The aim of this study was to answer two questions:

·Are all varicoceles hyperthermic?
·Is scrotal hyperthermia always associated with a varicocele?

We have collected 347 cases of men with abnormal spermiogram, oligospermia and/or asthenospermia and/or teratospermia. In each case we performed a complete clinical examination with a Valsalva test and a venous Doppler, as to diagnose the typical blood flow of the varicocele; then we performed a contact thermography. We have adopted the technique of Lewis and Harrison (1979) and modified it as described here : the man is lying on the examining table with the lower part of his body unclothed. After a 7- to 10-minute rest period the thermography is performed. This lapse of time is necessary for the equilibration of the body temperature with the room temperature of 20 to 22°C (Zorgniotti and MacLeod, 1973). The scrotum and the penis are isolated from the abdomen and the legs with an isolating plate as shown on Fig. 1. In this condition it is very easy to place the thermographic plate on the scrotum and penis so the contact with the cholesterolic liquid crystal plate is good and very selective.

Two photographs are always taken: in the first, the man is lying down and in the second, the man is standing up. During the time of recumbency, for 10 minutes at least, the venous blood of the varicocele flows back into the abdomen and is rewarmed at 37°C. When the man stands up, some warm blood flows back from the central veins into the varicose vein. Therefore, the varicocele is warmer in the standing position after this manoeuvre (Figs. 2 and 3). This method is not performed for an exact scrotal temperature evaluation, but it is a good technique for the anatomical analysis of the temperature distribution.

Generally the thermographic picture gives two indications:

(1) The possible presence of vascular hyperthermia (varicocele)
(2) The temperature of the total scrotal area.

The temperature of the scrotum is normally lower than the reference (our reference is the temperature of the proximal part of the penis). When the scrotal temperature is higher or equal to the reference, there is a total hyperthermia.

*Temperature and Environmental Effects on the Testis*
Edited by A. W. Zorgniotti, Plenum Press, New York, 1991

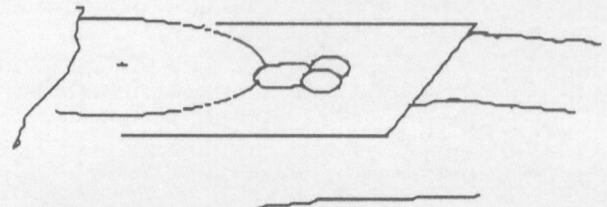

Figure 1. The scrotum placed on the isolating plate.

First question: Is there a relation between total hyperthermia and varicocele? Table 1 shows the result of 347 observations. We observed that:

(1) All hyperthermic cases were not associated with a varicocele. This has been recently suggested by Mieusset (1987): hyperthermia can be the consequence of other factors.

(2) The distribution of the total hyperthermia is not a result of chance. In the varicocele population the frequency of hyperthermia is very high (53 %) but in the population without varicocele the percentage is only 37.5 %. This difference is significant ($p < 0.01$): varicocele is an important factor of hyperthermia.

Table 1. Distribution of total hyperthermia in varicocele and non-varicocele group.

|  | Varicocele | |
|  | Present | Absent |
| --- | --- | --- |
| No hyperthermia | 107 (47%) | 75 (62.5%) |
| Total hyperthermia | 120 (53%) | 45 (37.5%) |
| Total | 227 | 120 |

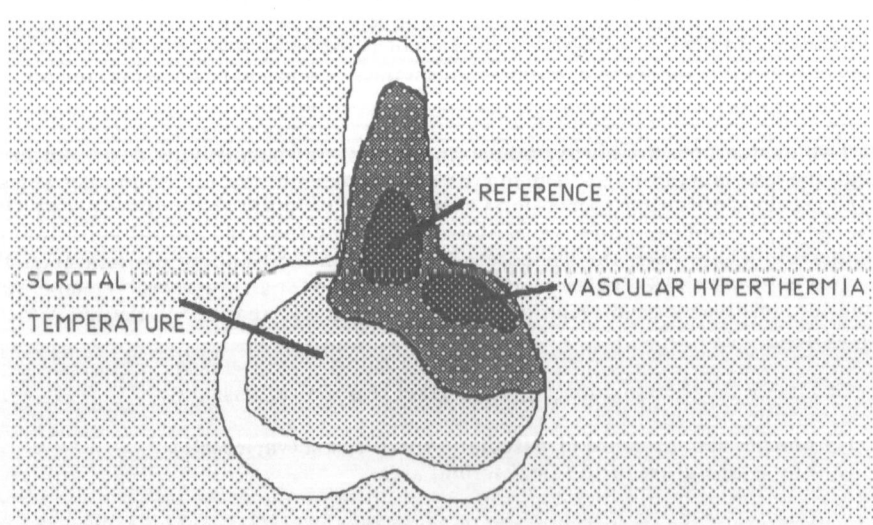

Figure 2. Thermography in recumbent position.

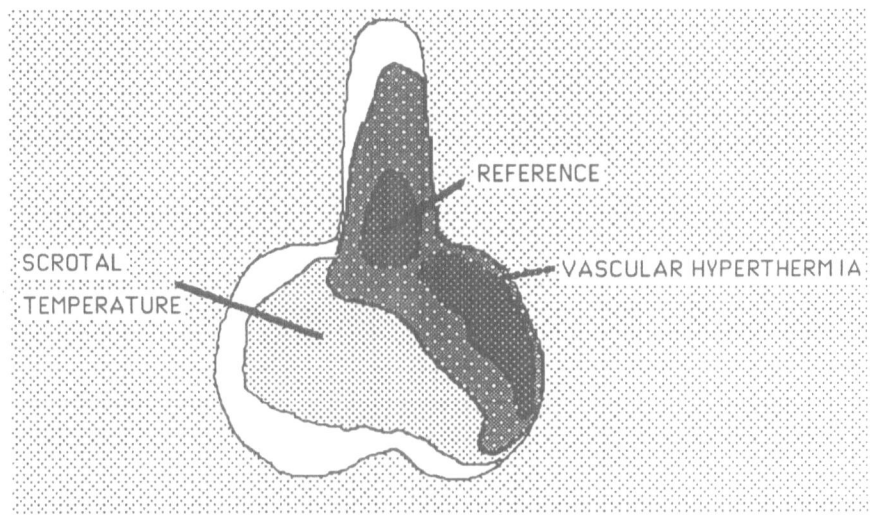

Figure 3. Thermography in standing position.

Second question: Is there a relation between the extension of varicocele and total hyperthermia? Our technique gives a good indication of the extension of the varicocele

Table 2. Distribution of total hyperthermia according to the length of the varicocele

|  | Varicocele Extension | | |
|  | Very Extended | Medium | Apical |
| --- | --- | --- | --- |
| No hyperthermia | 4 (22%) | 18 (78%) | 52 (88%) |
| Total hyperthermia | 14 (78%) | 5 (22%) | 7 (12%) |
| Total | 18 | 23 | 59 |

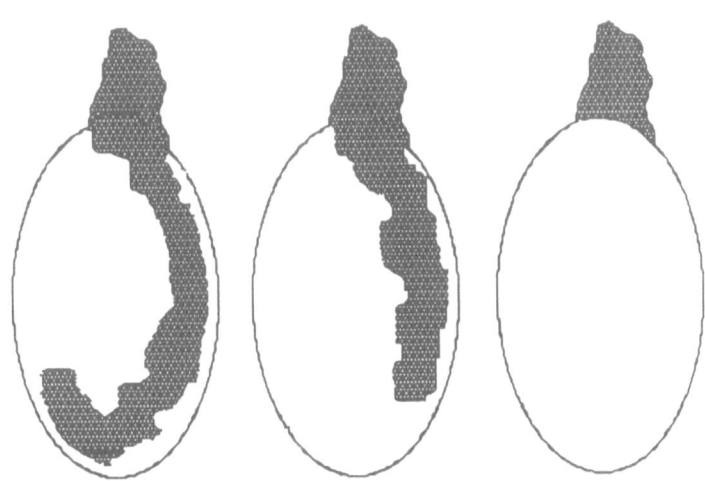

Figure 4. Varicocele extension observed by contact thermography.

on the testicular surface (Fig. 4). Table 2 shows the result of 100 varicoceles studied. The longer the varicocele, the higher is the frequency of total hyperthermia. For the very extended varicocele we found 78 % of total hyperthermia and in the apical varicocele the frequency was only 12 % (p < 0.001). There is a relation between the length of varicocele and the temperature of the scrotum.

In some cases after surgery of a varicocele, we observed that the vascular hyperthermia disappeared but sometimes total hyperthermia was still persisting.The relationship between these two factors is not clear but it is obvious that there is no simple relation between the vascular abnormality and the thermic perturbation.

The following are some examples of hyperthermia without varicocele:

·In the case of a testicular unilateral atrophy, the normal testis was cold and the atrophic testis hyperthermic. The volume of the testis can be the factor of hyperthermia. The contact between the testis and the scrotum is necessary for a good conduction of heat.
·In the case of treatment with Orciprenaline, a vasodilatator's drug, we found a total scrotal hyperthermia. This hyperthermia could be the result of vascular modification.
·In the case of an infection, epididymitis localized at the cauda epididymis, we found a local scrotal hyperthermia. The infection is the source of calories and vasodilatation. This can be responsible for total hyperthermia.


CONCLUSION

Total hyperthermia is more frequent in the group of varicocele patients. The size and extension of varicocele has an influence on scrotal temperature but the relation between varicocele and scrotal hyperthermia is not absolute. Hyperthermia can be the consequence of simultaneous actions of many different factors: varicocele, infection, hormones, neurological disease, etc.

In our point of view, thermography and temperature evaluation must be routine examinations for all sterility exploration.

REFERENCES

Comhaire, F., Monteyne, R. and Kunnen, M. 1976. The value of scrotal thermography as compared with selective retrograde venography of the internal spermatic vein for the diagnosis of "subclinical" varicocele. Fertil. Steril., 27: 694.

Lewis, R.W. and Harrison, R.M. 1979. Contact scrotal thermography: application to problems of infertility. J. Urol., 122: 40-42.

Lewis, R.W. and Harrison, R.M. 1980. Contact scrotal thermography. II. Use in the infertile male. Fertil. Steril., 34: 259-263.

Mieusset, R., Bujean, L., Mondinat, C., Mansat, A., Pontonnier, F., Grandjean, H. 1987. Association of scrotal hyperthermia with impaired spermatogenesis in infertile men. Fertil. Steril., 48: 1006-1011.

Monteyne, R. and Comhaire, F. 1978. The thermographic characteristics of varicocele: An analysis of 65 positive registrations. Br. J. Urol., 50: 118.

Zorgniotti, A.W. and MacLeod, J. 1973. Studies in temperature, human semen quality and varicocele. Fertil. Steril., 24: 854-863.

# THE SIGNIFICANCE OF ELEVATED SCROTAL TEMPERATURE

# IN AN ADOLESCENT WITH A VARICOCELE

Joseph A. Salisz, Evan J. Kass and
Bruce W. Steinert

William Beaumont Hospital
3601 West Thirteen Mile Road
Royal Oak, Michigan 48073

## ABSTRACT

Simultaneous measurements of left and right scrotal, and axillary skin temperatures were recorded in 58 consecutive adolescents (mean age 14.4 years) with a grade II-III left sided varicocele, and nine control adolescents without genital pathology (mean age 15.7 years). Left and right testicular volumes were determined in both groups. The adolescents with a varicocele had a significant bilateral elevation of the scrotal temperatures compared to the control subjects. This relative hyperthermia was present in both supine and standing positions. The mean left scrotal temperature of varicocele patients was significantly higher in the standing position than in the supine position, which may reflect the dependent venous filling of the varicocele. Those varicocele patients who maintained a left scrotal temperature at least 1.4°C cooler than axillary did not have significant left testicular volume loss, whereas those whose left temperature was approximately equal to axillary did have significant growth retardation of the left testis. Following successful varicocele surgery, left scrotal temperatures were significantly cooler, and statistically indistinguishable from controls. The left testicular volumes were also significantly improved with respect to corresponding right testicular volumes. These observations suggest that adolescents with a moderate to large left varicocele have a significant bilateral loss of testicular thermoregulation. In those individuals with a significantly warmer left hemiscrotum, there is a definite increased potential for left testicular volume loss. Varicocele surgery can reverse this process.

## INTRODUCTION

It has been known for some time that a varicocele first becomes clinically evident during early adolescence[1,2]. Some investigators have suggested that a varicocele may have its greatest effect upon the rapidly growing testis of the adolescent, because it has been demonstrated that a varicocele can be responsible for significant testicular growth retardation and that this effect upon testicular growth can be reversed by varicocele ligation[3,4,5]. Additionally, it has been reported that a left sided varicocele in a teenager can result in impaired testicular function[6] and abnormal testicular histology[7]. The exact mechanism by which a varicocele induces a testicular injury remains an enigma, although many theories have been proposed. An attractive theory supported by several investigators is that it produces a disturbance of the thermoregulatory system[8], which results in a loss of the normal abdominal-scrotal temperature difference, and this increase in scrotal temperature may be responsible for impaired spermatogenesis[9,10]. Zorgniotti and MacLeod[11] observed elevated scrotal temperatures in infertile men with a varicocele. Agger[12] suggested that there was a good correlation between the fall in the temperature of the left testis following varicocele correction and the observed improvement in spermatogenesis. Since scrotal skin temperatures have been shown to correlate well with intratesticular temperatures[13,14], we measured scrotal skin temperatures in 58 adolescents with a varicocele in order to determine if there is a significant temperature effect of the varicocele in this age group.

## MATERIALS AND METHODS

A total of 58 boys between the ages of 8 and 19 (mean = 14.4 years old) with a left sided varicocele were evaluated between July 1988 and October 1989. The majority of patients were referred by their primary physicians because a scrotal mass was identified during a routine school or athletic physical examination. Most of the patients with a varicocele denied prior knowledge of a scrotal abnormality and were unaware of the presence of the varicocele prior to their physical examination. None of these individuals presented for evaluation of infertility or scrotal pain. All patients had a palpable grade II-III varicocele; a Doppler ultrasound probe was not used to assist in the diagnosis of a varicocele in this study. Nine boys aged 12 to 17 (mean = 15.7 years old) who responded to a published request for volunteers, and who had a normal genital examination were included as control subjects.

Patients with bilateral varicoceles, or other intrascrotal pathology, such as hydroceles or hernias, were excluded from this study. No patient had a unilateral right sided varicocele.

All patients were examined in both the standing and supine positions, and the size of the varicocele was graded according to Steeno[1]. Testicular volumes were then determined with a standard orchidometer[15]. Next, using a small piece of silk tape, a skin surface temperature probe was attached to the subjects' right and left hemiscrota, in positions over the anterolateral aspects of the testes, and a third probe was attached to the patients' left axillary skin. Using a three-channel thermocouple sensor (Sensortek), simultaneous temperature measurements were recorded in the supine and standing positions. Since the probes measure instantaneous temperatures with a time constant of 0.15 sec, they were allowed to stabilize before recording, and each temperature recording was checked twice before accepting a value. Control patients had testicular volume measurements and temperatures determined in exactly the same manner. Room temperature remained constant.

Following successful varicocele ligation, 31 patients were reevaluated by testicular volumetrics, and 17 by temperature determinations; 12 had both post-operative temperature and volume measurements.

Table 1.  Left Scrotal Temperature

|  | Supine | Erect | Difference (Erect-Supine) |
|---|---|---|---|
| Varicocele | 34.3±1.2 | 35.3±1.0 | 1.0* |
| Control | 32.9±0.9 | 33.1±1.1 | 0.2 ¥ |
| Difference (Varicocele - Control) | 1.4 † | 2.2 † |  |

Temperatures in degrees centigrade
\* p<0.001 by Student's t test
† p<0.001 by Mann-Whitney Nonparametric test
¥ not significant

Table 2. Right Scrotal Temperature

| | Mean Supine ±SD | Mean Erect ±SD | Difference (Erect-Supine) |
|---|---|---|---|
| Varicocele | 33.3±0.9 | 33.3±1.0 | 0 [†] |
| Control | 32.6±0.5 | 32.6±0.6 | 0 [†] |
| Difference (Varicocele-Control) | 0.7* | 0.7* | |

Temperatures in degrees centigrade
* p<0.05 by Mann-Whitney Nonparametric test
[†] no significant difference on right with respect to position

## RESULTS

### Scrotal Temperatures

In adolescents with a varicocele, there is a significant bilateral increase in scrotal temperature. The left scrotal temperature was significantly higher (p < 0.008) in varicocele patients than the left scrotal temperature of control subjects in both the supine and standing positions (Table 1). The mean temperature of the right hemiscrotum in varicocele patients was also significantly warmer (p < 0.05) than in control subjects regardless of position (Table 2). The mean left scrotal temperature in those with a left varicocele was significantly higher (p < 0.0001) in the standing than in the supine position. The average elevation was 1.00°C. There was no significant positional change in control subjects (Table 1).

In varicocele patients, the left scrotum was significantly warmer than the right scrotum in both positions (p < 0.0001). In contrast, the left scrotal temperature did not statistically differ from the right in those without a varicocele.

In patients with a varicocele, the mean left scrotal temperature was significantly lower than the mean axillary temperature only in the supine position (p < 0.0002). In the

Table 3. Axillary - Left Scrotal
Temperature Differences

| Varicocele | Difference | p [†] |
|---|---|---|
| (Ax-L) Supine | 0.61 | <0.001 |
| (Ax-L) Erect | -0.22* | >0.1 |
| Control | | |
| (Ax-L) Supine | 1.83 | <0.001 |
| (Ax-L) Erect | 1.91 | <0.01 |

Temperatures measured in degrees centigrade
* Negative value indicates mean left scrotal temperature is warmer than mean axillary temperature in erect position.
[†] Significance calculated by Student's t test.

standing position, the mean left scrotal temperature increased significantly and was essentially equal to the mean axillary temperature. Again, the controls did not display this positional change in scrotal temperature, and on the contrary, they maintained a significant axillary-scrotal temperature difference in both supine and erect positions (Table 3).

## Testicular Volumes

In varicocele patients, the magnitude of elevation of the mean left scrotal temperature correlated with measurable left testicular volume loss. When left scrotal temperature in the standing position was at least 1.4°C cooler than corresponding axillary temperature, there was no significant left testicular volume loss when compared to control subjects (p > 0.07). However, varicocele patients with left scrotal temperatures equal to or warmer than simultaneous axillary temperatures had significantly greater (p < 0.0001) left testicular volume loss than controls, as well as those in whom left scrotal temperatures were cooler than axillary.

## Operative Results

Following varicocele correction, there was a significant decrease in the mean volume difference between the left and right testis, from 3.2 cc to 1.8 cc (p < 0.0009). Left scrotal temperatures also significantly decreased postoperatively (p < 0.001), and were not statistically distinguishable from controls (Table 4).

## DISCUSSION

In a wide variety of plants and animals there is a narrow temperature range that is suitable for gametogenesis. Temperatures above this optimal range tend to prevent gamete formation, and this temperature effect seems to be peculiar to the male of the species; female gametogenesis is not altered by similar degrees of temperature elevation. In mammals with scrotal testes the effect of heat on testicular function is similarly injurious whether it is whole body heat, heat applied directly to the scrotum or heat derived by placing the testis within the body cavity, as occurs with cryptorchidism[16].

A normal countercurrent heat exchange system was postulated by Dahl and Herrick[8] in which inflowing core temperature spermatic arterial blood is cooled by the outflowing cooler venous blood of the pampiniform plexus. Varicosities of this plexus may decrease the effectiveness of testicular cooling. Numerous investigators have demonstrated that there is a significant bilateral increase in both testicular blood flow and temperature following the creation of a unilateral varicocele in an experimental animal model[17-21]. These changes are reversible following varicocele ligation. The authors postulate that the

Table 4. Operative Results

| | Mean Temperature Difference (Ax-L Scrotal) Erect | Mean Testicular Volume Difference (R-L) |
|---|---|---|
| Pre-Op Varicocele | -0.2±1.1* | 3.2±3.0 |
| Post-Op Varicocele | 1.8±1.1 | 1.8±1.7 |
| Controls | 1.9±1.4 | -0.4±0.8† |

Temperatures in degrees centigrade
Volumes measured in cubic centimeters
* Negative value indicates mean left scrotal temperature was warmer than mean axillary temperature in erect position.
† Negative value indicates left testis was larger than right in some control patients.

increase in testicular blood flow interferes with the countercurrent heat exchange mechanism, resulting in an increased temperature and a secondary impairment of spermatogenesis.

The varicocele-induced increase in scrotal and testicular temperature has been implicated as a possible cause for the observed testicular changes in adult men with a varicocele and infertility. Histochemical studies of testis biopsy material from oligospermic men with a varicocele have demonstrated a decrease in Sertoli cell glycogen stores and an increase in phosphorylase activity[22]. These data suggest that the observed increase in testicular temperature induces an increase in phosphorylase activity, resulting in a subsequent reduction in glycogen stores due to the increased metabolic activity. The increase in metabolic activity may be the initial mechanism by which a varicocele induces testicular injury.

Previous studies have demonstrated that a unilateral left sided varicocele is associated with a bilateral elevation of the scrotal skin temperature in adult men with varicocele[11,14,23]. Our data in adolescents with a grade II-III varicocele demonstrate that similar alterations of scrotal temperature take place in adolescents with a unilateral left sided varicocele. This bilateral elevation in scrotal temperature in adolescents lends further support to the theory that elevated testicular temperatures play an important role in the pathophysiology of varicocele. Since the scrotal temperature increase in varicocele patients is significantly greater in the standing position and this positional temperature increase was not observed in controls, it is likely that retrograde filling of the dilated varicosities accounted for this loss of thermoregulation.

It is interesting to note that the magnitude of the scrotal temperature increase is directly related to the potential for testicular volume loss. Volume loss of the left testis was much more common in those individuals with a scrotal temperature equal to or greater than axillary, and was not observed in those adolescents who maintained a significantly cooler scrotum, despite the presence of a varicocele. Perhaps those individuals with a varicocele who maintain a significant abdomino-scrotal temperature difference constitute an unaffected varicocele group. The importance of the scrotal temperature increase is further supported by the observation that following varicocele correction a normal axillary-scrotal temperature difference is noted and, in individuals with volume loss of the left testis, this normalization of temperature is associated with a relative increase in the left testicular volume.

## CONCLUSION

Unilateral large left varicoceles significantly warm the entire scrotum. This is an important observation because unilateral varicoceles have been implicated in infertility, the pathophysiology of which would logically involve a bilateral problem.

In varicocele patients, we have observed highest left scrotal temperature recordings being nearly equal to axillary when the patient is standing and the varicocele is maximally filled. The mass of veins visually deflates and the scrotal temperature decreases when the subject reclines. The warming effect of the dependent position is presumably due to retrograde filling of the dilated veins, or to reduced outflow. The rapidity in which the change happens, however, would support reflux of warmer systemic blood into the spermatic venous system. Subjects without a varicocele did not have a positional variation of the scrotal temperature.

We have identified that an adolescent with a left varicocele has a tendency for left testicular growth retardation, and that higher left scrotal temperatures correlate with greater left testis hypotrophy. After varicocele ligation, scrotal temperatures and testicular volumes significantly improved. We have also observed that significant testicular volume loss did not occur in adolescent varicocele patients with cooler scrotal temperatures. Although data regarding fertility in our adolescent population will not be available for several years, this observed loss of thermoregulation induced by a varicocele may be related to infertility.

# REFERENCES

1. Steeno, O., Knops, J., Declerck, L. et al. 1976. Prevention of fertility disorders by detection and treatment of varicocele at school and college age. Androl., 8: 47-53.

2. Oster, J. 1971. Varicocele in children and adolescents. Scand. J. Urol. Nephrol., 5:27-32.

3. Mieusset, R., Bujan, L., Mondinat, C., Mansat, A., Pontonnier, F. and Grandjean, H. 1987. Association of scrotal hyperthermia with impaired spermatogenesis in infertile men. Fertil. Steril., 48(6): 1006-1011.

4. Lipshultz, L.I. and Corriene, J.N. 1977. Progressive testicular atrophy in the varicocele patient. J. Urol., 117: 175-176.

5. Kass, E.J. and Belman, A.B. 1987. Reversal of testicular growth failure by varicocele ligation. J. Urol., 137: 475-476.

6. Kass, E.J., Freitas, J.E. and Bour, J.B. 1989. Adolescent varicocele: objective indications for treatment. J. Urol., 579-582.

7. Kass, E.J., Chandra, R.S. and Belman, A.B. 1987. Testicular histology in the adolescent with a varicocele. Peds, 79: 996-998.

8. Dahl, E.V. and Herrick, J.F. 1959. A vascular mechanism for maintaining testicular temperature by counter-current exchange. Sur., Gyn. & Obs., 108(6): 697-705.

9. Procope, B.J. 1965. Effect of repeated increase in body temperature on human sperm cells. Int J. Fertil., 10: 333.

10. Robinson, P., Rock, J. and Menkin, M.F. 1968. Control of human spermatogenesis by induced changes of intrascrotal temperature. JAMA, 204: 290.

11. Zorgniotti, A.W. and MacLeod, J. 1973. Studies in temperature: human semen quality and varicocele. Fertil. Steril., 24: 854-863.

12. Agger, P. 1971. Scrotal and testicular temperature: its relation to sperm count before and after operation for varicocele. Fertil. Steril., 22: 286-297.

13. Kurz, K.R. and Goldstein, M. 1986. Scrotal temperature reflects intratesticular temperature and is lowered by shaving. J. Urol., 135: 290.

14. Goldstein, M. and Eid, J.F. 1989. Elevation of intratesticular and scrotal skin surface temperature in men with varicocele. J. Urol., 142: 743-745.

15. Takihara, H., Cosentino, M.J. and Cockett, A.T.K. 1983. Significance of a new orchidometer in andrology clinic: testicular atrophy in the varicocele patient and recovery after varicocelectomy. Presented at the annual meeting of the American Urological Association, Las Vegas.

16. Cowles, R.B. 1965. Hyperthermia, aspermia, mutation rates and evaluation. Quart. Rev. Biol., 40: 341.

17. Kay, R., Alexander, N.J. and Baugham, W.L. 1979. Induced varicoceles in rhesus monkeys. Fertil. Steril., 31: 195-199.

18. Zippe, C.D., Tomashefsky, P. and Nagler, H.M. 1987. Altered testicular microvascular bed function in the experimental varicocele. Surg. Forum, 38: 645-646.

19. Saypol, D.C., Howards, S.S. and Turner, T.T. 1981. Influence of surgically induced varicocele on testicular blood flow, temperature and histology in adult rats and dogs. J. Clin. Invest., 68: 39-45.

20. Hurt, G.S., Howards, S.S. and Turner, T.T. 1986. Repair of experimental varicosities in the rat: long-term effects on testicular blood flow and temperature on cauda epididymal sperm concentration and mobility. J. Androl., 7: 271.

21. Fussel, E.N., Lewis, R.W. and Roberts, J.A. 1981. Early ultrastructural findings in experimentally produced varicocele in the monkey testis. J. Androl., 2: 111-119.

22. Re M., Iannitelli, M. and Cerasaro, A. 1983. Histochemical study of glycogen and phosphorylase activity on bilateral biopsies of oligospermic men with varicocele. Arch. Androl., 10(1): 79-83.

23. Green, K.F., Turner, T.T. and Howards, S.S. 1984. Varicocele: reversal of testicular blood flow and temperature effects by varicocele repair. J. Urol., 131: 1208.

# DEEP BODY TEMPERATURE MEASUREMENT FOR THE

# NONINVASIVE DIAGNOSIS OF VARICOCELE

Hiroshi Takihara, Masatoshi Yamaguchi, Kazuhiko Ishizu,
Takuya Ueno and Jisaburo Sakatoku

Department of Urology
Yamaguchi School of Medicine
Ube, Yamaguchi, Japan

## INTRODUCTION

Since varicocele treatment is one of the few effective treatments of male infertility, it is important to make the diagnosis correctly so as to administer surgical treatment where it is appropriate[1,2]. We reported that deep body temperature is a practical index of the intrascrotal temperature and is useful in the clinical examination of patients with varicocele[3]. In light of those observations, the purpose of this study is to investigate the relationship between grading of varicocele and the intrascrotal deep body temperature before and after varicocelectomy.

## MATERIALS AND METHODS

Forty-four patients were evaluated by deep body temperature measurement before and after varicocelectomy in our Andrology Clinic. These patients were classified into three groups on the basis of physical examination. The first group consisted of 15 patients with grade I varicocele; the second group consisted of 12 patients with grade II varicocele and the third group consisted of 17 patients with grade III varicocele. All patients were examined and operated by one author (H.T.).

The deep body temperature measurement was performed using the Terumo deep body temperature system according to the procedure previously reported[3,4]. This system consisted of a deep body temperature monitor, Coretemp CTM204, Terumo recorder TFR-102 (Terumo Corporation, Tokyo). All patients were asked to expose their genitalia for 10 minutes while supine. An aircap mask was placed between scrotum and thighs to eliminate the influence of heat from the thighs. (An aircap mask is a small device containing air to protect electric appliances, etc. from shock.) The bilateral intrascrotal temperatures were recorded in the supine position for 15 minutes and then monitored continuously in the standing position, with the environmental temperature at 24°C to 25°C.

We measured the maximum temperatures in both the supine and standing positions in all patients. Results were expressed as mean ± standard errors and were analyzed for statistical significance using t test for differences among groups and using paired t test with each patient serving as his own control.

*Temperature and Environmental Effects on the Testis*
Edited by A. W. Zorgniotti, Plenum Press, New York, 1991

## RESULTS

### Deep Body Temperature of the Left Side

When mean maximum temperature of the left side before varicocelectomy were compared among groups, the mean deep body temperatures were not significantly different in the supine position. In the standing position, differences among groups were significant in proportion to the grade of varicocele. The grade III varicocele's mean deep body temperature was 1.2°C higher than the grade I varicocele's (Fig. 1).

Then we studied the difference in temperature ($\triangle T$) due to the postural change from supine to standing, in order to exclude the influence of basal temperature changes in each patient. Before varicocelectomy, the mean $\triangle T$ showed more pronounced correlation with grade of varicocele than the basal deep body temperature at standing position (Figs. 1 and 2). The mean $\triangle T$ of grade II varicocele was 1.1°C higher than grade I varicocele's, while the mean $\triangle T$ of grade III varicocele was 0.8°C higher than grade II varicocele's. In contrast, after varicocelectomy, the mean $\triangle T$ of all grades showed negative values; thus, there were no significant differences among groups (Fig. 2). Our preliminary results show a temperature decline in all normospermatic men without varicocele (ten fertile volunteers), which may be due perhaps to the scrotum being distant from the trunk.

### Deep Body Temperature of the Right Side

When mean maximum temperatures of the right side before varicocelectomy were compared among groups, no significant differences were observed in contrast to that of the left side, in both supine and standing positions (Fig. 3). When the difference in

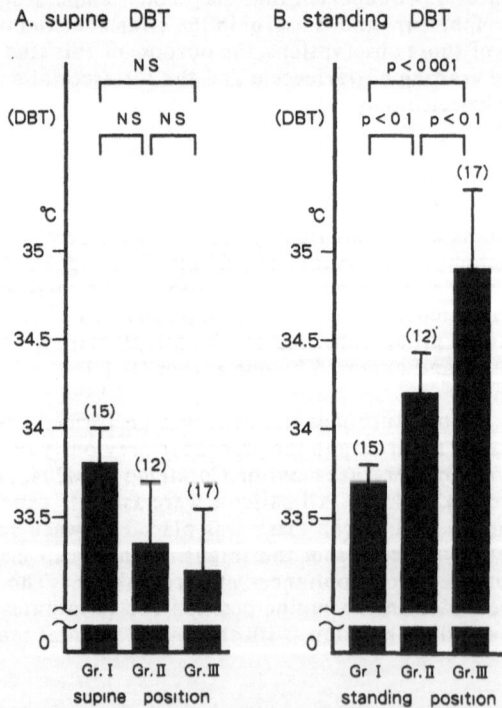

Figure 1. Left deep body temperature (DBT) of preoperation group. (A) Supine DBT; the mean DBT among each group is not significant; (B) Standing DBT; the mean DBT of the grade III group is higher than that of the other two groups.

temperature due to postural change is measured at the right side, the mean $\triangle$T of grade III varicocele alone showed a positive value of +0.30°C, whereas the mean $\triangle$T of grades I and II varicocele were around -0.4°C before varicocelectomy. After varicocelectomy, the mean $\triangle$T of all grades of varicocele showed negative values; thus, there were no significant differences among groups (Fig. 4).

## The Comparison of $\triangle$T before and after Varicocelectomy

When $\triangle$T changes of both the left and right sides in grade I varicocele were compared before and after varicocelectomy, the mean $\triangle$T of the left side showed a significant decline of 0.86°C after varicocelectomy, and postoperative $\triangle$T revealed negative values in all cases except one. The mean $\triangle$T of the right side did not show a significant decline after varicocelectomy because the tendency of each case was not constant (Fig. 5). In grade II varicocele, the mean $\triangle$T of the left side showed a significant decline of 2.0°C after varicocelectomy, and postoperative $\triangle$T showed negative values in all cases. In contrast, $\triangle$T of the right side revealed no constant tendency (Fig. 6). In grade III varicocele, the mean $\triangle$T of the left side showed a dramatical drop of 2.7°C after varicocelectomy, and $\triangle$T of all cases in grade III varicocele were positive before varicocelectomy, while $\triangle$T of all cases after varicocelectomy were negative. In addition, the mean $\triangle$T of the right side also revealed a significant decline of 0.79°C after varicocelectomy, which is a marked difference from grades I and II varicocele (Fig. 7). Although improvement of the semen data was observed in the majority of cases after varicocelectomy, some patients showed no improvement of the semen data in spite of the clear decline of the deep body

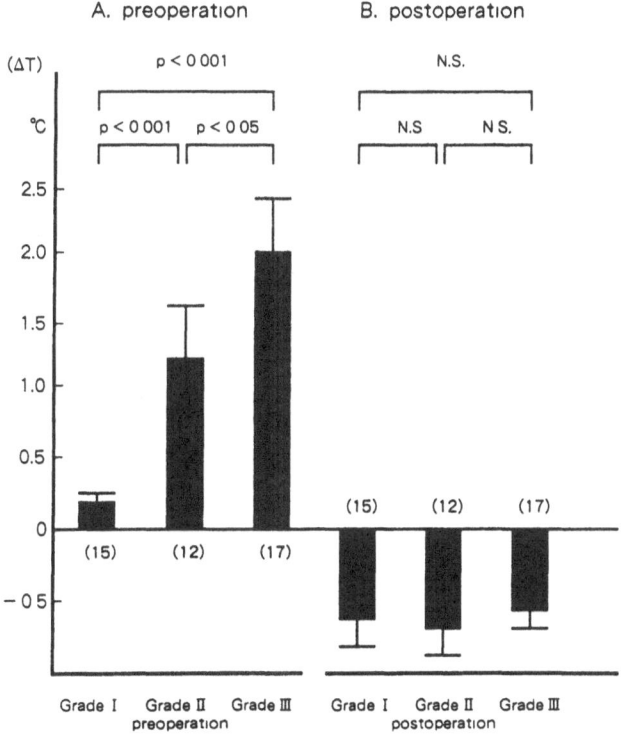

Figure 2. Difference in temperature ($\triangle$T) of the left side. (A) $\triangle$T of preoperation group; its difference is more remarkable than the left standing DBT. (B) $\triangle$T of postoperation group; the mean $\triangle$T among each group is not significant.

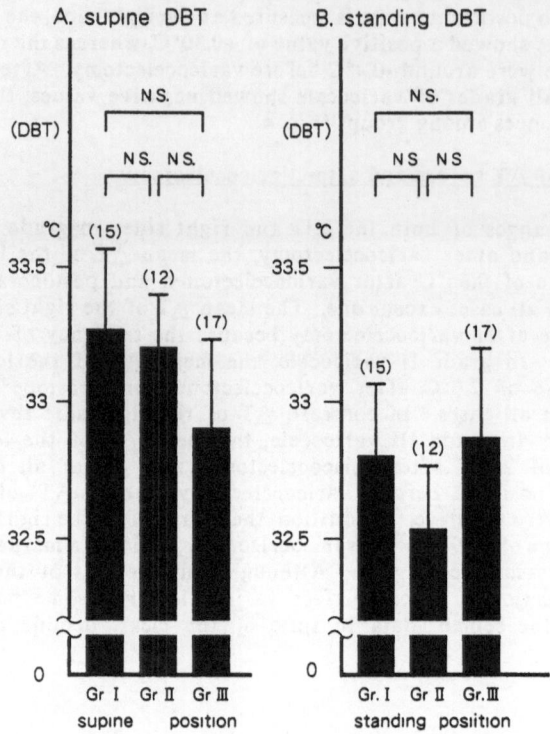

Figure 3. Right deep body temperature (DBT) of preoperation group. (A) Supine DBT; the mean DBT among each group is not significant. (B) Standing DBT; the mean DBT among each group is not significant.

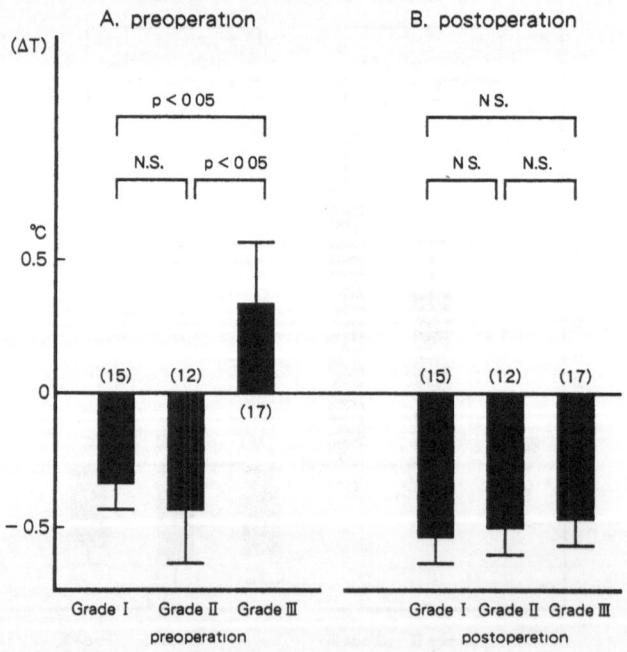

Figure 4. Difference in temperature ($\triangle$T) of the right side. (A) $\triangle$T of preoperation group; the mean $\triangle$T is +0.30°C in grade III varicocele group, whereas it is approximately -0.4°C in the other two groups. (B) $\triangle$T of postoperation group; the mean $\triangle$T among groups is not significant.

temperature, which might suggest that other factors besides temperature had also to be considered as affecting semen findings.

DISCUSSION

The method of measuring deep body temperature was first developed by Fox et al.[5] and improved by Togawa[6]. The temperature of body surface is generally lower than that of deep body because of the effect of the air. It is reported that the temperatures of body surface and deep body become equal if body surface is covered with heat insulation material to prevent the effect of the air temperature[5,6]. The measurement of deep body temperature, which is practically applied to clinical use, such as diagnosis and follow-up of skin diseases and peripheral circulatory disturbance caused by angiopathy and neuropathy, is regarded as useful in each field; however, there has been no report of the application of the deep body temperature measurement for the diagnosis of varicocele.

The scrotum itself and funicular vasculature play an important role to lower both intratesticular and intrascrotal temperatures by 2-4°C than intraperitoneal temperature. If thermoregulatory structure was disturbed or the temperature around the testicle was high, as animal experiments proved, some disturbance in spermatogenesis occurred[8]. Agger[9] investigated the semen before and after varicocelectomy and measured testicular

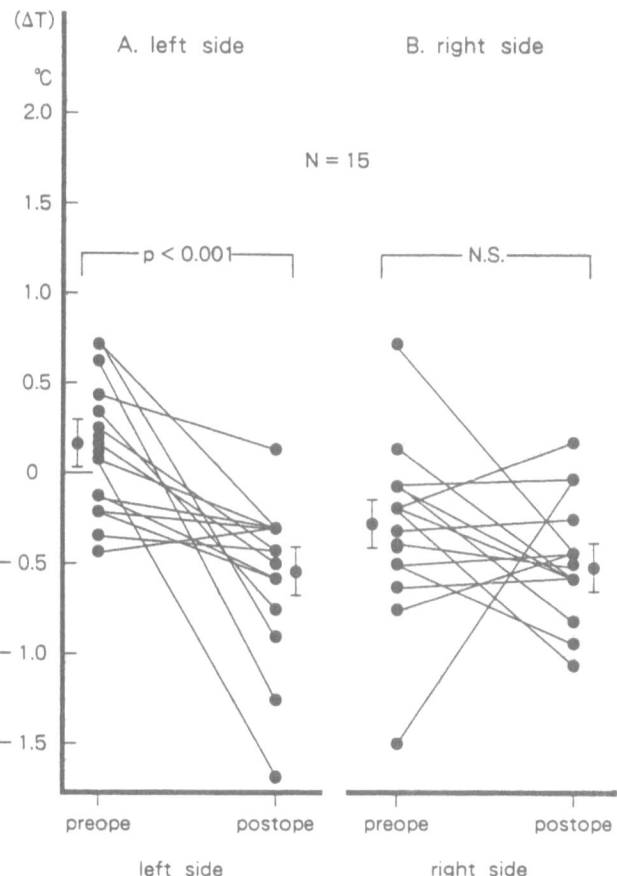

Figure 5. Difference in temperature ($\triangle$T) of grade I varicocele group. (A) $\triangle$T of left side; the mean $\triangle$T is +0.16°C in the preoperation group, whereas it is -0.70°C in the postoperation group. (B) $\triangle$T of right side; the mean $\triangle$T between preoperation and postoperation group is not significant.

temperature. The left testicular temperature of the patients with varicocele was 0.1°C higher than the right side before the operation, and the postoperative temperature of the left side was 0.2°C lower than the right side. Thus, the lower the temperature dropped, the better the results of the semen obtained. He concluded that "some effect of temperature rise cannot be excluded". The conventional method of the measurement of testicular temperature, however, was the needle puncture of the testicle by a probe under anesthesia, and this method could not be clinically applied especially for the postoperative observation. Zorgniotti and MacLeod[7] tried to solve this problem by inventing a new method of measurement. They reported that a change in position from supine to standing resulted in a drop of 0.6°C in intrascrotal temperature[7,9], and that there was no difference in right versus left intrascrotal temperatures except in standing varicocele subjects in whom the left intrascrotal temperature was significantly higher (0.3°C). In the present study the deep body temperature of the left testicle was high before varicocelectomy and that at the standing position, this tendency was more definitely shown. The change in temperature soon after standing ($\triangle$T) was positive only in the left testicle of patients with varicocele before operation. Our data confirm the previous reports[7,9], which were based on different measurement techniques from ours,

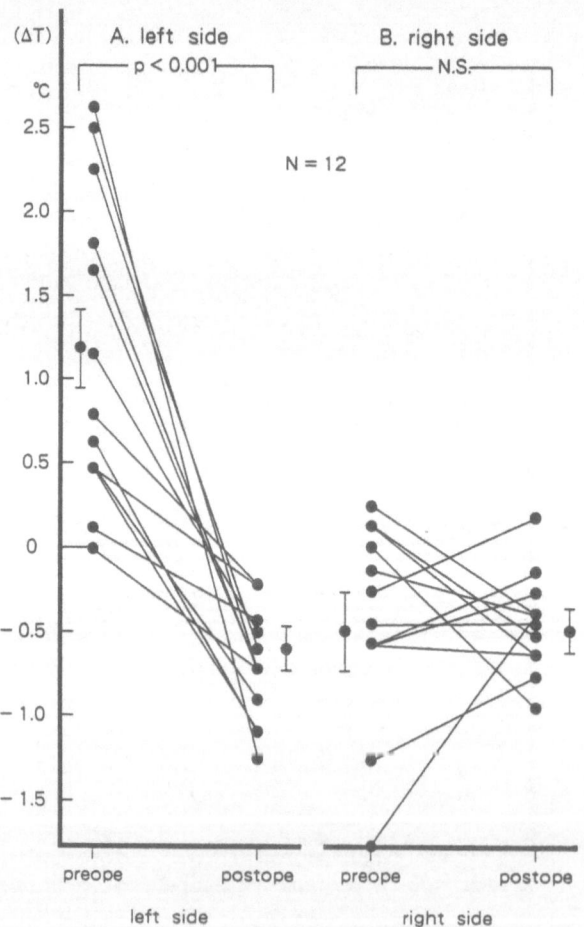

Figure 6. Difference in temperature ($\triangle$T) of grade II varicocele group. (A) $\triangle$T of left side; the mean $\triangle$T is +1.31°C in the preoperation group, whereas it is -0.71°C in the postoperation group. (B) $\triangle$T between preoperation and postoperation group is not significant.

thus indicating the usefulness of this deep body temperature measurement for the screening of varicocele. Comhaire et al.[11] conducted the comparative study of temperature according to each grade of varicocele using thermography and reported that the temperature difference was clearly greater in cases with grades II and III varicocele as compared with subclinical or grade I varicocele. Our results of $\triangle T$ according to each grade showed satisfactory correlation, which proved the efficiency of this method for the diagnosis according to each grade.

Preoperative $\triangle T$ of the right testicle showed positive only in grade III varicocele, while it showed negative in grades I and II varicocele, which might suggest the possible effect of the conspicuous rise of the left testicular temperature by varicocele on the right side. After varicocelectomy, $\triangle T$ at all grades of varicocele showed negative and the difference in temperature according to grade was not recognized. Thus, the measurement of intrascrotal deep body temperature is an effective examination not only for the diagnosis but also for the postoperative management of varicocele at the outpatient male infertility clinic. It is noninvasive, rapid and practical.

Though the measurement of deep body temperature is useful as described in the above discussion, it has the following problems. When we measure the intrascrotal temperature, we have to eliminate the influence of heat from the thighs. We succeeded in the femoral insulation by the placement of an adiabatic aircap mask between the scrotum and thighs. In thermography, a technique of insulation using adiabatic plate was

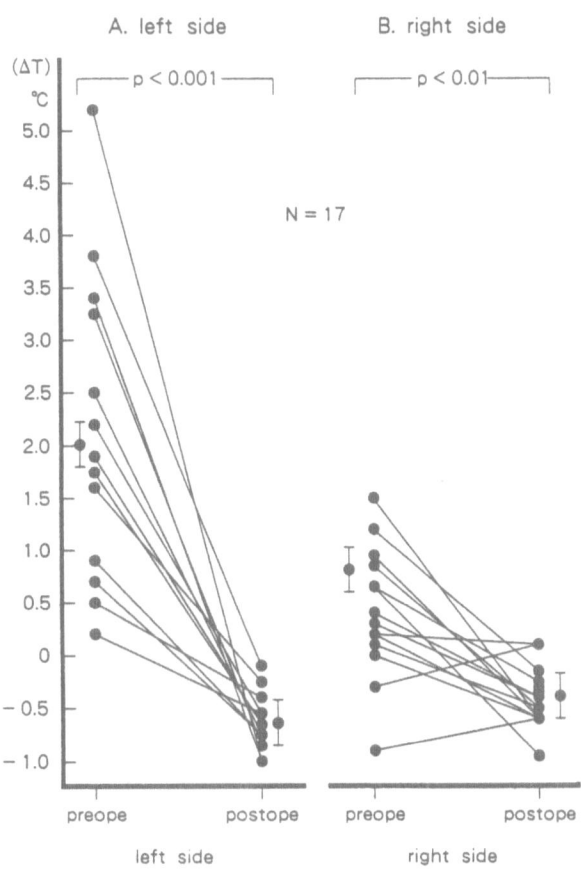

Figure 7. Difference in temperature ($\triangle T$) of grade III varicocele group. (A) $\triangle T$ of left side; the mean $\triangle T$ is +2.08°C in the preoperation group, whereas it is -0.61°C in the postoperation group. (B) $\triangle T$ of right side; the mean $\triangle T$ is +0.30°C in the preoperation group, whereas it is -0.49°C in the postoperation group.

reported[12], but it is not applicable to our method which requires change in the position of the patient during examination. The following cases should be excluded from the measurement because of high temperature: inflammatory disease in the scrotum and the high position of the testicle which is close to the trunk. In the case of thick pubis, we have to confirm whether the probe is attached in the appropriate position because we sometimes record low temperature.

In conclusion, the measurement of intrascrotal deep body temperature is useful for both the diagnosis of varicocele according to each grade and the postoperative follow-up. More investigation is necessary concerning the change in testicular temperature of the patients before and after varicocelectomy and the differences in the results of their seminal findings.

## ACKNOWLEDGMENTS

The authors wish to thank Dr. Naotoshi Murakami, Professor and Chairman of the 2nd Department of Physiology, Yamaguchi University School of Medicine for his helpful discussions and instructions.

## REFERENCES

1. H. Takihara, T. Ueno, K. Ishizu, M. Yamaguchi, R. Isoyama, Y. Baba and J. Sakatoku. 1989. The varicocele. Path. Clin. Med., 7: 311.

2. A.T.K. Cockett, M.J. Cosentino and H. Takihara. 1984. The varicocele. Fertil. Steril., 41: 5.

3. M. Yamaguchi, J. Sakatoku and H. Takihara. 1989. The application of intrascrotal deep body temperature measurement for the noninvasive diagnosis of varicocele. Fertil. Steril., 52: 295.

4. H. Takihara, M. Yamaguchi, Y. Baba and J. Sakatoku. 1990. Deep body intrascrotal thermometer: theory and methodology, In "Temperature and Environmental Factors on the Testis", A.W. Zorgniotti (ed.) Plenum Press, New York.

5. R.H. Fox and A.J. Solman. 1970. A new technique for monitoring the deep body temperature in man from the intact skin surface. J. Physiol., 212: 8.

6. T. Togawa. 1973. Medical thermometer making use of zero heat flow method. Rep. Inst. Med. Dent. Engng., 7: 75.

7. A.W. Zorgniotti and J. MacLeod. 1973. Studies in temperature, human semen quality and varicocele. Fertil. Steril., 24: 854.

8. E. Steinberger and W.J. Dixon. 1969. Some observations on the effect of heat on the testicular germinal epithelium. Fertil. Steril., 10: 578.

9. P. Agger. 1971. Scrotal and testicular temperature: its relation to sperm count before and after operation for varicocele. Fertil. Steril., 22: 286.

10. A.W. Zorgniotti, A. Toth and J. MacLeod. 1979. Infrared thermometry for testicular temperature determinations. Fertil. Steril., 32: 347.

11. F. Comhaire, R. Monteyne and M. Kunnen. 1976. The value of scrotal thermography as compared with selective retrograde venography of the internal spermatic vein for the diagnosis of "subclinical" varicocele. Fertil. Steril., 27: 694.

12. R.H. Gold, R.M. Ehrlich, B. Samuels, A. Dowdy and R.T. Young. 1977. Scrotal thermography. Radiology, 122: 129.

# A COMPARATIVE STUDY OF THE DIAGNOSTIC VALUE OF TELETHERMOGRAPHY

# AND CONTACT THERMOGRAPHY IN THE DIAGNOSIS OF VARICOCELE

Veljko Vlaisavljevič

Hospital Maribor
Department of Gynecology, Reproductive Unit
Maribor, Yugoslavia

## INTRODUCTION

Varicocele is the most frequent operable cause of male infertility. It is often difficult to diagnose small lesions merely by clinical examination. The fact that a measurable rise of scrotal temperature occurs in a certain number of infertile patients has helped to extend thermography also to the detection of such cases (Comhaire et al., 1976; Zorgniotti and MacLeod, 1973).

## METHODS

The objective of our study was to compare the efficiency of telethermography (T) and contact thermography (CT) in the diagnosis of varicocele in patients of the Outpatient Station for Infertile Couples. T was performed by means of the AGA Thermovision 680 system. The patients were examined after a 15-minute cooling period, standing upright at room temperature of 20 ± 1°C.

For CT, four thermographic plates with a temperature range of 28.1 - 32.3°C, 29.0 - 33.0°C, 30.6 - 35.1°C and 32.2 - 36.6°C, respectively, were used. The flexible 300 by 80 mm plates were made at the Jozef Stefan Institute (Ljubljana, Yugoslavia).

The examination was performed immediately after the T examination. Those thermograms of the scrotum in which the measured temperature differences between the symmetrical areas were greater than 1.5°C were defined as pathological. All thermograms

Table 1. Correlation between findings obtained by means of contact thermography and telethermography

| Contact thermography | Telethermography Abnormal | Normal |
|---|---|---|
| Abnormal (n = 69) | 54 | 15 |
| Normal (n = 174) | 16 | 158 |
| Total          243 | 70 | 173 |

*Temperature and Environmental Effects on the Testis*
Edited by A. W. Zorgniotti, Plenum Press, New York, 1991

Table 2. Absolute temperature values above the left pampiniform plexus measured by contact thermography and telethermography

| Diagnosis | N | Telethermography | | Contact thermography | |
|---|---|---|---|---|---|
| | | Left | Right | Left | Right |
| C[a] + T[b] normal | 100 | 30.50 ± 1.13 | 30.44 ± 1.12 | 30.21 ± 1.07 | 30.21 ± 1.07 |
| C normal + T abnormal | 33 | 32.84 ± 1.75 | 30.42 ± 1.08 | 32.23 ± 1.59 | 30.32 ± 1.17 |
| Varicocele + T abnormal | 28 | 33.75 ± 1.90 | 30.39 ± 1.74 | 32.96 ± 1.77 | 30.67 ± 1.32 |
| T normal[c] | 13 | 30.74 ± 1.14 | 30.53 ± 1.20 | 30.64 ± 0.97 | 30.31 ± 0.67 |
| T normal[d] | 23 | 30.56 ± 1.24 | 30.58 ± 1.25 | 30.34 ± 0.85 | 30.34 ± 0.85 |

a: clinical findings
b: thermography
c: suspicious clinical findings
d: control after surgery

with established temperature differences between the two hemiscrota greater than 1.5°C were considered pathological CTs (Vlaisavljević, 1985a).

To confirm the existence of unpalpable circulatory disturbances in the scrotum of thermographically positive patients, we applied dynamic angioscintigraphy of the scrotum. Isotope scanning was performed only in oligoasthenospermatic patients. Expected positive findings were an indication for venographic exploration. After injecting 10 mCi[99m] Tc pertechnetate into the elbow vein, we observed the perfusion of the scrotum. Scintigrams with increased perfusion over the hyperthermic hemiscrotum with scintigraphic index exceeding 15% of activity were thought pathological (Vlaisavljević, 1985b).

RESULTS

Using both methods, we examined 246 patients. In three patients the scrotal temperature after cooling was lower than can be measured by CT plates. In the group were 34 patients with clinically evident varicocele established by the first clinical examination. The relation between abnormal and normal thermographic findings of both examinations is shown in Table 1.

The results of CT and T did not correlate in 16/70 (22.8%) abnormal T findings and in 15/173 (8.7%) normal T findings. In 87.2% of examined patients, the abnormal and normal findings of both examinations correlated.

Comparison of absolute values of the measured temperatures in patients with identical results (normal or abnormal) of both examinations gave no significant differences in the use of various methods of measurement. Absolute temperatures were measured above the left pampiniform plexus in patients with normal clinical and thermographic findings, in patients with normal clinical and abnormal thermographic findings, and in patients with abnormal clinical and abnormal thermographic findings, suspicious clinical and normal thermographic findings and after surgical correction of varicocele (Table 2).

Of the 213 patients with normal clinical findings, 65 had an abnormal thermogram (T). In 4 patients we established the cause for the abnormal T by inspection (skin fold, anatomic abnormality, inflammation (2 x)). In 19 patients perfusion of the scrotum was evaluated by scintigraphy immediately after T. Increased activity of the isotope above the hyperthermal area was recorded in 17 cases (89.5%). The remaining patients in whom no further investigation by means of angioscintigraphy was used to disclose the cause of abnormal T, were not taken into account (n = 42). The diagnostic value of T and CT in the diagnosis of varicocele for our group of patients is shown in Table 3.

Table 3. Correlation between thermographic and final diagnosis

| Diagnosis | Telethermography | | Contact thermography | |
| | Normal | Abnormal | Normal | Abnormal |
| --- | --- | --- | --- | --- |
| Varicocele n = 51 | 8 | 43 | 10 | 41 |
| No varicocele n = 153 | 147 | 6 | 133 | 20 |
| Sensitivity | 0.843 | | 0.804 | |
| Specificity | 0.961 | | 0.869 | |
| Accuracy | 0.931 | | 0.853 | |

By changing the level of the cut-off point, which means using various rigorous criteria of separation between abnormal and normal thermograms, the ratio between false positive and true positive results also changed (Table 4). The curve of ratio between true positive and false positive results of T and CT for different thresholds between abnormal and normal findings (1°C, 1.5°C, 2°C and 2.5°C) is shown.

## DISCUSSION

The efficiency of T in the diagnosis of varicocele has enabled better diagnosis of subclinical varicoceles (Comhaire et al., 1976). CT was first applied in order to enable a wider use of thermographic diagnosis in andrology (Amiel et al., 1976; Sillo-Seidl, 1976). This method also proved successful in the diagnosis of subclinical varicoceles (Lewis and Harrison, 1980). In literature known to us there are no data about a study which would estimate the comparative value of CT and T.

A comparison of both methods has shown that their diagnostic value is similar. A comparison of findings obtained by the study of the entire group of patients with those in which we actually established varicocele or circulatory disturbances has shown minimal discordance between CT and T in the last group of patients. This implies that the two methods differ in the evaluation of the threshold value (temperature difference of about 1.5°C). Pronounced hyperthermal findings obtained by T in patients with verified subclinical varicocele appeared as such in CT as well. In 12.8% of patients CT and T findings did not correspond. This can be explained by the possible influence of the compression exerted by the thermographic plate upon the vascular complex during the examination.

An abnormal T associated with negative clinical findings is due to the changes in perfusion of the hyperthermal area, which was proved by scintigraphy in 89.5% of cases (Vlaisavljević, 1985b).

The comparison of two methods with regard to threshold criteria between abnormal and normal thermogram has shown that the optimal value of temperature difference is 1.5°C. At this value both methods give a relatively large number of true positive findings with the least number of false positive findings.

Table 4. Correlation of success of telethermography and contact thermography on chosen cut-off point value between normal and pathologic findings. All patients (n = 204) were investigated by both methods simultaneously

| Cut-off point value T (°C) | Telethermography | | Contact thermography | |
|---|---|---|---|---|
| | TPF | FPF | TPF | FPF |
| 0.9 | 0.86 | 0.08 | 0.82 | 0.10 |
| 1.4 | 0.84 | 0.04 | 0.80 | 0.14 |
| 1.9 | 0.79 | 0.01 | 0.80 | 0.13 |
| 2.4 | 0.59 | 0.006 | 0.39 | 0.01 |

TPF: true positive findings
FPF: false positive findings

## ACKNOWLEDGMENTS

This study was supported by the Slovenian Research Foundation (09-3177-334).

# REFERENCES

Amiel, J.P., Vignalou, L., Tricoire, J., Jamain, B., and Ravina, J.H. 1976. Thermographie du testicule. J. Gynecol. Obstet. Biol. Reprod. (Paris), 5: 917.

Comhaire, F., Monteyne, R. and Kunnen, M. 1976. The value of scrotal thermography as compared with selective retrograde venography of the internal spermatic vein for the diagnosis of "subclinical" varicocele. Fertil. Steril., 27: 694.

Lewis, R.W. and Harrison, R.M. 1980. Contact scrotal thermography. II. Use in the infertile male. Fertil. Steril., 34: 259.

Metz, C.E. 1978. Basic principles of ROC analysis. Semin. Nucl. Med., 8: 283.

Sillo-Seidl, G. 1976. Platten-thermographische Untersuchungen an den männlichen äusseren Geschlechtsorganen, Urologe A., 15: 126.

Vlaisavljević, V. 1983. Thermography in the evaluation of scrotal circulatory disturbances. Br. J. Sex Med., 10: 9.

Vlaisavljević, V. 1985a. Thermographic characteristics of the scrotum in the infertile male, in: Recent advances in medical thermology. E.F.J. Ring and B. Phillips (eds.), Plenum Press, New York, p. 415.

Vlaisavljević, V. 1985b. The value of angioscintigraphy in detection of varicocele (croat). Lijec Vjestn, 107: 511.

Zorgniotti, A.W. and MacLeod, J. 1973. Studies in temperature, human semen quality and varicocele. Fertil. Steril., 24: 854.

# SCROTAL THERMOGRAPHY IN VARICOCELE

Frank Comhaire

State University Hospital
Department of Internal Medicine
Ghent, Belgium

## INTRODUCTION

In a cross-cultural, multicentre study on 8504 men coming to consultation as partner of an infertile couple, varicocele was found to be present in between 0% and 47% of cases with abnormal semen quality (W.H.O., 1987). This very high level of variability of varicocele frequency between different centres could not be attributed to regional differences, nor to differences in coincidental or additional causes of infertility in either the male or the female partner. The difference could only be explained by the expertise of the investigator to detect varicocele upon clinical investigation. Clearly, centres which included trained andrologists or urologists in the team of investigators reported significantly higher frequencies of varicocele than those centres lacking such team members.

It has been shown that blood reflux in the internal spermatic vein, being the pathogenic factor in testicular impairment associated with varicocele, can be present without palpable distention of the pampiniform plexus in the so-called subclinical varicoceles. It is estimated that the latter constitute 40% of men in whom the presence of spermatic venous reflux can be documented by means of retrograde venography of the internal spermatic vein (W.H.O., 1985).

Several methods are available to detect spermatic venous reflux among which scrotal thermography, Doppler ultrasonography and isotope scanning after injection of technetium pertechnetate have been advocated for screening purposes.

## TECHNICAL DEVELOPMENTS IN SCROTAL THERMOGRAPHY

Originally the presence of scrotal hyperthermia associated with varicocele was demonstrated by thermometry of the skin and infrared photography (Zorgniotti and MacLeod, 1973) and telethermography (Kormano et al., 1970; Gasser et al., 1973). Using the latter method, it was assessed that a temperature difference of more than 0.5 °C between both hemiscrota was highly suggestive of unilateral varicocele, whereas bilateral temperature increase was associated with bilateral reflux. The thermographic detection of bilateral reflux was, however, rather inaccurate (Comhaire et al., 1976).

The availability of thermosensitive liquid crystals initiated the development of devices for contact thermography (Fochem and Pflanzer, 1975). Initially the liquid crystals were applied onto rigid plates, similar to those used for contact thermography of the breast (Sillo-Seidl, 1976). Later on, flexible devices were developed in which the

Figure 1. Varicoscreen.

crystals were applied on a cellophane or plastic film (Lewis and Harrison, 1980) (Clark Topical Thermograph, Clark Research and Development Inc., New Orleans, Louisiana).

All these devices were more or less impractical to use because of their poor compliance with the male anatomy. Indeed, the region of the scrotal skin most probable to present hyperthermia is located near the scrotal neck and is difficult to reach with stiff and even flexible strips because of the presence of the penile shaft.

For this reason we have developed a specific device for contact thermography of the scrotum, presenting a peculiar shape similar to inverted spectacles (Fig. 1) (Varicoscreen, Promedex Inc., 7 E. Shore Drive, Kamfe Lake, Bloomingdale, NJ 07403). This particular design allows for accurate and complete investigation of the scrotal skin including the cranial quadrants outside of the penile shaft.

In addition, the choice of the composition of the high performance liquid crystals was such that colour changes occur at intervals of 0.6 °C and that these changes are easily detectable in the critical range between 31.3 °C (88.3 °F) and 35.3 °C (95.5 °F) using one single device.

Finally, the film and liquid crystals are coated to allow for the device to be disinfected. Thanks to these developments; the screen can accurately be used for over 500 examinations without losing its reliability.

## TECHNIQUE OF SCROTAL THERMOGRAPHY

Scrotal thermography should be performed in subjects with abnormal spermatozoa and presenting no clinical abnormalities on urogenital examination, or in whom the presence of varicocele is suspected upon palpation or because of unilateral decrease of testicular volume. The patient has to stand undressed for 5 to 10 minutes in a room where the temperature does not exceed 22 °C. Alternatively, the scrotal skin can be cooled using a handheld ventilator or cold air blower. With the patient standing, the investigator brings the scrotum forward with both hands and the Varicoscreen is applied. In a normal man, the temperature of the scrotal skin is symmetrically distributed and does not exceed 33 °C. Any abnormalities in the distribution of the temperature of the scrotal skin should be recorded (Comhaire et al., 1976; Lewis and Harrison, 1979; Pochaczevsky et al., 1986).

## COMPARISON BETWEEN CONTACT THERMOGRAPHY AND TELETHERMOGRAPHY

The performance of telethermography necessitates expensive equipment, whereas contact thermography using flexible strips such as Varicoscreen is a cheap and easy office-method for the detection of scrotal hyperthermia. We have compared the findings of tele- and contact thermography on 108 patients consulting for infertility. The results of the two investigations were evaluated and recorded by three independent clinicians who were unaware of the clinical data. The results were also compared with the outcome of subsequent retrograde venography of the internal spermatic vein. The results of tele- and contact thermography were identical in all cases. False-positive thermographic findings occurred in 8% of cases (Comhaire and de Thibault de Boesinghe, 1982).

## COMPARISON BETWEEN CONTACT THERMOGRAPHY AND OTHER DIAGNOSTIC METHODS

Using the result of retrograde venography of the internal spermatic veins as gold standard, the accuracy of different screening methods for varicocele was assessed in a multicentre study organized by the World Health Organization involving 141 men (W.H.O., 1985). Isotope scanning has a high false-negative rate in cases with small varicoceles, including Valsalva positive and subclinical varicoceles. In addition, the findings were commonly inconclusive in cases with palpable varicoceles, making this method unsuitable for diagnosis. Both Doppler ultrasonography and contact thermography had a low false-negative rate. However, the frequency of false-positive findings was higher with Doppler ultrasonography than with scrotal contact thermography. This is probably due to confounding occurrence of signals caused by contraction and relaxation of the cremasteric muscle during the Valsalva manoeuvre, which is required for the Doppler test. Hence, scrotal thermography was found to be the most accurate test for the detection of small and subclinical varicoceles, while this technique was highly confirmative in cases with clinically palpable varicoceles.

## CONCLUSION

Among all cases with pathological blood reflux in the internal spermatic vein, large varicoceles are easily detected by simple inspection in 13% of cases, or palpation without Valsalva manoeuvre in 25% of cases. In 22% of cases the venous reflux can only be palpated during Valsalva manoeuvre. In the latter cases, and in the remaining 40% of patients with impalpable, so-called subclinical varicoceles, the diagnosis is difficult if not impossible without technical help. Hence, it can be estimated that the diagnosis of reflux in the internal spermatic vein is overlooked in about 50% of cases (Marsman, 1985).

Among different methods for detecting reflux filling of the pampiniform plexus, contact thermography was found to be the most accurate and simple as well as cost-effective, while not requiring any participation of the patient. The development of a cheap, robust, and easily applicable as well as readable device (Varicoscreen) has made systematic scrotal contact thermography accessible to all centres involved in the investigation and management of the infertile couple.

## REFERENCES

Comhaire, F.H., Monteyne, R. and Kunnen, M. 1976. The value of scrotal thermography as compared with selective retrograde venography of the internal spermatic vein for the diagnosis of 'subclinical' varicocele, Fertil. Steril., 27:69.
Comhaire, F. and de Thibault de Boesinghe, L. 1982. Thermostrip (R) detection of varicocele, J. Androl., 3:32 (abstract).
Fochem, K., and Pflanzer, K. 1975. Indikationsmöglichkeiten der Plattenthermographie, Röntgen-Berichte, 4:169.

Gasser, G., Strassl, R. and Pokieser, H. 1973. Thermogramm des Hodens und Spermiogram, Andrologie, 5:127.

Kormano, M., Kahanpaa, K., Svinhufvud, U. and Tahti, E. 1970. Thermography of varicocele, Fertil. Steril., 21:558.

Lewis, R.W. and Harrison, R.M. 1979. Contact scrotal thermography: application to problems in infertility, J. Urol., 122:40.

Lewis, R.W. and Harrison, R.M. 1980. Contact scrotal thermography. II. Use in the infertile male, Fertil. Steril., 34:259.

Marsman, J.W. 1985. Clinical versus subclinical varicocele: venographic findings and improvement of fertility after embolization, Radiology, 155:635.

Pochaczevsky, R., Lee, W.J. and Malett, E. 1986. Management of male infertility: Roles of contact thermography, spermatic venography, and embolization, AJR, 147:97.

World Health Organization. 1985. Comparison among different methods for the diagnosis of varicocele, Fertil. Steril., 43:575.

World Health Organization. 1987. Towards more objectivity in diagnosis and management of male infertility, Int. J. Androl., supplement 7.

Zorgniotti, A.W. and MacLeod, J. 1973. Studies in temperature, human semen quality and varicocele, Fertil. Steril., 24:854.

# THE ROCC ANALYSIS OF FIVE DIFFERENT METHODS

# IN THE DIAGNOSIS OF VARICOCELE

Veljko Vlaisavljević

Hospital Maribor
Department of Gynecology, Reproductive Unit
Yugoslavia

## INTRODUCTION

The optimistic results of surgical treatment of varicocele and publications on the role of subclinical varicocele in male infertility were the reasons for the study of different diagnostic methods of evaluation of this condition. For more than 15 years thermography is used routinely in the evaluation of circulatory disturbances in infertile patients who consulted our Reproductive Unit at the Department of Gynecology.

Telethermography is a method which can explore the thermoregulatory mechanisms under very physiological conditions. The equipment used for investigation is AGA Thermovision. We investigate the scrotum under standard temperature conditions (20°C) after the patient was asked to undress and remain in upright position for 15 minutes. During this period the cooling rate of the scrotum is constant and symmetrical in both hemiscrota. In patients with reflux/varicocele, the temperature over the affected area will be constantly high or it will increase during the period of cooling (Vlaisavljević 1984a).

In 1064 patients examined because of the possible presence of circulatory disturbances, we found 292 (27.3%) thermograms which were not symmetrical and a temperature difference over the hemiscrotum exceeding 1.5°C. Such positive findings can be registered in the group with clinically manifest varicocele and in clinically normal patients without clinically evident reflux (Table 1).

The most interesting finding was that the quality of ejaculate can be related to the thermographic manifestations of thermoregulatory mechanisms over the scrotum. We found statistically significant relationships between the sperm count and the motility of the sperm and the rate of abnormal thermograms in 360 clinically normal patients (Vlaisavljević 1984b; Table 2).

The recognition that a small varicocele has an equally damaging effect on spermatogenesis as a large one (Dubin and Amelar, 1970) even while it is still in its "subclinical" unpalpable form (Comhaire et al., 1976), presented new demands on the diagnosis of circulatory scrotal disturbances. In literature we can find numerous reports on the diagnosis of these changes not only by clinical investigation but also by telethermography, contact thermography, scintigraphy, venography and Doppler stethoscope. Unfortunately, the conditions of the investigation, the evaluation of findings and the characteristics of the population investigated in these studies are usually not comparable with each other (Comhaire, 1983).

*Temperature and Environmental Effects on the Testis*
Edited by A. W. Zorgniotti, Plenum Press, New York, 1991

Table 1. Incidence of abnormal thermograms in 1064 patients

| Thermographic findings | Temperature difference (°C) | N | % |
|---|---|---|---|
| Normal | < 1 | 740 | 69.5 |
| Abnormal | 1 - 1.4 | 32 | 3.0 |
| | 1.5 - 1.9 | 73 | 6.8 |
| | > 2 | 219 | 20.5 |

The aim of the present study was to investigate the validity of the diagnostic tests on the basis of medical decision-making theory.

On the basis of specificity and sensitivity values, the probability of the presence (or absence) of varicocele/reflux was determined if the result of investigation was abnormal (or normal). For the calculation of this posterior (post test) probability of disease, Bayes' theorem was used, changing the values of disease prevalence (pre-test probability) from 1 to 99 (McNeil and Adelstein, 1976).

## MATERIALS AND METHODS

The study included 491 patients who consulted our Division for Male Infertility because of infertile marriage. After clinical investigation the genital status was determined. The result of the first clinical investigation is considered as the admission diagnosis. The correct final clinical diagnosis was established if an independent second examiner confirmed the admission diagnosis. After the ejaculate was analyzed, all patients were examined by means of telethermography (AGA Thermovision system, Sweden).

For the determination of the optimal level between normal and abnormal findings of the tested methods, a receiver operating characteristic curve (ROCC) analysis was made (Metz, 1978).

Table 2. Comparison of semen quality and rate of abnormal scrotal thermograms registered in 360 clinically normal patients

| Sperm concentration ($10^9$/L) | Sperm motility (%) | Number of patients | Abnormal thermograms | Rate |
|---|---|---|---|---|
| < 20 | < 60 | 56 | 20 | 0.357[a] |
| < 20 | > 60 | 50 | 15 | 0.300 |
| > 20 | < 60 | 40 | 8 | 0.200 |
| > 20 | > 60 | 214 | 34 | 0.159 |
| > 40 | > 60 | 144 | 21 | 0.145 |
| > 60 | > 60 | 86 | 9 | 0.104 |

[a]$p < 0.005$ when compared with the higher semen quality

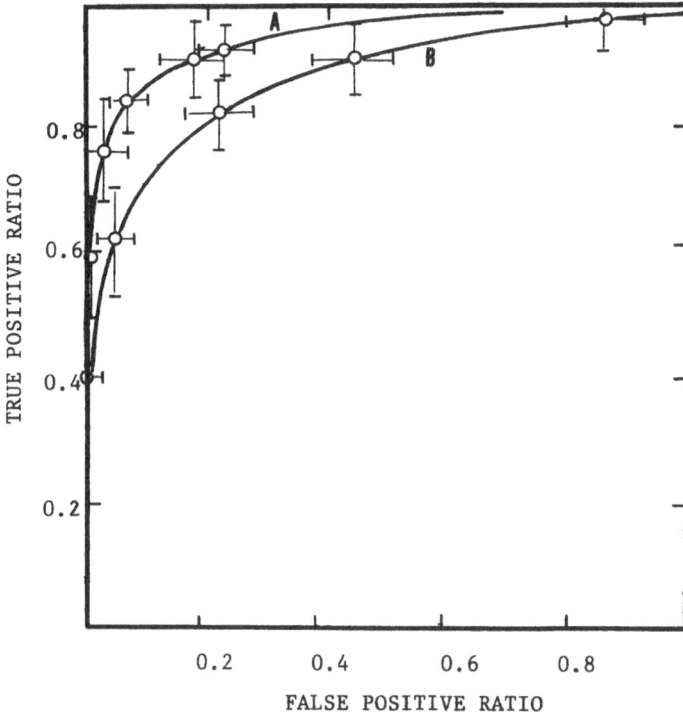

Fig. 1. The decision fractions resulting from the data obtained from thermographic measurements of absolute temperature values and maximal temperature difference, plotted as points in a ROC (receiver operating characteristic) space (±2 standard deviation), with possible ROC curves on which these points could lie. Curve A: Cut-off points from left to right are: 2.5°C, 2.0°C, 1.5°C and 0.5°C. Curve B: Cut-off points from left to right are: 33.4°C, 32.4°C, 31.4°C, 30.4°C and 29.4°C.

On the basis of the final diagnosis we tried to analyze which thermographic diagnostic criteria are better for the detection of clinically not manifest reflux and visualization of clinically manifest varicocele: measurements of absolute temperatures over the testis or measurements of maximal temperature differences between both hemiscrota. The results of thermographic examinations were plotted as points in ROC (receiver operating characteristic) space (±2 standard deviation) with a possible ROC curve on which these points can lie. The decision fractions result from the data obtained from thermographic measurements of absolute temperature values and maximal temperature differences over the scrotum (Fig. 1).

We tried to compare the results obtained by using more sophisticated and expensive telethermographic equipment with those obtained by using the cleaner and simpler method of contact thermography. For contact thermography, four thermographic plates with different temperature ranges were used.

A group of 204 randomly selected patients were examined by contact thermography using flexible reusable plates (J. Stefan Institute, Yugoslavia; Vlaisavljevič and Levstik, 1984).

The optimal cut-off point between a normal and abnormal thermogram was decided at 1.5°C temperature difference between both hemiscrota, regardless of whether we use telethermography or contact thermography (Vlaisavljevič, 1983).

The biggest problem in the investigation of methods was the group of patients with clinically normal findings and thermographically abnormal image, and vice versa. A

Table 3. Comparison of thermographic, scintigraphic and venographic findings

| Temperature difference (°C) | 0 – 1.4 | 1.5 – 2.9 | 3.0 – 4.4 | 4.5 – 5.9 |
|---|---|---|---|---|
| Investigated scintigraphically | 34 | 16 | 11 | 5 |
| Abnormal scan (%) | 4 (12) | 9 (56) | 8 (73) | 5 (100) |
| Mean activity | 13.72 | 19.05 | 85.12* | 55.03* |
| ± SEM | ±3.54 | ±4.98 | ±21.63 | ±17.03 |
| Median (Range) | 6.5 (0.3 – 61.8) | 11.5 (0.6 – 75.9) | 51.5 (2.3 – 197.3) | 46.0 (14.4 – 117.6) |
| Investigated venographically | 4 | 7 | 7 | 3 |
| Presence of reflux (%) | 1 (25) | 4 (57) | 6 (86) | 3 (100) |

*Values significantly different from control at p < 0.001

medical indication for the application of more invasive methods (radionucleotide angioscintigraphy and/or venography) was only given in the group of patients with abnormal semen analysis in which some benefit from correct diagnosis and therapy can be expected.

The suspicious and thermographically disconcordant findings obtained with previously mentioned methods were researched by radionucleotide angiography and/or retrograde venography as described by Nahoum et al. (1980) and Comhaire et al. (1976). This group of patients included those with negative clinical and positive telethermographic findings, and vice versa. From this group of patients examined by isotope scanning, a smaller group was selected for venographic investigation on the basis of ejaculate analysis. Only oligoasthenospermatic patients (WHO standard) were examined venographically because therapy was planned on the basis of venographic findings (Table 3).

Radionuclide angioscintigraphy was performed in 80 infertile patients. After the injection of a bolus of 185 MBq of 99m Tc pertechnetate into the cubital vein, we observed the perfusion of the scrotum and dynamic changes over both hemiscrota by means of a Picker-Dina 4/15 gamma camera. The rise of activity over the hemiscrota was registered as a curve for the left and right side. The summary of the 16 sequentional scintigrams was expressed as a scintigraphic index (Vlaisavljevič, 1985).

The analysis of ROC curves has shown that the method is useful in the diagnosis of only clinically manifest lesions. The through-cut-off point between the positive and negative findings lies between 10 and 15% of activity. The size of varicocele influences the scintigraphic expression of the lesion. In small lesions, the scintigraphic index was lower than in the bigger ones. In subclinical varicocele/reflux, angioscintigraphy used as an independent method is not adequate for the visualization of the lesion (Fig. 2).

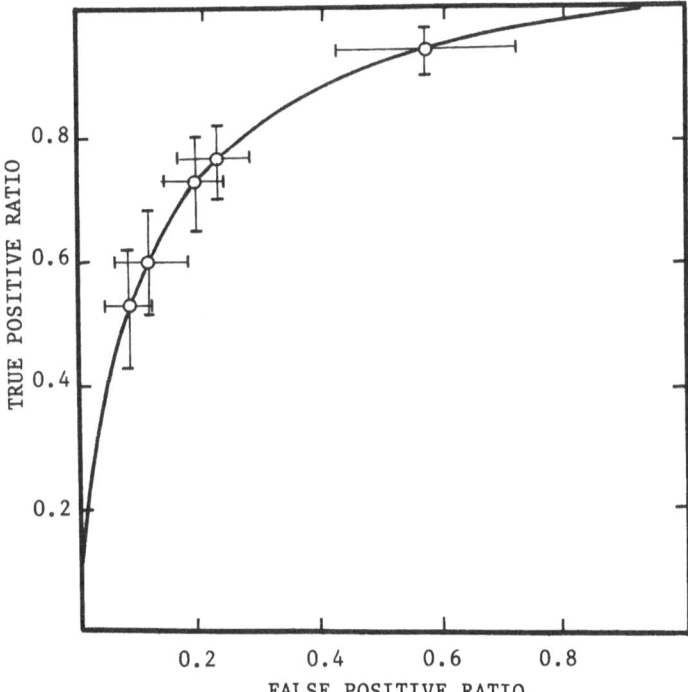

Fig. 2. The decision fractions from the data obtained from scrotal angioscintigraphy. Values of scintigraphic index plotted as points in ROC space (±2 standard deviation) with ROC curve on which these points could lie. Cut-off points from left to right are: 25%, 20%, 15%, 10% and 5%.

Table 4. The results of screening for varicocele/reflux in 491 patients

| Diagnostic method | Number of patients | Result of investigations | Varicocele or reflux Present | Varicocele or reflux Absent | Spec. (%) | Sen. (%) |
|---|---|---|---|---|---|---|
| Clinical examination | 491 | positive negative | 73 39 | 27 352 | 92.8 | 65.2 |
| Teletothermography | 491 | positive negative | 101 11 | 27 352 | 92.8 | 90.2 |
| Contact thermography | 204 | positive negative | 46 5 | 26 127 | 83.0 | 90.2 |
| Scintigraphy | 79 | positive negative | 38 5 | 3 33 | 91.7 | 83.7 |
| Venography | 35 | positive negative | 27 1 | - 7 | 100.0 | 96.4 |

RESULTS

The diagnostic value of the method describes its capability to give a positive result when a disease is in fact present and to exclude disease (give a negative result) in the healthy person. The results of the success of each tested method for the detection of varicocele/reflux are shown in Table 4.

The estimation of validity was calculated from this diagnostic matrix. The obtained values are inserted into the ROC space and thus the curves on which these values lie are determined. The dependence of the final posterior probability on the initial probability of the disease after obtaining positive (or negative) findings of the investigation is shown for every tested method (Fig. 3).

The post test probability difference curve (Fig. 4) represents the difference between the post test probability of varicocele/reflux with an abnormal test and the post test probability of the disease with a normal test. The peak of this curve indicates the prevalence with which the diagnostic test most effectively discriminates between varicocele/reflux presence or absence.

DISCUSSION

The testing of several methods only on the basis of its sensitivity and specificity is simple in cases when one test gives doubtlessly higher values than the other. In cases when the differences are minimal or the value of one parameter for the evaluation of validity is equal in both tests, but different in the second parameter, it is more difficult to decide. Since the value of the investigation does not depend on its specificity and sensitivity, but also on the frequency of the disease in the investigated group, we evaluated the post test probability in the presence of the disease changing its prevalence. Such an approach was advocated in the works of Turner (1978), Metz (1978), Swets (1979) and others. The major problem for the evaluation of previous methods of screening/ diagnosis of varicocele/reflux is in the invasive nature of venography and scintigraphy used as a detection test. This is the reason why the prevalence of clinically evident and subclinical varicocele is lower in the first group (clinical examination and thermography) and higher (but only subclinical form) in the venography and scintigraphy group.

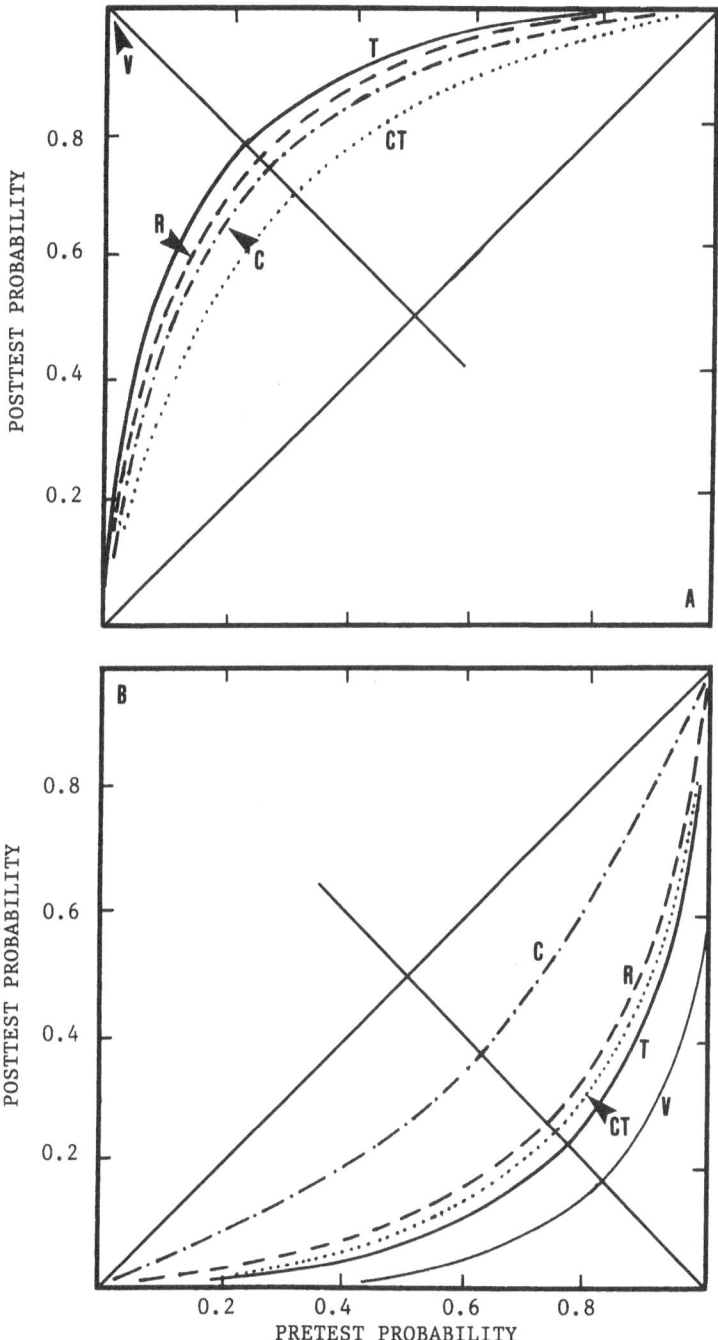

Figs. 3. The pretest probability or prevalence of varicocele/reflux versus the post test probability of disease for an abnormal test result (A) and a normal test result (B). T: telethermography; CT: contact thermography; R: radionucleotide angiography; C: clinical investigation; V: venography.

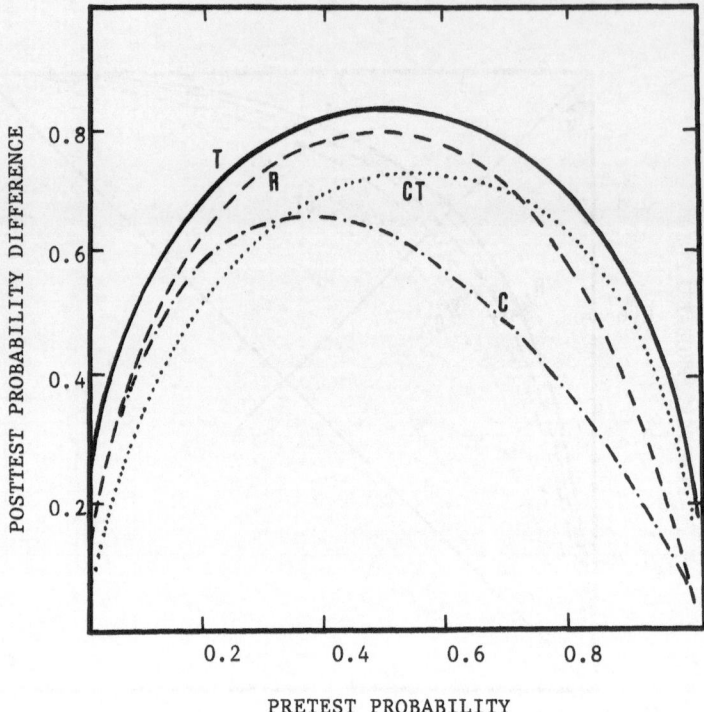

Fig. 4. The post test probability difference curves. From the graph it is apparent which test provides the best discrimination between the disease and the absence of disease based on the pretest probability of disease (prevalence). T: telethermography; CT: contact thermography; R: radionuclide angiography; C: clinical investigation.

The curves generated in the ROC space showed that the value of the method used independently for diagnosing circulatory changes in the scrotum decreases in this order: venography, telethermography, radionulceotide angiography, clinical examination and contact thermography. Because the prevalence of the disease greatly varies in the group submitted to the different diagnostic methods, from 22.8% in the total population, over 54.4% in the scintigraphy group and 80.0% in the venography group, the validation will automatically favor the latter diagnostic methods if comparison is made only on the basis of its sensitivity and specificity. Different prevalence of clinically evident varicocele (excluded in the last two groups) and subclinical varicocele may influence the results of testing. Comparison of methods on the point of theoretically chosen prevalence of disease describes its value in routine use.

Today a method with which we could express the statistical significance of the difference between two curves is still not known. The difference between them is described with the incline of the curve (larger incline - higher specificity) and with graphic analysis of the sections on the straight line crossing the curves, and set vertically on the line denoting 50% of disease probability. In cases when the initial probability of the disease is 20%, positive findings of clinical investigation will increase the probability of reflux/varicocele for 3.5 times, radionucleotide angiography the same, telethermography for 3.8 times and contact thermography for 2.8 times. From the graph it is evident that if we wish to attain the same value of the posterior probability of varicocele/reflux, regardless of whether we use telethermography, then the disease prevalence in the examined group must be higher when we use contact thermography. But these same methods differ in the value of the negative result (Fig. 3B). The steeper the curve is (nearer to the bottom right corner of ROC space), the more sensitive is the method and thus more capable of excluding healthy patients. Thus the negative findings of clinical investigation reduce the disease probability less effectively than telethermography, contact thermography or radionucleotide angiography. This means that the analyzed tests are more appropriate for the elimination of the disease than its detection.

Varicocele/reflux prevalence in the tested group is the decisive factor which influences the clinical value of the chosen method. Its value influences not only the clinical value of the chosen method, but also the prognostic value of positive or negative findings. Because of this, the test with high specificity and low sensitivity gives better results in evaluating the presence of disease or excluding it in cases when the prevalence of the sought disease is low (test adequate for screening). The counter situation is seen in tests with high sensitivity and low specificity which give better results in high disease prevalences (test adequate for use in selected population). This dependence is also presented graphically for our diagnostic methods (Fig. 4). The analysis of the curves showed that clinical investigation alone is not sufficient for a successful diagnosis of clinical varicocele and the detection of subclinical varicocele. Because of its uninvasiveness and high specificity, telethermography is a method which is adequate for selecting patients for venographic investigation. Where telethermography is not accessible, we can use contact thermography. Because of the good sensitivity of this method it can be used for excluding the existence of circulatory disturbances in the scrotum . In such a group of patients selected by thermography there is a high prevalence of patients with varicocele/reflux but additional testing by radionucleotide angiography will not bring any profit to diagnostic procedure. Since thermography has a low false negative rate and since the incidence of false positive findings will not be impressively reduced by isotope scanning, retrograde venography and embolization should be the next step in diagnosis and therapy of varicocele/reflux.

REFERENCES

Comhaire, F., Monteyne, R. and Kunnen, M. 1976. The value of scrotal thermography as compared with selective retrograde venography of the internal spermatic vein for the diagnosis of "subclinical" varicocele. Fertil. Steril., 27: 694.

Comhaire, F. 1983. Varicocele infertility an enigma. Int. J. Androl., 6: 401.

Dubin, L. and Amelar, R.D. 1970. Varicocele size and results of varicocelectomy in selected subfertile men with varicocele. Fertil. Steril., 21: 606.

McNeil, B.J. and Adelstein, S.J. 1976. Determining the value of diagnostic and screening tests. J. Nucl. Med., 17: 439.

Metz, C.E. 1978. Basic principles of ROC analysis. Semin. Nucl. Med., 8: 283.

Mieusset, R., Bujan, L., Mondinat, C., Mansat, A., Pontonnier, F. and Grandjean, H. 1987. Association of scrotal hyperthermia with impaired spermatogenesis in infertile men. Fertil. Steril., 48: 1006.

Nahoum, C.R.D., De Almeida, A.S. and Flores, E. 1980. Scrotal scan in the diagnosis of varicocele. Fertil. Steril., 34: 287.

Swets, J.A. 1979. ROC analysis applied to the evaluation of medical imaging techniques. Invest. Radiol., 14: 109.

Vlaisavljević, V. 1983. Thermography in the evaluation of scrotal circulatory disturbances. Br. J. Sex Med., 95(10): 9.

Vlaisavljević, V. 1984a. Thermographic characteristics of the scrotum in the infertile male, in: "Recent advances in the medical thermology", pp. 415-420, ed. E.F.J. Ring and B. Phillips, Plenum Press, New York.

Vlaisavljević, V. 1984b. The influence of unpalpable circulatory disturbances on semen quality. Diab Croat, 13: 43.

Vlaisavljevič, V. and Levstik, I. 1984. The diagnosis of scrotal circulatory disturbances by contact thermography (slov), <u>Zdrav Vestn</u>, 53: 443.

Vlaisavljevič, V. 1985. The value of scrotal angioscintigraphy in detection of varicocele (croat). <u>Lijec Vjestn</u>, 107: 511.

# THE PATHOGENESIS OF EPIDIDYMO-TESTICULAR DYSFUNCTION IN

# VARICOCELE: FACTORS OTHER THAN TEMPERATURE

Frank Comhaire

State University Hospital
Department of Internal Medicine
Ghent, Belgium

## INTRODUCTION

Among factors accepted to cause impairment of sperm quality, clinical as well as subclinical varicoceles are the most common since they are detected in approximately one third of subfertile men. Several hypotheses have been proposed to explain the mechanism of epididymo-testicular dysfunction in these men. An important problem with the majority of these theories resides in the fact that they do not account for the observation that varicocele may have such a variable influence on testicular function, as to leaving it apparently unaltered in some cases and causing a total arrest of spermatogenesis and/or deficient androgen production in others.

Increased temperature of the testes has been the favourite among hypotheses. In our view, this mechanism does not explain the observed alterations in all cases. Indeed, testicular temperature at the non-affected side is commonly normal whereas cases with subclinical varicocele may also present normal testicular temperature in spite of increased temperature over the pampiniform plexus.

## ALTERNATIVE HYPOTHESES

### Accumulation of $CO_2$ and other Substances in Testicular Tissue

Another theory suggests that impaired venous reflux results in the accumulation of $CO_2$ in testicular tissue. Together with hypercapnia and hypoxia, the accumulation of lactic acid would cause metabolic disturbances which could impair spermatogenesis (Donohue and Brown, 1969; Netto et al., 1977). Lactic acid is, however, an essential constituent of the natural milieu for spermatogenesis and can hardly be imagined to be severely harmful. Besides, measurement of blood gases and testicular fluid pH in patients with varicocele-associated infertility did not reveal any significant abnormalities. On the other hand, raised concentrations of sodium, potassium and urea have been recorded in spermatic venous blood of men with varicocele and these may, among other factors, be involved in the pathogenesis of impaired spermatogenesis (Adamopoulos et al., 1987).

### Reflux of Adrenal Steroids

MacLeod (1965) has suggested that reflux of adrenal steroids may cause deficient sperm production, similar to the patterns seen after stress. However, measurement of the concentration of cortisol and dehydro-epiandrosterone sulphate in the refluxing spermatic venous blood did not reveal any increase over values in peripheral venous

Table 1. Catecholamine concentrations in spermatic venous blood

| | Adrenaline | Noradrenaline |
| --- | --- | --- |
| | (ng per dl) | |
| Peripheral venous blood | 15.8 | 17.1 |
| Spermatic venous blood | | |
|    recumbent | 14.9 | 28.0 |
|    standing | 17.4 | 44.0 |

blood (Koumans et al., 1969; Agger, 1971). On the contrary, except for a few cases with an adreno-gonadal by-pass, the cortisol concentration in refluxing blood tended to be lower than that in peripheral blood, which is similar to the situation in renal venous blood (Comhaire and Vermeulen, 1974).

Elevation of the Concentration of Catecholamines

Significant elevation of the concentration of catecholamines has been documented in refluxing testicular venous blood (Comhaire and Vermeulen, 1974; Cohen et al., 1975). This was demonstrated both in blood samples obtained during surgery and in samples aspirated during selective catheterization of these veins (Table 1). The mean concentration of noradrenaline in the refluxing blood was almost three times higher than that in peripheral blood, mimicking the make-up of renal venous blood.

We hypothesize that part of this noradrenaline is exchanged from the veins to the testicular artery at the level of the pampiniform plexus. Such counter-current exchange

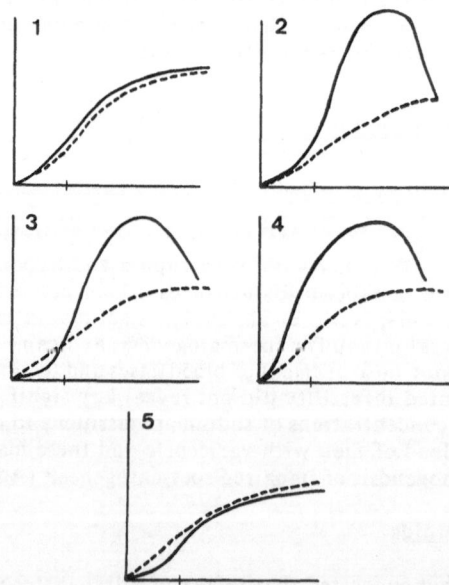

Fig. 1. Time-activity curves as registered in the regions of interest over the left and right hemiscrotum using the method of rapid sequence scintigraphy after injection of $^{99}$Tc pertechnetate (for explanation of different curves see text).

has been documented for testosterone (Amann and Ganjam, 1976; Fleet et al., 1982), and it could be accentuated through the increased hydrostatic pressure in the scrotal veins (Shafik and Bedeir, 1980).

The noradrenaline may originate directly from the adrenal gland and reach the pampiniform plexus via an adreno-gonadal by-pass, in which case the cortisol concentration will also be increased. Alternatively, it may be forced from the renal vein into the spermatic vein by the increased renal venous pressure resulting from the 'nut-cracker phenomenon' (Comhaire and Vermeulen, 1974). In the latter situation a mixture of renal and adrenal blood will reflux into the spermatic vein, and the make-up of this blood may depend on the localization of the adrenal venous outlet as related to that of the spermatic vein (Johnstone, 1957; Ahlberg et al., 1965; Clegg, 1970; Volter et al., 1975).

In our hypothesis arterial testicular blood of certain men with varicocele will contain an enhanced noradrenaline concentration which would cause constriction of the intratesticular arterioles. The latter has indeed been documented in electron-micrographs of testicular biopsies taken from men with varicocele-associated infertility (Terquem and Dadoune, 1981).

Long-lasting exposure of the testicular arterioles to the vasoactive catecholamines may cause endothelial hyperplasia with permanent impairment of testicular perfusion (Terquem and Dadoune, 1981; Hoffmann et al., 1982).

As a result of either functional or anatomical restriction of blood supply progressive deterioration of the Sertoli cells may occur, which present vacuolization and finally release of the spermatogenic cells before maturation is complete (Etriby et al., 1967; Dubin and Hotchkiss, 1969; Cameron et al., 1980; Terquem and Dadoune, 1981).

Disturbance of Testicular Arterial Blood Supply

In an effort to study the blood supply of the testes we have used the technique of rapid sequence scintigraphy after injection of the radioisotope $^{99}$Technetium pertechnetate (Comhaire et al., 1983). After rapid injection of a bolus of the radionuclide, the level of radioactivity at "first passage" of the tracer through a well defined organ is determined by the arterial bloodflow. This principle has been applied routinely to study brain and kidney perfusion. Similarly, rapid sequence scintigraphy can be used to study the time activity curve over the left as compared to the right hemiscrotum yielding interesting data concerning the arterial blood supply of the testes (Comhaire and Simons, 1982) and of the venous reflux (Mali et al., 1984). The latter was found to be related to the severity of the 'nut-cracker' constriction of the renal vein and to the pressure gradient from the lateral to the proximal segment of the renal vein. The rapidity of spermatic venous reflux was also correlated with the degree of venous pooling and the clinical stage of varicocele (Comhaire and Simons, 1982; Mali et al., 1984).

The arterial blood supply to the testes can be estimated from the radioactivity occurring respectively over the left and right hemiscrotum within the first 7 to 10 sec after the appearance of radioactivity in the abdominal aorta. In normal men, as in the majority of patients with idiopathic testicular failure, the ratio of activity of both hemiscrota is close to one (0.9 - 1.2), since the regions of interest within which the activity is measured show a strictly symmetrical localization. Five different patterns of arterial and venous blood circulation can be observed (Fig. 1) (Comhaire et al., 1983):

1. A pattern with symmetrical activity in both the arterial and venous phases was registered in 85 per cent of patients with idiopathic testicular failure, in 36 per cent of patients with subclinical or grade I varicocele, and in 17 per cent of men with larger varicoceles. Half of the men treated for varicocele presented this normal type of curve, which was usually seen in cases of rapidly recovering spermatogenesis (Fig. 1, 1);

2. Symmetrical arterial activity with significant venous pooling occurred in only 14 per cent of patients with grade I varicocele. This pattern was never found in idiopathic testicular failure, post-treatment registrations or subclinical varicoceles (Fig. 1, 2);

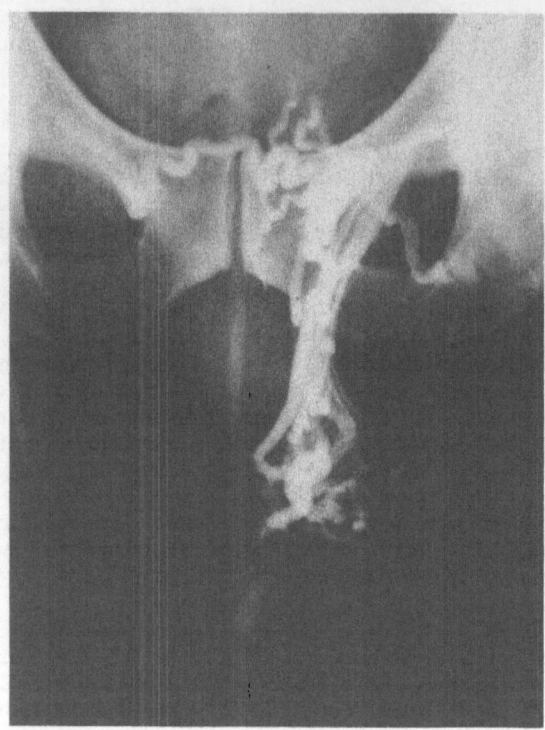

Fig. 2. Retrograde venogram of the internal spermatic vein exemplifying the presence of left-to-right shunting of spermatic venous blood in a patient with left sided grade II varicocele.

3. Decreased arterial perfusion of the testis at the side of the varicocele together with venous pooling occurred in 17 per cent of grades II and III varicoceles and was not seen in other situations (Fig. 1, 3);

4. The pattern of early reflux with increased activity already occurring during the first 10 sec of the recording, was always found together with extensive venous pooling and indicated a very rapid and intense reno-gonadal flow. This type of recording was found in 17 per cent of grades II and III varicoceles, but not in other cases (Fig. 1, 4);

5. The last and most remarkable pattern consists of a decreased arterial supply without signs of venous pooling. It was found in 15 per cent of patients with idiopathic testicular failure, 57 per cent of men with subclinical varicocele, 14 per cent of patients with grade I varicocele and 30 per cent of cases with grades II or III varicoceles. Half of the patients present this pattern immediately after treatment. It may either disappear within the first month after treatment, or persist corresponding with poor recovery of spermatogenesis (Fig. 1, 5).

It seems reasonable to accept that the noradrenaline may pass from the varicocele side to the opposite hemiscrotum causing subsequent deterioration of the contralateral testicle (Fig. 2). Hence, it may be impossible to demonstrate the difference between the left and the right side in cases where impairment of testicular perfusion is bilateral. This could explain the negative scintigraphic findings in some cases.

Disturbances in Fluid Handling

Experiments performed on monkeys in whom an artificial varicocele was induced by restriction of the renal venous outlet, have shown disturbances of fluid handling by the testicular tissue (Fussell et al., 1981; Harrison et al., 1983). The latter could result in impaired supply of nutrients to the tubular tissue and in Sertoli cell dysfunction. In addition, stagnation of blood in micro-circulatory vessels has been documented, which may also contribute to disturbed fluid handling (Chakraborty et al., 1985).

## Testicular Hormonogenesis

Some authors suggest the primary cause of in varicocele-associated impairment of testicular function to be situated in the Leydig cells (Raboch and Starka, 1971; de la Torre et al., 1978; Pirke et al., 1983; Adamopoulos et al., 1984; Ando et al., 1984). They suggest testosterone biosynthesis to be deficient with increased production of 17-hydroxyprogesterone and impaired Leydig cell response to stimulation with hCG. The question remains whether these hormonal abnormalities are the cause or rather the result of tubular dysfunction (Pirke et al., 1983; Santiemma et al., 1983). There are good reasons to believe that Leydig cell dysfunction occurs in parallel with spermatogenic impairment and may rather result from a longlasting defect in testicular perfusion (Fabbrini et al., 1981).

## Additional Factors

Theories holding the reflux of prostaglandins (Ito et al., 1982) or serotonin responsible for the testicular dysfunction need further studies and suffer from major methodological problems.

Nistal et al. (1984) suggested partial obstruction of the efferent ductules by dilated testicular veins to be involved in the pathogenesis of oligozoospermia in some patients with large varicoceles.

Finally, it has been reported that an immunological factor may play a role in varicocele-associated infertility since antisperm antibodies were more commonly present in these patients than men with infertility due to other causes (Golomb et al., 1986). The latter may be related to an increased prevalence of epididymal pathology among subfertile men with varicocele (Gerris et al., 1988).

## CONCLUSION

Several mechanisms can be held responsible for epididymo-testicular dysfunction in patients with varicocele and poor semen quality. Cases with reversible impairment of testicular arterial blood supply, and without complicating pathology such as epididymal degeneration, male accessory gland infection, formation of antisperm antibodies, or decrease of testicular volume have an up to 80 per cent probability of fertility repair after interruption of venous reflux to the pampiniform plexus. If, however, arterial perfusion defects become permanent or any of the complicating conditions have occurred, fertility prognosis after treatment is poor. Therefore, early detection and treatment of varicocele at pubertal or adolescent age is recommended (Sayfan et al., 1988).

## REFERENCES

Adamopoulos, D., Lawrence, D.M., Vassilopoulos, P., Kapolla, N., Kontogeorgos, L., and McGarrigle, H.H.G. 1984. Hormone levels in the reproductive system of normospermic men and patients with oligospermia and varicocele. J. clin. Endocrin. Metab., 59:447.

Adamopoulos, D., Kontogeorgos L., Abrahamian-Michalakis, A., Timoleon, T., and Vassilopoulos, P. 1987. Raised sodium, potassium, and urea concentrations in spermatic venous blood: an additional causative factor in the testicular dysfunction of varicocele? Fertil. Steril., 48:331.

Agger, P. 1971. Plasmacortisol in the left spermatic vein in patients with varicocele. Fertil. Steril., 22:270.

Ahlberg, N.E., Bartley, O., and Chidekel, N. 1965. Retrograde contrast filling of the left gonadal vein. Acta radiol. Diagn., 3:385.

Amann, R.P. and Ganjam, V.K. 1976. Steroid production by the bovine testis and steroid transfer across the pampiniform plexus. Biol. Reprod., 15:695.

Ando, S., Giachetto, C., Beraldi, E., Panno, M.L., Lombardi, A., Sposato, G., and Golpi, G. 1984. The influence of age on Leydig cell function in patients with varicocele. Int. J. Androl., 7:104.

Cameron, D.F., Snydle, F.E., Ross, M.H., and Drylie, D.M. 1980. Ultrastructural alterations in the adluminal testicular compartment in men with varicocele. Fertil. Steril., 33:526.

Chakraborty, J., Sinha Hikim, A.P., and Jhunjhunwala, J.S. 1985. Stagnation of blood in the microcirculatory vessels in the testes of men with varicocele. J. Androl., 6:117.

Clegg, E. 1970. The terminations of the left testicular and adrenal veins in men. Fertil. Steril., 21:36.

Cohen, M.S., Plaine, L., and Brown, J.S. 1975. The role of internal spermatic vein plasma catecholamine determination in subfertile men with varicocele. Fertil. Steril., 26:1243.

Comhaire, F. and Simons, M. 1982. Testicular arterial blood supply in varicocele patients. J. Androl., 3:13 (abstract).

Comhaire, F., Simons, M., Kunnen, M., and Vermeulen, L. 1983. Testicular arterial perfusion in varicocele: the role of rapid sequence scintigraphy with technetium in varicocele evaluation. J. Urol., 130:923.

Comhaire, F. and Vermeulen, A. 1974. Varicocele sterility: cortisol and catecholamines. Fertil. Steril., 25:88.

de la Torre, B., Noren, S., Hedman, M., and Diczfalusy, E. 1978. Studies on the relationship between sperm count and steroid levels in the spermatic and cubital veins of patients with varicocele. Int. J. Androl., 1:297.

Donohue, R.E. and Brown, J.S. 1969. Blood gases and pH determination in the internal spermatic vein of men with varicocele. Fertil. Steril., 20:365.

Dubin, L. and Hotchkiss, R.S. 1969. Testis biopsy in subfertile men with varicocele, Fertil. Steril., 20:50.

Etriby, A., Girgis, S.M., Hefnawy, H., and Ibrahim, A.A. 1967. Testicular changes in subfertile males with varicocele. Fertil. Steril., 18:666.

Fabbrini, A., Santiemma, V., Francavilla, S., Moscardelli, S., Francavilla, F., Incorvati, N., De Martino, C., and Bellocci, M. 1981. Leydig cell morphology and function in varicocele, in: "Oligozoospermia: Recent progress in andrology", Frajese, G., Hafez, E.S.E., Conti, C., and Fabbrini, A., E., eds., Raven Press, New York.

Fleet, I.R., Noordhuizen-Stasses, E.N., Setchell, B.P., and Wensing, C.J.G. 1982. The flow of blood from artery to vein in the spermatic cord of the ram with some observations on reactive hyperaemia in the testis and the effects of adenosine and noradrenaline. J. Physiol., 322:44.

Fussel, E.N., Lewis, R.N., Roberts, J.A., and Harrison, R.M. 1981. Early ultrastructural findings in experimentally produced varicocele in the monkey testis, J. Androl., 2:111.

Gerris, J., Van Nueten, J., Van Camp, C., Gentens, P., Van De Vijver, I., and Van Camp, K. 1988. Clinical aspects in the surgical treatment of varicocele in subfertile men. II. The role of the epididymal factor. Eur. J. Obstet. Gynecol. Reprod. Biol., 27:43.

Golomb, J., Vardinon, N., Homonnai, Z.T., Braf, Z., and Yust, I. 1986. Demonstration of antispermatozoal antibodies in varicocele-related infertility with an enzyme-linked immunosorbent assay (ELISA). Fertil. Steril., 45:397.

Harrison, R.M., Lewis, R.W., and Roberts, J.A. 1983. Testicular blood flow and fluid dynamics in monkeys with surgically induced varicoceles. J. Androl., 4:256.

Hoffmann, N., Hilscher, B., Passia, D., Hilscher, W., and Haider, S.G. 1982. Histological, morphometrical, and enzyme histochemical studies on varicocele orchiopathy, in: "Varicocele and male infertility. Recent advances in diagnosis and therapy", Jecht, E.W., and Zeitler, E., eds., Springer Verlag, Berlin.

Ito, H., Fuse, H., Minagawa, H., Kawamura, K., Murakami, M., and Shimazaki, J. 1982. Internal spermatic vein prostaglandins in varicocele patients. Fertil. Steril., 37:281.

Johnstone, F. 1957. The supra renal veins. Am. J. Surg., 94:615.

Koumans, J., Steeno, O., Heyns, W., and Michielsen, J.P. 1969. Dehydroepiandrosterone sulphate, androsterone sulphate and corticoids in spermatic vein blood of patients with left varicocele. Andrologia, 1:87.

MacLeod, J. 1965. Seminal cytology in the presence of varicocele. Fertil. Steril., 16:735.

Mali, W.P.T.M., Arndt, J.W., Coolsaet, B.L.R.A., Kremer, J., and Oei, H.Y. 1984. Haemodynamic aspects of left-sided varicocele and its association with so-called right-sided varicocele. Int. J. Androl., 7:297.

Netto, N.R., Lemos, G.C., and de Goes, G.M. 1977. Varicocele: relation between anoxia and hypospermatogenesis. Int. J. Fert., 22:174.

Nistal, M., Paniagua, R., Regadera, J., and Santamaria, L. 1984. Obstruction of the tubuli recti and ductuli efferentes by dilated veins in the testes of men with varicocele and its possible role of causing atrophy of the seminiferous tubules. Int. J. Androl., 7:309.

Pirke, K.M., Vogt, H.J., Sintermann, R., and Spyra, B. 1983. Testosterone in peripheral plasma, spermatic vein and testicular tissue under basal conditions and after hCG stimulation in patients with varicocele. Andrologia, 15:637.

Raboch, J. and Starka, L. 1971. Hormonal testicular activity in men with a varicocele. Fertil. Steril., 22:152.

Santiemma, V., Francavilla, S., Bruno, B., Casasanta, N.N Rosati, P., and Fabbrini, A. 1983. Leydig cell hyperplasia in varicocele: Is a local regulatory mechanism involved? in: "Recent advances in male reproduction: Molecular basis and clinical implications", D'Agata, R., Lipsett, M.B., Polosa, P., and van der Molen, H.J., eds., Raven Press, New York.

Sayfan, J., Soffer, Y., Manor, H., Witz, E., and Orda, R. 1988. Varicocele in youth. Ann. Surg., 207:223.

Shafik, A. and Bedeir, G. 1980. Venous tension pattern in cord veins: I. In normal and varicocele individuals. J. Urol., 123:383.

Terquem, A. and Dadoune, J.P. 1981. Morphological findings in varicocele: An ultrastructural study of 30 bilateral testicular biopsies. Int. J. Androl., 4:515.

Volter, D., Wurster, J., Aeikens, B., and Schubert, G.E. 1975. Untersuchungen zur Struktur und Function der Vena spermatica interna. Ein beitrag zur Atiologie der Varicozele. Andrologia, 7:127.

# VARICOCELECTOMY: EFFECT ON FERTILITY

David P. Gentile and Abraham T.K. Cockett

The University of Rochester
Department of Urology
Rochester, New York

The varicocele is the single most commonly identified, correctable form of male infertility[1]. It occurs in approximately 15% of the normal male population and in 40% of those presenting for the evaluation of infertility[1.] A marked left sided predominance exists. This is most probably secondary to a combination of factors. First, the spermatic vein on the left side is 8-10 cm longer than the right, predisposing it to greater hydrostatic pressures, particularly in the erect position. Second, the entry of the spermatic vein into the renal vein on the left compared to its entry into the inferior vena cava on the right leads to relatively elevated pressures on the left for two reasons. The pressure gradient at the spermaticorenal junction is lower than at the spermaticocaval junction, in so far as the pressure in the renal vein is higher than in the vena cava[2]. In addition, the Bernoulli principle of fluid dynamics dictates that in any horizontally moving fluid, pressure increases as the velocity of flow decreases. The lower flow rate through the renal vein as compared to the vena cava produces less "forward draw" on blood traveling in the left spermatic venous system. The end result is elevated pressures in the left system. Therefore, man due both to anatomic considerations as well as evolutionary, in that he has assumed an erect posture, is predisposed to the formation of varicoceles.

The etiology of male infertility associated with the varicocele remains controversial, although spermatogenic dysfunction with poor sperm motility seems to be the final common denominator. Numerous theories still predominate, including increased intrascrotal temperature, increased venous pressure with hypoxia, and retrograde flow of substances toxic to the testes such as catecholamines, serotonin, and prostaglandins, testifying to the fact that the pathophysiology remains incompletely understood. The role of serotonin has been emphasized by our group. Increased urinary levels of 5-HIAA have been associated with abnormal testicular biopsies and oligospermia[3-5]. Complicating matters is the fact that the majority of patients with a varicocele will have normal fertility. In addition, those with impaired fertility and a varicocele, will not necessarily have impaired fertility on the basis of the varicocele alone.

Predicting who will respond to varicocele surgery unfortunately remains an inexact science. This is in part due to the absence of objective criteria in selecting patients for varicocele surgery. Fertility rates following varicocelectomy vary widely in the literature. Pregnancy occurs in 20% to 60% of couples. Intervention in the male patient with a varicocele is indicated when it is determined to the greatest extent possible that the couple's inability to conceive is based on the husband's impaired spermatogenesis.

Of the above listed proposed mechanisms for varicocele-induced spermatogenic dysfunction, elevated intrascrotal temperature is most compelling. It is well documented

*Temperature and Environmental Effects on the Testis*
Edited by A. W. Zorgniotti, Plenum Press, New York, 1991

**289**

## Table 1

| Surgery Performed | Nª Patients | F/U Available | F/U > 1 Year |
|---|---|---|---|
| Bilateral | 290 (67.6%) | 229 | 203 |
| Left | 133 (31.0%) | 78 | 75 |
| Right | 5 ( 1.2%) | 3 | 2 |
| Unknown | 1 ( 0.2%) | 1 | 1 |
| Totals | 429 | 311 | 281 |

that for normal testicular function, testis temperature should be 1.5-2°C below normal core body temperature. It has also been documented that, although at times associated with normothermia, the varicocele often leads to elevated intrascrotal temperatures[6]. In canine experiments by Cockett et al., an artificially induced left varicocele was created by partially ligating the left renal vein proximal to the site where the left spermatic vein emptied into the renal vein[7]. After a sufficient recovery period, semen samples and scrotal temperatures were obtained and compared to preoperative base line controls. An increase in the number of immature and tapered spermatozoa was seen on semen analysis. In addition, bilateral elevation of testicular temperature was noted. Kay et al. in similar experiments induced left varicoceles in rhesus monkeys[8]. They reported a significant decrease in sperm concentration as well as elevated intratesticular temperature bilaterally. Turner, reporting his results in rats and dogs after artificially creating a left varicocele, noted an increase in testicular blood flow bilaterally with a resultant elevation in testicular temperature[9].

Over a fifteen-year period at our institution, from August 1975 to the present, we have seen 5,000 men in consultation for the evaluation of infertility. Of these men, 2,000 (40%) had varicoceles upon entry. This is consistent with reports of others that 35-40% of males presenting with infertility will on examination be found to have varicoceles. Of those 2,000 patients, approximately 20% (429) had semen analysis showing abnormal spermatogenesis: poor sperm motility and an increased percentage of tapered and immature forms. This finding, too, supports the fact that most patients with varicoceles will have unimpaired fertility, here demonstrated by the fact that 80% of our varicocele patients had normal spermatogenesis. The varicocele was found to be bilateral in 67.6%,

## Table 2

| Parameter Tested | Nª Showing Improvement | % With Follow-Up |
|---|---|---|
| Motility | 51 | 18.1 |
| Tapered Forms | 48 | 17.1 |
| Immature Forms | 41 | 14.6 |
| Sperm Count | 62 | 22.1 |
| Live Forms | 53 | 18.9 |
| Motility + Tapered Forms | 15 | 5.3 |
| Motility + Immature Forms | 22 | 7.8 |
| Tapered + Immature Forms | 12 | 4.3 |
| Motility, Tapered forms + Immature Forms | 6 | 2.1 |
| Any of the Above Parameters | 145 | 51.6 |

Table 3

| Surgery Performed | F/U > 1 Year | Pregnancy Achieved | Pregnancy Rate (%) |
|---|---|---|---|
| Bilateral | 203 | 107 | 52.7 |
| Left | 75 | 53 | 70.7 |
| Right | 2 | 0 | 0.0 |
| Unknown | 1 | 1 | 100.0 |
| Totals | 281 | 161 | 57.3 |

unilateral on the left in 31%, and unilateral on the right in 1.2%. High ligation of one or both internal spermatic veins was carried out using a high retroperitoneal approach (Table 1).

We have come to believe that due to venous communications through crossover vessels at the prepubic and scrotal levels, bilateral ligation is necessary. We perform high ligation because at this level, usually one and, at most, three veins are found on each side. We employ an entirely retroperitoneal approach. Our patients typically are discharged one to two days following surgery.

Postoperative semen analyses on our patients are obtained after a sufficient recovery period. Many parameters are evaluated, including sperm motility, the % tapered forms, % immature forms, total sperm count, and % live forms. Of the 281 patients available for follow-up, 145 (51.6%) showed improvement in one or more of the above parameters. Improvement was judged by at least a 10% reduction in tapered or immature forms and a 10% improvement in motility score, total sperm count, and percent live forms. Table 2 shows the number and percentages of patients with improvement in one or more semen parameters.

A pregnancy rate of 57.3% was achieved. Patients were deemed evaluable if, at the time of this report, surgery had been performed at least one year previously (Table 3).

With these encouraging pregnancy results, we attempted to analyze the time course postoperatively within which our patients achieved pregnancy. We found that in the first year post-varicocelectomy, 52.2% of those who would achieve pregnancy had done so (Table 4). In addition, the likelihood of achieving pregnancy drops quite precipitously as the time interval between varicocele surgery increases. Using these results, clinicians can perhaps better counsel their patients about the probability for achieving a pregnancy post-varicocelectomy.

One final aspect to our work has been evaluating the effect of varicocelectomy on testicular volume. Since the majority of the testis is composed of tissue solely devoted to the production of spermatozoa, it seemed logical that if sperm number and quality increased postoperatively, then perhaps testicular volume did as well. Postoperative increases in testicular volume following varicocelectomy have certainly been well documented in the pediatric age group.

Of 89 patients in whom we had obtained pre- and postoperative testicular volumes, there was an increase in volume on both sides, regardless of whether unilateral or bilateral varicocelectomy was performed. Of note, right testicular volume increased more than left, an average of 4.76 ccs on the right compared to an average of 3.63 ccs on the left. When broken down by whether pregnancy had or had not been achieved, no correlation with postoperative testicular volume changes could be appreciated. In both sets of patients in whom testicular volumes increased postoperatively, again right sided enlargement was more prevalent than left sided. It appears clear that although the

## Table 4

| Year(s) Post Op | Nº Pregnant | 6 Month Increment | % of Total |
|---|---|---|---|
| 1.0 | 84 | - | 52.2 |
| 1.5 | 105 | 21 | 65.2 |
| 2.0 | 118 | 13 | 73.3 |
| 2.5 | 128 | 10 | 79.5 |
| 3.0 | 134 | 6 | 83.2 |
| 3.5 | 138 | 4 | 85.7 |
| 4.0 | 141 | 3 | 87.6 |
| 4.5 | 144 | 3 | 89.4 |
| 5.0 | 146 | 2 | 90.7 |
| 10.0 | 150 | (4) | 93.2 |
| >10.0 | 161 | (11) | 100.0 |

symptomatic varicocele results in reduced testicular volumes, increased testicular volume postoperatively does not predict a successful pregnancy.

In conclusion, the varicocele results in decreased fertility through impairment of spermatogenesis in approximately 20% of patients with varicoceles. Although the exact mechanism of impaired spermatogenesis has yet to be conclusively established, elevated intrascrotal temperature is certainly a key factor. In couples who have impaired fertility with no other identifiable cause, varicocelectomy at our institution has resulted in a pregnancy rate of 57.3%. If pregnancy is going to occur post-varicocelectomy, it will likely occur within three years. Testicular volume can be expected to increase post-varicocelectomy, but this is not predictive of improved fertility.

## ACKNOWLEDGEMENT

The authors would like to acknowledge Sue Yells for her assistance in the preparation of this manuscript.

## REFERENCES

1. Cockett, A.T.K., Urry, R.L. and Dougherty, K.A. The varicoceles and semen characteristics, J. Urol., 121:435 (1978).

2. Shafik, A., Moftakh, A., Olfat, S., Mohi-El-Din, M. and El Sayed, A. Testicular veins: Anatomy and role in varicocelogenesis and other pathological conditions. Urol., 35(2):175 (1990).

3. Segal, S., Sadovske, E., Palti, Z., Pfeifer, Y. and Polishuk, W.Z. Serotonin and 5-Hydroxyindole acetic acid in fertile and subfertile men. Fertil. Steril., 26: 314 (1975).

4. Urry, R.L., Dougherty, K.A. and Cockett, A.T.K. Correlation between follicle stimulating hormone, luteinizing hormone, testosterone, and 5-Hydroxyindole acetic acid with sperm cell concentration. J. Urol., 116:322 (1976).

5. Urry, R.L. and Cockett, A.T.K. Elevated urinary levels of 5-Hydroxyindole acetic acid and its relationship among levels of plasma follicle stimulating hormone, testosterone, and testicular pathology in patients with severe oligospermia and/or azoospermia. J. Urol., 118:591 (1977).

6.  Zorgniotti, A.W. and MacLeod, J. Studies in temperature, human semen quality, and varicocele. Fertil. Steril., 24:854 (1973).

7.  Cockett, A.T.K., Al-Juburi, A., Altebarmakian, V., Vergamini, R.F. and Caldamone, A.A. The varicocele: New experimental and clinical data. Urol., 15(5):492 (1980).

8.  Kay, R., Alexander, N.J. and Baugham, W.L. Induced varicoceles in Rhesus monkeys, Fertil. Steril., 31:195 (1979).

9.  Turner, T.T. Varicocele: Still an enigma. J. Urol., 129:695 (1983).

10. Nagler, H.M. Varicocele: Adult. In: "Common Problems in Infertility and Impotence" Chapt. 7, pp. 51-59. A.J. Wein and J. Rajfer (eds), Year Book Medical Publishers, Chicago (1990).

APPENDIX A:

VARICOCELE IN THE ADOLESCENT

Omer P. Steeno

Department of Internal Medicine
University Hospital Gasthuisberg and
Youth Health Care Center
University of Leuven, Belgium

DEFINITION

The word varicocele is derived from the Latin "varix" which means dilated vein and the Greek "kele" meaning a tumor or swelling (Horner, 1960; El-Gohary, 1984). In the first century, Paulus Cornelius Celsus (cited by Saypol, 1981 and by Allemann and Jenny, 1980) defined the condition of varicocele as "swollen veins, twisted over the testicle, which becomes smaller than its fellow, in as much as its nutrition has become defective". In 1550, Ambroise Pare (cited by Ivannissevich, 1960) considered a varicocele as "a compact pack of vessels quite filled with melancholic blood". Bennett (1889) defined it as "a condition of varicosity of the veins of the spermatic cord, of congenital origin, resulting in or associated with a deficient development or functional imperfection of the corresponding testis in the majority of cases".

Ivannissevich (1918), who performed the most operations for varicocele, gave as definition: "Varicocele is an anatomoclinical syndrome. Anatomically it is characterized by varices inside the scrotum; clinically, by venous reflux, i.e., valvular insufficiency of the spermatic cord" (cited by Ivanissevich, 1960). Later definitions are in agreement with the previously given (Tulloch, 1952; Ribeiro, 1956; Saypol, 1981; Schickedanz and Kleinteich, 1982, 1987; Sawczuk et al., 1985).

HISTORY

Although varicocele was already known as a clinical symptom for many centuries, it took some time to challenge the traditional attitude that varicocele is an innocuous condition. Marriage and the passage of years were considered to usually bring about a diminution in size of the varicocele (Skinner, 1941; Turner, 1943; Avid, 1949: cited by Russell, 1954), so that the incidence would likely decrease with age. Still, in a 1956 textbook of surgery edited by Turner and Rogers, it was pointed out that in the majority of patients with varicocele, "the condition is purely psychological and will either disappear in marriage or continue without producing symptoms (Harrison, 1966; Verstoppen and Steeno, 1977a).

Idiopathic varicocele in a boy under 12 yr of age may result, according to Campbell (1944), from chronic excessive masturbation. That varicocele as a congenital condition was rarely observed prior to puberty, was already known to Bennett (1889). Horner (1960) pointed out that varicocele starts at puberty following the increased vascularity of the testicles at that time.

It was difficult for most authors in the past to explain why a lesion which is generally one-sided should affect spermatogenesis in the other testis. Russell (1954) was the first to put forward as possible explanation a slight but prolonged alteration in temperature within the scrotum, affecting both testes. Although Tulloch (1952) is cited by most authors as the first author to have proven a relationship between varicocele and subfertility, the suggestion of such a relationship was already made by Curling (1856, cited by Saypol, 1981 and by Gattuccio et al., 1988) ("a decrease in the secreting powers of the gland"); by Bennett (1889), who mentioned that the production of spermatozoa may be arrested ("sperm quality became better after the surgical correction of a varicocele"); by Macomber and Sanders (1929, cited by Saypol, 1981), who described a man with bilateral varicocele and oligozoospermia who became normozoospermic following varicocele repair and whose wife ultimately conceived; by Wilhelm (1937, cited by Saypol, 1981), who stated that the importance of varicocele as cause of male infertility probably is underestimated although he believed that varicocele as etiology of partial or absolute sterility would only be accepted in cases of bilateral varicosity in the scrotal sac (Rost et al., 1975); by Hamman (1944, cited by Saypol, 1981) who identified varicoceles in 11.9% of 291 patients attending his infertility clinic and recorded sperm counts of less than 60 million in 92% of the patients with varicocele, and by Hotchkiss (1944) who often found large varicoceles to be associated with a soft testis and presumed that extensive damage to the spermatic tissue would result. According to Hotchkiss, testicular biopsies would ultimately prove or disprove this contention.

Influence of the condition of varicocele on testicular size and consistency in teenagers was already noted by Barwell (1885), who found that following ligation of the varicocele "the testicle became harder and regained some of its dimensions". A few years later, Bennett (1889) wrote that after surgical care, the testis "assumed its natural size and consistency" and "the secretion of seminal fluid improved in character".

The higher incidence of varicocele in patients attending a subfertility clinic than in husbands whose wives were attending an antenatal clinic was shown by Russell (1954). Dubin and Amelar (1971) found in their private urological praxis 39% of their patients presenting clinical symptoms of varicocele.

In the general population the prevalence of varicocele was checked in military men (Lewis, 1950; Appleby, 1955; Clarke, 1966; Johnson et al., 1970; Bergman and Wahlqvist, 1974).

In school populations prevalence studies were done by Horner (1960) in the United Kingdom, by Oster (1970, 1971) in Denmark and by Steeno et al. (1975, 1976) in Belgium. Our group was the first to study the influence of varicocele on the testicles (volume and/or consistency) according to the degree of varicocele in a large population (N: 4,067) of school boys and college students.

Operations for varicocele in children and adolescents were performed already for a long time in Russia (Arbuliev, 1967; Isakov and Arbuliev, 1969).

Varicocele repair as a prophylactic measure against testicular damage was first proposed by Hotchkiss (1944). At that time he considered the surgical correction of a varicocele probably effective only as a prophylactic measure against damage which might ensue with passing years. According to him it would be unlikely that the operation has little other than cosmetic value in long-standing cases of a small, soft testis. Johnson et al. (1970) advised operation in unmarried men only in cases of sperm disturbances. Our group started with operative treatment of varicocele in adolescents for purely preventive reasons around 1970 (Steeno, 1972). Operation was considered to be indicated in cases of varicocele associated with testicular volume loss or diminishment in consistency, either unilateral or bilateral. Initially many reactions against this preventive attitude had to be faced in Belgium as well as in the Netherlands. Our preventive attitude was supported in the ensuing years, initially in Germany (Günther et al., 1974; Rost et al., 1975).

In situations where operation was seen as the only possible form of treatment, post-operative hydrocele was the only complication which caused concern. Therefore,

Bergman and Wahlqvist (1974) could consider early operation to prevent subfertility only if post-operative hydrocele could be avoided.

## ETIOLOGY

The abnormal dilatation of the veins of the pampiniform plexus may be due to several factors:

-the erect posture of man (Buch and Cromie, 1985; Steeno, 1972; Vorontsov et al., 1979) which makes the outflow of venous blood through the right-angle entry of the left spermatic vein into the high-pressure venous system of the left renal vein difficult (Appleby, 1955; Arbuliev, 1968; Buch and Cromie, 1985; Erokhin et al., 1980; Ivanissevich, 1960; Sawczuk et al., 1985; Schickedanz and Kleinteich, 1982; Verstoppen and Steeno, 1977a);
-the long hydrostatic column, especially in tall boys and particularly during the growth spurt at the time of puberty (Allemann and Jenny, 1980; Buch and Cromie, 1985; El-Gohary, 1986a,b; Erokhin et al., 1980; Ivanissevich, 1960; Lindner, 1982; Lyon et al., 1982; Oster, 1970, 1971; Pozza et al.; 1983; Sawczuk et al., 1985; Schickedanz and Kleinteich, 1982; Verstoppen and Steeno, 1977a; Vorontsov et al., 1979);
-deficient number, absent or inefficient anti-reflux valves, resulting in blood stasis (Allemann and Jenny, 1980; Appleby, 1955; Audry et al., 1986, 1987; Diamond and Ravitz, 1975; El-Gohary, 1986a,b; Isakov et al., 1977; Ivanissevich, 1960; Kass, 1988; Palomo, 1949; Pozza et al., 1983; Sawczuk et al., 1985; Schickedanz and Kleinteich, 1982; Verstoppen and Steeno, 1977a);
-retrograde flow of blood from the left renal vein into the left spermatic vein and to the pampiniform plexus (Arbuliev, 1968; Audry et al., 1986, 1987; Dubin and Amelar, 1971; Isakov and Arbuliev, 1969; Verstoppen and Steeno, 1977a);
-compression of the left renal vein between the aorta posteriorly and the superior mesenteric artery anteriorly (nutcracker phenomenon) (Buch and Cromie, 1985; El-Gohary, 1986a,b; Ivanissevich, 1960; Kass, 1988; Schickedanz and Kleinteich, 1982; Verstoppen and Steeno, 1977a);
-venous dilatation caused by the loss of the muscular pump mechanism due to the lack of supporting structures (El-Gohary, 1986a,b; Schickedanz and Kleinteich, 1982), cremasteric muscle atrophy (Buch and Cromie, 1985), hereditary weakness of the veins (Erokhin et al., 1980; Schickedanz and Kleinteich, 1982), and degenerative changes in the pampiniform plexus (Sawczuk et al., 1985);
-venous renal hypertension caused by reduction of the blood outflow from the renal into the vena cava inferior (Isakov et al., 1977);
-anomalies in the number of veins above the iliac crest, anomalies in the caliber (ectasia), presence of anastomotic branches, irregular junction with the left renal vein (Pozza et al., 1983);
-blood stream turbulences since the left suprarenal vein enters the left renal vein juxta-opposite to the left internal spermatic vein (Schickedanz and Kleinteich, 1982) causing in 10% of the cases with varicocele retrograde blood flow with high concentrations of adrenal corticosteroids (Steeno, 1972);
-increased vascularity at the time of puberty and testicular development (Allemann and Jenny, 1980; Horner, 1960; Isakov et al., 1977; Lindner, 1982; Lyon et al., 1982).
-pressure of the ileal segment (filled with feces) on the trunk of the spermatic vein (Ivanissevich, 1960);
-influence of sports (Ristic et al., 1986);
-secondary causes: retroperitoneal tumors, renal vein thrombosis, hydronephrosis (Diamond and Ravitz, 1975).

## PREVALENCE

### In Infertility Clinics

In infertility clinics the number of cases of varicocele recognized is related to the discipline and the interest of the clinician and is also determined by the number of patients referred by family physicians and specialists in other disciplines. Hadziselimovic (1983) gives a number of 50%. Dubin and Amelar (1971) in 1,294 consecutive cases seen in their private urological praxis in New York, found in 39% of cases varicocele as cause of male subfertility. Risser and Lipshultz (1984) found 37% in their literature search on 2,121 subfertile men. Schickedanz and Kleinteich (1987) noticed an increased incidence according to the year of publication (data derived from 31 publications).

| Period | Number of patients | Varicocele (%) |
|--------|--------------------|----------------|
| 1950 - 1959 | 3,057 | 9.06 |
| 1960 - 1969 | 23,463 | 8.73 |
| 1970 - 1979 | 8,460 | 23.58 |
| 1980 - 1984 | 6,713 | 32.59 |

Steeno et al. (1975, 1976) also found an increase over the years.

| Period | Varicocele (range in %) |
|--------|-------------------------|
| 1955 - 1965 | 4.6 - 8.1 |
| 1965 - 1970 | 9.4 - 22.6 |
| 1970 - 1975 | 14.7 - 39.0 |

## In Recruits (Military Men)

In Table 1 the prevalence of varicocele in 7 populations of recruits is given. The prevalence ranges from 4.1% to 30.7%.

## Prevalence in the General Population

Lipshultz and Corriere (1977) examined 83 healthy volunteers and found a varicocele in 13 of them (15.85%). Thomason and Fariss (1979), during a routine physical

Table 1. Prevalence of Varicocele in Young Men of Army Age

| Authors | Country | Number of Recruits | Varicocele (%) |
|---------|---------|--------------------|----------------|
| Lewis (1950) | U.S.A. | 1,500 | 16.5 |
| Appleby (1955) | U.S.A. | 4,000 | 10.0 |
| Clarke (1966) | U.S.A. | 189 | 8.0 |
| Johnson (1970) | U.S.A. | 1,592 | 9.5 |
| Bergman (1974) | Sweden | 122 | 4.1 |
| Wutz (1977) | Germany | 3,490 | 5.2 |
| Thomason (1979) | U.S.A. | 909 | 30.7 |
| Alcalay (1986) | Israel | 1,479 | 9.1 |
| All Populations | | 13,281 | 10.6 |

Table 2. Prevalence of Varicocele in School Boys and Adolescents

| Authors | Country | Number of Boys | Vari-cocele (%) | Moderate to severe grades (%) |
|---|---|---|---|---|
| Horner (1960) | U.K. | 1,211 | 15.9 | 10.7 |
| Oster (1971) | Denmark | 837 | 16.2 | 7.8 |
| Steeno (1976) | Belgium | 4,067 | 14.7 | 5.3 |
| Garnier (1978) | France | 494 | 8.3 | |
| Erokhin (1979) | U.S.S.R. | 10,000 | 12.4 | 8.4 |
| Berger (1980) | U.S.A. | 586 | 9.0 | |
| D'Ottavio (1982) | Italy | 5,177 | 25.8 | 12.9 |
| Kleinteich (1983) | E.Germany | 1,262 | 18.0 | |
| Risser (1984) | U.S.A. | 423 | 13.7 | |
| Ristic (1986) | Yugoslavia | 776 | 20.6 | |
| El-Gohary (1986) | U.A.E. | 1,546 | 13.4 | 6.9 |
| Gattuccio (1988) | Sicily | 1,449 | 16.0 | |
| All Populations | | 27,828 | 16.1 | 8.8 |

examination of army men, aged 17 to 42 yr (mean age 22 yr), noticed 30.7%. Kleinteich and Schickedanz (1983a) in a literature search on 3,009,495 adult men obtained a percentage of 16.97%.

Prevalence in School Boys and Adolescents

In general, the prevalence of varicocele in adolescent boys is already equivalent to that in the general male population (Buch and Cromie, 1985). Literature search revealed an overall prevalence of 12.4 to 16.2% (5.3 to 10.7% for moderate to marked forms) (El-Gohary, 1984) and 13.4% for 9,000 young men (Risser and Lipshultz, 1984). In Table 2 the percentages are given for the school populations studied.

D'Ottavio et al. (1981, 1983, 1987) found no significant differences between the school populations of different regions in Italy (range 18.9 to 21.3%). Before puberty almost no varicocele could be detected (Table 3).

Ethnic Distribution

In the past Goulart (1935, cited by Alcalay and Sayfan, 1985) reported less incidence of varicocele in black men because of the postulated existence of more competent valves in the left spermatic vein of black people. Alcalay and Sayfan (1985) did not find any

Table 3. Prevalence of Varicocele before Puberty

| Authors | Country | Number of Boys | Age Range | Varicocele (%) |
|---|---|---|---|---|
| Horner (1960) | U.K. | 132 | 3-11 | 0.0 |
| Oster (1971) | Denmark | 188 | 6-9 | 0.0 |
| Garnier (1978) | France | ? | <10 | 0.0 |
| Ristic (1986) | Yugoslavia | 160 | 7-9 | 3.1 |

ethnic difference (13.7% varicocele in black adolescents). Risser and Lipshultz (1984) noticed 13.7% varicocele in 423 black adolescent males (against 14 - 15% in white adolescents).

Alcalay et al. (1986) in Israel recorded the country of origin in the populations studied and could not find any significant difference. El-Gohary (1986b) did the same in the United Arabian Emirates and also found no significant differences between the males of different origin.

SYMPTOMATOLOGY

Varicocele in adolescents is mostly asymptomatic and the diagnosis is made chiefly during a routine clinical examination at school or at the army recruitment. According to Allemann and Jenny (1980), the diagnosis is made by chance in 90% of the cases. Most patients are not aware of the presence of a scrotal mass (only 8.6% in the epidemiological study of Risser and Lipshultz, 1984; 10% in the series of Berger, 1980).

Varicocele is seldom painful (Buch and Cromie, 1985). In the groups examined in a juvenile court detention center, Berger (1980) noted that 10 out of 59 adolescents with varicocele reported pain or a sense of fullness. Mothes (1986) recorded only two cases with pain (out of 197 adolescents with varicocele), Wutz (1977) only in 5.19%.1 Scrotal pain was present in 19% of the adolescents referred for operation to Audry et al. (1986), in only one out of 23 patients operated by El-Gohary (1984), and in 6 out of 22 boys in the series of Hienz et al. (1980).

Problems (scrotal discomfort) may arise during increased effort or during basic training (Alcalay and Sayfan, 1984; Appleby, 1955; Risser and Lipshultz, 1984).

The diagnosis may also be made at the time of a posttraumatic scrotal hematoma (13% in the series of El-Gohary, 1984) or during examination for other groin pathology, such as undescended testes or inguinal hernia (17% in the series of El-Gohary, 1984).

Testicular growth failure as a symptom has become more evident in the last years because at school examination more attention is paid to this condition (29% in the series of Reitelman et al., 1987; Ivanissevich, 1960).

Increased information to the parents about possible fertility problems in the future also increases referral for operation because of concern for fertility (Levitt et al., 1987; Steeno, 1972, 1975, 1976).

Symptoms are generally only present in cases with varicocele degree II (medium) or III (large varicoceles) (Appleby, 1955; Kleinteich and Schickedanz, 1983b).

Two studies on the presenting symptoms and signs in adolescent patients, referred for operation, are especially valuable (Levitt et al., 1987; Okuyama et al., 1988) (Table 4).

CLINICAL AND TECHNICAL EXAMINATIONS

Clinical Examination

The clinical examination must be performed in the upright position (Allemann and Jenny, 1980; Ivanissevich, 1960; Oster, 1971; Steeno, 1972). Venous reflux can be evaluated by means of a Valsalva maneuver (Steeno, 1972). The varicocele will be less pronounced and even disappear in the lying or Trendelenburg position (Allemann and Jenny, 1980; Ivanissevich, 1960).

Reduced testicular volume or arrested testicular growth, which may be associated with varicocele, can be evaluated by means of orchidometry (Buch and Cromie, 1985; Kass and Belman, 1987; Lipshultz and Corriere, 1977). Orchidometry facilitates the early detection of testicular growth failure. Normally there should be no more than 2 cc volume difference between testes (Kass, 1988). There exists a clear relationship between

Table 4.  Presenting Symptoms and Signs in Adolescents with Varicocele

| Authors | Levitt et al. | Okuyama et al. |
|---|---|---|
| Number of patients | 26 | 40 |
| Symptoms | | |
| Progressive increase in varicocele size | 11 | 22 |
| Pain or discomfort | 2 | 17 |
| Pain and progressive increase in varicocele size | 4 | |
| Testicular asymmetry | 3 | 12 |
| Testicular asymmetry and increase in varicocele size | 1 | |
| Torsion of the appendix testis | 1 | |
| Inguinal swelling | 1 | |
| Concern for fertility | 3 | |

the duration of the varicocele and the degree of testicular "atrophy" of the left (or both) testicle(s). Audry et al. (1986) found a smaller testicle at the left side in 71% of the cases operated in their series; Kass and Belman (1987) in 70%; Lyon et al. (1982) in 77%; Wyllie (1985) in 90%. In an epidemiological study, we found an influence on the left or both testicles in 34.4% of cases with grade II and in 81.2% in cases with grade III (Steeno et al., 1976). This is in contrast with the statement of Lyon et al. (1982) who postulated that there exists no consistent relationship between the varicocele grade and the degree of testicular growth arrest. The loss of testicular mass is most striking in the eight to thirteen year group, suggesting that it is in the years of rapid growth that varicocele has its most damaging effects (Lyon et al., 1982).

Catch-up growth after treatment can also be objectively evaluated by orchidometry (Kass and Belman, 1987).

Not only changes in testicular volume but also testicular consistency are important as signs of an already existing testicular damage, but the clinical appraisal of softer testicles may be difficult (D'Ottavio et al., 1982; El-Gohary, 1986b).

Total situs inversus should be considered whenever right-sided varicocele is solitary or predominant (Wilms et al., 1988).

The development of varicocele is related to the physical changes which occur during puberty (Lyon et al., 1982). Boys with varicocele tend to be generally taller than others in the control group but the differences in most studies are not significant (El-Gohary, 1986b; Horner, 1960; Oster, 1970, 1971). The mean weight of boys with varicocele was noted to be significantly lower than that of their colleagues (El-Gohary, 1986b) but body weight in the age-group of 14-15 yr was greater (but not significant) in the epidemiological study of Oster (1971) and Horner (1960). Pubertal stage evaluation revealed that puberty is more advanced in boys with varicocele (D'Ottavio et al., 1982; El-Gohary, 1986b). The differences are apparent in 13 year-olds and most prominently in 14 year-olds, but the difference however is not significant (Oster, 1970, 1971).

The clinical symptoms are important but do not allow for a prediction on the dangers to fertility in the future (Mothes, 1986). The duration of the varicocele is more important for testicular damage and risk for infertility (Mothes, 1988). The appearance of varicocele at the start of puberty, however, increases the chance of testicular growth arrest (Lyon et al., 1982).

Mothes (1988) noted that in 4% of the cases a wrong diagnosis was made at school examination. In these cases the condition of varicocele was misdiagnosed as hernia, torsion, tumor or hydrocele.

## Hormonal Data

The importance of hormonal plasma data (testosterone, FSH, LH, LH-RH test) is very relative at pubertal age, that is in adolescent boys (Allemann and Jenny, 1980). It is difficult to evaluate hormonal data according to the ages of the boys and their pubertal development. In fact, no early detection of damage of the germinal epithelium is possible (Allemann and Jenny, 1980).

Kass et al. (1988) and Okuyama et al. (1981) found no significant differences in basal values and response of serum LH to LH-RH. But with the progress of pubertal stage, both the basal value and response of serum FSH to LH-RH became higher in comparison with normal boys. The exaggerated response pattern was especially abnormal in the groups of varicocele with testicular volume loss. They found that early testicular failure is presented in ca. 50% of adolescents with a moderate to large size varicocele.

## Radiographical Explorations

A urography may exclude the existence of a secondary varicocele, especially in prepubertal boys (Allemann and Jenny, 1980; Audry et al., 1986; Campbell, 1944; Diamond and Ravitz, 1975; Hecker, 1973; Kraeft et al., 1981). At prepubertal age a varicocele may be the first symptom of a kidney tumor, a congenital renal anomaly or a disturbance of the venous flow (Hecker, 1973). The presence of a solitary right varicocele may be a clinical sign of a retroperitoneal tumor (Gunther et al., 1974; Berger, 1980) or the presence of situs inversus (Berger, 1980; Wilms et al., 1988). A urography can be replaced by an echography of the abdomen.

A preoperative or intraoperative venography is advised by several authors (Allemann and Jenny, 1980; Arbuliev, 1968; Erokhin et al., 1980; Gorenstein et al., 1986;

Table 5. Localization of Varicocele in Operated Cases of Boys and Adolescents

| Authors | Country | Localization of Varicocele | | |
| | | Left | Right | Bilateral |
| --- | --- | --- | --- | --- |
| Allemann (1980) | Switzerland | 27 | 1 | 0 |
| Audry (1986) | France | 31 | 0 | 0 |
| Boeckx (1985) | Belgium | 51 | 0 | 2 |
| El-Gohary (1984) | U.K. | 22 | 0 | 1 |
| Hienz (1980) | W.Germany | 20 | 1 | 1 |
| Janneck (1979) | W.Germany | 43 | 0 | 0 |
| Kawamura (1987) | Japan | 19 | 0 | 1 |
| Mothes (1988) | E.Germany | 223 | 0 | 25 |
| Okuyama (1988) | Japan | 33 | 2 | 5 |
| Palomo (1949) | Guatemala | 36 | 0 | 2 |
| Reitelman (1987) | U.S.A. | 40 | 0 | 0 |
| Schürholz (1978) | W.Germany | 21 | 1 | 1 |
| Wyllie (1985) | Australia | 10 | 0 | 0 |
| Total | | 576 | 5 | 38 |
| Percentage | | 93 | 0.8 | 6.2 |

Isakov and Arbuliev, 1969; Pozza et al., 1983; Schickedanz and Kleinteich, 1985; Vorontsov et al., 1985). Pozza et al. (1983) found in 68.6% one or more anomalies of the left spermatic vein (number of veins above the iliac crest, anomalies in caliber, presence of anastomotic branches, irregular junction with the left renal vein). Schickedanz and Kleinteich (1985) advise an antegrade transscrotal phlebography of the internal spermatic vein preoperatively to avoid the risk of recurrence or persistence of the varicocele postoperatively (they consider a retrograde phlebography to be difficult in boys). Mothes (1986) only advocates a phlebography in cases of recurrence or persistence after operation. Levitt et al. (1987) perform a routine post-ligation venography for the detection of residual communicating veins between the pampiniform plexus and the internal spermatic veins (present in 39% of their series). With this strategy they have been able to reduce the risk of recurrence to 3.6%.

With the introduction of sclerotherapy, phlebography is now routinely performed in all centers, where this kind of therapy is available.

## Thermography and Scintigraphy

Preoperative and postoperative thermography allow for the evaluation of temperature influence of the varicocele on the testicles (Allemann and Jenny, 1980; Buch and Cromie, 1985; Ponchietti et al., 1986).

Radionuclide varicocele scans may also be helpful, especially in doubtful cases and in patients with clinical varicocele but without temperature increase at thermography (Buch and Cromie, 1985).

## Spermiogram

Spermiograms are not possible in pubertal boys and adolescents (Allemann and Jenny, 1980; Buch and Cromie, 1985). Furthermore, no normal data are known for these age groups. The variation in pubertal development makes it also impossible to interpret data of spermiograms before adult age.

Spermiograms were performed on 97 army recruits with varicocele (age range 17 - 24 yr; mean age 19.2 yr) by Johnson et al. (1970). In 27% the sperm count was less than 20 million sperm cells per ml without relation to the size of varicocele. Sperm motility was reduced in 56%, also not related to the size of varicocele. The morphology was disturbed in the oligozoospermic groups. They could indicate that left-sided varicocele is detrimental to spermatogenesis and that small varicoceles require as much attention as large ones.

Sayfan et al. (1988) could show that the disturbance of spermatogenesis is less severe in adolescents and young adults (16 - 23 yr) than in the "older" (25 - 38 yr) varicocele groups. In younger patients only the total sperm count was significantly depressed. They conclude that their data prove the progression of damage to the testicles.

## Other Exploration Tests

Ultrasonographic examination is used to determine testicular volume more accurately (Sayfan et al., 1988) and to exclude retroperitoneal tumors (Schickedanz and Kleinteich, 1987). An echography can also replace an urography (Mothes, 1986). Echo-doppler evaluation is promoted by Buch and Cromie (1985) and by Ponchietti et al. (1986).

Phlebotonometry with measurement of the pressure gradient was advocated by Russian groups (Arbuliev, 1968; Erokhin et al., 1980; Isakov and Arbuliev, 1969; Isakov et al., 1977).

Testicular biopsies for light or electron microscopy will show that the disturbance of spermatogenesis in the affected testes becomes prominent with the progress of puberty, while changes in the testis of the opposite side are mild. On the opposite side the first sign of damage is slight thickening of the tubular wall (Okuyama et al., 1981).

## Table 6. Localization of Varicocele at Army Age

| Authors | Country | Total Recruits | Localization of Varicocele | | |
| --- | --- | --- | --- | --- | --- |
| | | | Left | Right | Bilateral |
| Lewis (1950) | U.S.A. | 1,500 | 246 | 0 | 1 |
| Appleby (1955) | U.S.A. | 4,000 | 396 | 0 | 4 |
| Clarke (1966) | U.S.A. | 189 | 15 | 0 | 0 |
| Wutz (1977) | Germany | 3,490 | 181 | 0 | 0 |
| Total | | 9,179 | 838 | 0 | 5 |
| Percentage | | | 99.4 | 0 | 0.6 |

## LOCALIZATION

In the groups of patients referred for operation, almost all had a left-sided varicocele, as can be seen in Table 5.

The data obtained in epidemiological studies on the incidence of varicocele in the general population are given in Tables 6 and 7.

A literature study of 75 publications (total cases: 13,641) revealed a left-sided varicocele in 95.13%, a right-sided varicocele in 1.63% and a bilateral varicocele in 3.24% (Schickedanz and Kleinteich, 1985).

We believe that the higher percentage of bilateral cases in adolescents with varicocele, referred for treatment, is due to the fact that bilateral varicocele attracts the attention of referring doctors to a greater degree and was the reason for greater concern for fertility in the future.

## CLASSIFICATION OF VARICOCELES: METHODS FOR GRADING OF VARICOCELE

Varicoceles are classified or graded although there exists a gradual transition from the minor to the most severe situation (Verstoppen and Steeno, 1977b). No clear relationship between the grade of varicocele, the severity of sperm disturbance and the result of the treatment could be shown (Steeno, 1972).

The first grading in three classes had been proposed by Lewis (1950):

-class I (severe form): symptomatic: pain, dragging sensation, noticeable testicular atrophy;
-class II: small asymptomatic;
-class III: moderate or small symptomatic.

This classification is not in a gradually increasing order, but almost all following classifications proposed are restricted to three classes, degrees or grades.

The classification of Horner (1960) was applied later in the studies of Oster and El-Gohary (1984). He proposed three grades A, B and C in an increasing order:

-mild or varicocele A: palpable but no visible prolongation, dilatation and/or convolution of the veins around the testis;
-moderate or varicocele B: varicocele is both palpable and visible;
-marked or varicocele C: mass of tortuous veins so pronounced as to impede examination of the testis, or subjective symptoms are present.

Table 7. Localization of Varicocele in School Populations

| Authors | Country | Localization of Varicocele | | |
| | | Left | Right | Bilateral |
| --- | --- | --- | --- | --- |
| Horner (1960) | U.K. | 192 | 1 | 0 |
| Oster (1971) | Denmark | 136 | 0 | 0 |
| Risser (1984) | U.S.A. | 55 | 0 | 2 |
| | | | | |
| Total | | 384 | 1 | 2 |
| Percentage | | 99.1 | 0.3 | 0.6 |

In this classification not only clinical signs but also subjective symptoms are taken into account.

In the same year Kiszka and Cowart (1960) proposed a more simple clinical classification, which was also used by Johnson et al. (1970):

-small (grade I): not obvious on inspection, but contained a few tortuous and palpable enlarged veins;
-medium (grade II): veins approximately 1 cm in diameter;
-large (grade III): findings exceed the latter classification.

Uehling (1968, cited by Steeno 1972, Steeno et al., 1975, 1976) classified the small and moderate cases in grade I and made a division in more severe forms:

-grade I: palpable scrotal varicosity ("grape-sized mass of veins");
-grade II: varicocele mass as big as the testis;
-grade III: venous mass fills the left scrotum and is already visible at simple inspection.

The simple classification of Dubin and Amelar (1970, cited by Steeno et al., 1976) was applied by most authors (Audry et al., 1986, 1987; D'Ottavio et al., 1982; Alcalay and Sayfan, 1984; Lyon et al., 1982; Pozza et al., 1983; Reitelman et al., 1987; Risser and Lipshultz, 1984; Steeno, 1972; Thomason and Fariss, 1979):

-grade I (small size): barely palpable varicocele, detected only by Valsalva maneuver; diameter of the vas less than 1 cm;
-grade II (moderate size): veins clearly palpable as cords; diameter of the mass 1 to 2 cm;
-grade III (large size): massive distention of venous system, visible before palpation; diameter of the mass more than 2 cm.

Steeno et al. (1975, 1976) combined the classification of Uehling and those of Dubin and Amelar to the following degrees:

-grade I: palpable scrotal varicosity; diameter of the vein < 1 cm, but with reflux at the Valsalva maneuver;
-grade II: pronounced varicocele mass; diameter of the vein 1 to 2 cm;
-grade III: venous mass fills the scrotum; diameter of the vein > 2 cm with positive reflux easily visible at a distance.

This classification was followed by Boeckx and Vereecken (1985), El-Gohary (1984), Gattuccio et al. (1988) and Kass (1988).

The classification of Steeno et al. was also applied by Rost et al. (1975) but in reverse order (grade I of Rost = grade III of Steeno).

Table 8. Results of Grading of Varicocele in Clinical Young Patients

| Authors | Country | Grade I (%) | Grade II (%) | Grade III (%) |
|---------|---------|-------------|--------------|---------------|
| Pozza (1983) | Italy | 2.8 | 22.9 | 74.3 |
| Dias (1985) | Belgium | 4.0 | 24.0 | 72.0 |
| Audry (1986) | France | 6.5 | 71.0 | 22.5 |

The grading of Kleinteich and Schickedanz (1983 a,b; see also Schickedanz and Kleinteich, 1982, 1987) is also simple and clinically practicable:

-grade I: enlarged veins, less in volume than the ipsilateral or contralateral testis;
-grade II: venous mass equal to the testicular volume;
-grade III: venous mass exceeds testicular volume.

In their publication of 1982, Lyon et al. followed the classification of Dubin and Amelar, but in 1983 they give a grading in two forms:

-grade I: veins clearly palpable as cords with the patient in the upright position;
-grade II: massive distention of the venous system in the scrotum.

They also grade the testicular mass (Lyon et al., 1983) as:

-grade I: testis clearly smaller than its counterpart;
-grade II: testis 1/4 or less than the size of the right one.

Hennebert et al. (1985) proposed also grades IV and V but gave no clear definition of these latter grades.

CLASSIFICATION OR GRADING OF VARICOCELES: RESULTS

To interpret the results of the classification according to the degree of varicocele, distinction must be made between the data obtained in clinics (boys referred for operation) (Table 8), in military men (recruits) (Table 9) and in school populations (Table 10). Only the classifications in increasing order are of severity are included in the tables.

In Table 11 the prevalence (%) of varicocele in the school populations according to the degree of severity is given.

From these data it is clear that the results of grading are dependent on the subjective interpretation of the clinical status of varicocele by the different authors. In

Table 9. Results of Grading of Varicocele in Military Men (Recruits)

| Authors | Country | Total Men | % Recruits with Varicocele | Grade I (%) | Grade II (%) | Grade III (%) |
|---------|---------|-----------|----------------------------|-------------|--------------|---------------|
| Appleby (1955) | U.S.A. | 4,000 | 9.1 | 64.0 | 35.0 | 1.0 |
| Thomason (1979) | U.S.A. | 909 | 30.7 | 45.5 | 31.9 | 22.5 |
| Alcalay (1984) | Israel | 1,479 | 9.1 | 44.0 | 40.0 | 16.0 |

Table 10. Results of Grading of Varicocele in School Populations

| Authors | Country | Total Boys | % Boys with Varicocele | Grade I (%) | Grade II (%) | Grade III (%) |
|---|---|---|---|---|---|---|
| Horner (1960) | U.K. | 1,450 | 14.2 | 32.6 | 45.1 | 22.2 |
| Oster (1971) | Denmark | 837 | 16.2 | 52.2 | 44.1 | 3.7 |
| Steeno (1976) | Belgium | 4,067 | 14.7 | 63.9 | 24.4 | 11.7 |
|  |  | 894 | 13.5 | 61.5 | 28.9 | 9.6 |
| D'Ottavio (1981) | Italy | 2,792 | 20.4 | 44.9 | 32.4 | 22.9 |
| (1983) |  | 5,177 | 25.8 | 48.0 | 27.5 | 24.5 |
| Risser (1984) | U.S.A. | 423 | 13.7 | 45.0 | 29.0 | 27.0 |

boys referred to specialists for treatment, almost 95% have a moderate or severe form of varicocele (70 to 75% severe form); in military recruits with varicocele, 55% (ca. 20% severe form); in school populations 40 to 55%. The population of school boys "at risk" for possible later fertility problems (grades II and III) amounts to 5.3 to 12.9% (Table 11).

In a literature study, Kleinteich and Schickedanz (1983a,b) and Schickedanz and Kleinteich (1987) show that in adults the percentage of patients with a moderate or severe degree of varicocele is higher than in children and adolescents (Table 12). These results prove the progressive character of the condition.

## INFLUENCE ON THE TESTICLES ACCORDING TO THE GRADING OF VARICOCELE

Ristic et al. (1986) found smaller and softer testicles at the left side or bilaterally in 2.93% of boys (7 - 18 yr) without varicocele, but in 36.8% of boys with a varicocele.

The influence on the testicles (affected testis smaller and/or softer than contralateral one or both testicles smaller and/or softer than expected according to the pubertal stage) according to the grading of the varicocele was studied by Steeno et al. (1975, 1976) and El-Gohary (1986a,b) (Table 13).

From these data it can be concluded that moderate forms of varicocele have a clinical influence on the ipsilateral or both testicles in one third of the boys with varicocele; severe forms, however, in more than 80% of the boys with varicocele. The influence on the testicle(s) increases with the duration of the condition (to almost 98%).

Table 11. Prevalence of Varicocele in School Boys according to the Degree of Severity

| Authors | Country | Total Boys | Grade I (%) | Grade II (%) | Grade III (%) | Grade I+II+III (%) | Grade II+III (%) |
|---|---|---|---|---|---|---|---|
| Horner (1960) | U.K. | 1,450 | 4.3 | 6.0 | 3.9 | 14.2 | 9.9 |
| Oster (1971) | Denmark | 837 | 8.4 | 7.2 | 0.6 | 16.2 | 7.8 |
| Steeno (1976) | Belgium | 4,067 | 9.4 | 3.6 | 1.7 | 14.7 | 5.3 |
| Erokhin (1979) | U.S.S.R. | 10,000 | 4.0 |  |  | 12.4 | 8.4 |
| D'Ottavio (1983) | Italy | 5,177 | 12.4 | 7.1 | 5.8 | 25.3 | 12.9 |
| El-Gohary (1986) | U.A.E. | 1,823 | 6.5 |  |  | 13.4 | 6.9 |

Table 12. Results of Grading of Varicocele in Children and Adolescents versus Adult Men

| Authors | Age Group | Degree I (%) | Degree II (%) | Degree III (%) |
|---------|-----------|--------------|---------------|----------------|
| Kleinteich (1983) | 6 - 17 yr | 58.44 | 34.69 | 6.87 |
|  | adults | 41.39 | 29.88 | 28.73 |
| Schickedanz (1987) | 6 - 17 yr | 62.20 | 28.74 | 9.06 |
|  | adults | 42.94 | 31.02 | 26.04 |

## PATHOGENESIS OR MECHANISM OF INJURY

Many theories on the deleterious effects of varicocele on the testicular development and/or function have been proposed. The mechanism of injury certainly is multifactorial. The proven mechanisms and the hypothetical ones can be summarized as follows:

1. Increase in scrotal temperature: increase in retroflow of venous blood results in a bilateral increase in testicular temperature, especially at the affected side, with loss of the abdomino-scrotal thermic gradient (Allemann and Jenny, 1980; Audry et al., 1986; Buch and Cromie, 1985; Kass, 1988; Kleinteich and Schickedanz, 1981; Steeno, 1972; Verstoppen and Steeno, 1978; Wutz, 1977).

2. Stasis in the testicular veins disturbs the testicular circulation with venous hyperpressure, hypoxia of the germinal epithelium, extra- and intracellular acidosis and nutrition disorders with loss of testicular mass (Allemann and Jenny, 1980; Audry et al., 1986; Hienz et al., 1980; Kleinteich and Schickedanz, 1981; Lipshultz and Corriere, 1977; Schickedanz and Kleinteich, 1982; Steeno, 1972; Verstoppen and Steeno, 1978; Wutz, 1977).

3. Toxic damage by high concentrations of known and unknown substances of renal and adrenal origin (catecholamines, renin, cortisone, hydrocortisone, DHEA, androsterone, serotonin, prostaglandins) (Allemann and Jenny, 1980; Audry et al., 1986; Buch and Cromie, 1985; Kleinteich and Schickedanz, 1981; Steeno, 1972; Verstoppen and Steeno, 1978; Wutz, 1977).

4. Impact on the Sertoli and Leydig cell function, resulting in a decrease in intratesticular testosterone (Kass, 1988; Ponchietti et al., 1987). The involvement of biosynthetic pathway of testosterone production adversely affects the functional maturation of the testis and results in impaired spermatogenesis (Ponchietti et al., 1987) with premature sloughing of immature cells in the seminiferous tubules (Dubin and Amelar, 1971).

5. Possible existence of immunological reactions responsible for disturbance in spermatogenesis (Isakov et al., 1977).

Table 13. Influence on the Testicles (%) according to the Grading of Varicocele

| Authors | Country | Age Range of Boys | Total # | Grade II (%) | Grade III (%) |
|---------|---------|-------------------|---------|--------------|---------------|
| Steeno | Belgium | 12 - 25 | 4,067 | 34.4 | 81.2 |
|  |  | 18 - 25 | 2,118 | 34.1 | 97.6 |
| El-Gohary | U.A.E. | 9 - 17 | 1,823 | 35.0 | 78.0 |

## HISTOLOGICAL OBSERVATIONS

Testicular biopsies showed a variable degree of damage (Bento et al., 1984). The disturbance of spermatogenesis in the affected testis becomes prominent with the progress of puberty, while changes in the testis of the opposite side are milder (Okuyama et al., 1981; Ponchietti et al., 1986). The first sign in the opposite side is slight thickening of the tubular wall (Okuyama et al., 1981). Kawamura et al. (1987), however, found no significant difference between the right and left testis. The histological changes are identical as in adult males with varicocele (D'Ottavio et al., 1982; Hadziselimovic, 1983; Hienz et al., 1980; Jenny and Hadziselimovic, 1982; Ponchietti et al., 1986; Pozza et al., 1983; Schurholz and Weissbach, 1978). Histological changes worsen with passing of time (D'Ottavio et al., 1982). The problem remains unanswered whether histological changes improve following varicocele ligation (Kass et al., 1987).

The degenerative changes can be summarized according to the testicular compartment as follows:

1. tubular development:
   -delay of maturation of tubules in respect to chronological age (D'Ottavio et al., 1982);
   -reduction of tubular diameter (Hennebert et al., 1985);
2. changes in the tubular wall:
   -thickening of tubular walls (D'Ottavio et al., 1982; Jenny and Hadziselimovic, 1982; Ponchietti et al., 1986);
   -slight hyalinization of the walls of individual tubules (Hienz et al., 1980);
   -tubular sclerosis due to collagen deposition (Jenny and Hadziselimovic, 1982, 1987; Jones et al., 1988; Ponchietti et al., 1986; Schurholz and Weissbach, 1978);
3. regressive and degenerative lesions of the seminiferous epithelium:
   -arrest of spermatogenesis or hypospermatogenesis (Cesaroli et al., 1985; D'Ottavio et al., 1982; Hennebert et al., 1985; Hienz et al., 1980; Jenny and Hadziselimovic, 1982);
   -desquamation of germinal epithelium in lumen (Hadziselimovic, 1983; Hienz et al., 1980; Jones et al., 1988; Schurholz and Weissbach, 1978);
   -reduction of percentage of germ cells present as late stage forms (secondary spermatocytes, spermatids and spermatozoa (Jones et al., 1988);
   -spermatogonia show a diffuse vacuolization and disarrangement in the junctional system with the Sertoli cells (Ponchietti et al., 1986);
4. Sertoli cells:
   -involution (D'Ottavio et al., 1982);
   -reduction of mean germ cell/Sertoli cell ratio (Jones et al., 1988);
   -large dilatation in the endoplasmic reticulum of the Sertoli cells, involving the entire cytoplasm (Ponchietti et al., 1986);
   -big vacuoles in cytoplasm (Jenny and Hadziselimovic, 1982);
5. Interstitium:
   -interstitial edema (Hienz et al., 1980; Schurholz and Weissbach, 1978);
   -focal interstitial fibrosis (Hienz et al., 1980; Jenny and Hadziselimovic, 1987);
   -collagenization (Jenny and Hadziselimovic, 1987; Ponchietti et al., 1986; Schurholz and Weissbach, 1978);
   -presence of many mast cells (Hadziselimovic, 1983; Jenny and Hadziselimovic, 1982);
   -degenerative alterations of Leydig cells (Ponchietti et al., 1987);
   -diffuse vascular degeneration of the cytoplasm (Ponchietti et al., 1986; Jenny and Hadziselimovic, 1982);
   -decrease in organelle number;
   -decrease in lipid droplets;
   -large amounts of lipofuscin pigment present;
   -membrane system of smooth endoplasmic reticulum enlarged;
   -cisternae of Golgi apparatus dilated with decreased membrane-bound vesicles;
   -mitochondria exhibit a less dense matrix with fragmentation of tubular cristae;
   -Reinke crystalloids often observed;

6. alterations of the vascular compartment:
   -arterial lumina markedly reduced with protrusion of endothelial cells which
   have many bundles of microfilaments (Hienz et al., 1980; Ponchietti et al.,
   1987);
   -capillaries present thickening of the basement membrane with a great number
   of pinocytotic vesicles (Hienz et al., 1980; Jones et al., 1988; Ponchietti et al.,
   1987; Schurholz and Weissbach, 1978);
   -venous walls are fibrotic and thickened (Ponchietti et al., 1987; Schurholz and
   Weissbach, 1978).

## STRATEGY: TO TREAT OR NOT TO TREAT?

Most authors advise an early treatment in boys and adolescents for several reasons:

-because of the relationship between varicocele and subfertility and in order to
avoid problems of subfertility or infertility in adult age (Allemann and Jenny, 1980;
Bento et al., 1984; Berger, 1980; Cesaroli et al., 1985; Dias et al., 1985, 1987a,b; El-Gohary,
1986a,b; Gorenstein et al., 1986; Janneck, 1979; Kiszka and Cowart, 1960; Kleinteich and
Schickedanz, 1983a,b; Lipshultz and Corriere, 1977; Mothes, 1986; Rost et al., 1975;
Steeno, 1972; Wyllie, 1985;
-since no early recognition of damage to the germinal epithelium is possible
(Allemann and Jenny, 1980);
-since the deleterious effects of varicocele on testicular mass and spermatogenesis
are more important in boys and adolescents than in adults (Audry et al., 1986, 1987; Lyon
et al., 1982; Wyllie, 1985);
-because of the relationship between the duration or evolution of the varicocele and
the degree of testicular involvement (progressive damage) (Audry et al., 1986, 1987;
D'Ottavio et al., 1981, 1982; Erokhin, 1979b; Garnier and Szatkovnik-Plot, 1978; Hienz et
al., 1980; Holschneider et al., 1978; Hotchkiss, 1944; Jones et al., 1988; Kass, 1988;
Kawamura et al., 1987; Lipshultz and Corriere, 1977; Lyon et al., 1982; Mothes, 1986;
Okuyama et al., 1981; Petrenko et al., 1988; Pozza et al., 1983; Sayfan et al., 1988);
-since patients treated before adult age show a better sperm picture at adult age than
those operated later on (Boeckx and Vereecken, 1985; Hosli, 1988; Kass, 1988; Okuyama
et al., 1988);
-since bilateral testicular involvement becomes more important with increasing age
(Cesaroli et al., 1985; Ponchietti et al., 1986);
-since catch-up growth can be obtained if operation occurs before adult age (Hosli,
1988; Kass and Belman, 1987);
-since timely treatment reverses some of the histological changes (Ponchietti et al.,
1987).

There is a general agreement that adolescent males with a moderate (grade II) or
severe (grade III) form of varicocele must be treated (El-Gohary, 1986a,b; Hennebert et
al., 1985; Holschneider et al., 1978; Steeno, 1972; Steeno and Adimoelja, 1975; Steeno et
al., 1975, 1976).

Patients with grade II or III and/or underdevelopment of the underlying testis
should be considered as high-risk group regarding future subfertility (El-Gohary,
1986a,b).

For adolescents with a mild (grade I) form of varicocele a yearly follow-up
examination is indicated. Changes in size of the varicocele and the corresponding testis
should be checked (El-Gohary, 1986a,b; Holschneider et al., 1978; Steeno and Adimoelja,
1975).

Other authors even believe that treatment during childhood is necessary as soon as
possible after diagnosis, regardless of the degree of severity, the presence or absence of
symptoms and phlebographic criteria (Gunther et al., 1974; Hienz et al., 1980; Johnson et
al., 1970; Mothes, 1986; Sayfan et al., 1988; Thon et al., 1989).

The argument that infertility does not necessarily result in each case of varicocele is not relevant due to the uncertainty involved in an individual case (Hienz et al., 1980; Sayfan et al., 1988).

The ultimate goal of treatment is to reduce future fertility problems (Bento et al., 1984; Steeno et al., 1975, 1976; Wyllie, 1985).

The high risk of later infertility should be compared with the low risk of surgery or other kinds of treatment during childhood (Audry et al., 1986, 1987; Hienz et al., 1980). The patient has nothing to lose but much to win (Rost et al., 1975).


INDICATIONS FOR TREATMENT

Already in 1889, Bennett considered operation to be <u>expedient</u> in the following cases:

1. cases in which varicocele has led to the rejection of candidates for military service on account of physical imperfection;
2. cases in which the mere size of the varicocele is sufficient to lead to risk of injury, or causes enough inconvenience to make the patients seek relief;
3. cases of double varicocele with marked impairment of testicular function in which the cure may be undertaken for the purpose of allowing the development of proper physiological activity of the testes.

According to Bennett, the operation may be rendered <u>necessary</u> by the following symptoms:

1. pain;
2. tenderness;
3. rapid increase in the size of varicocele;
4. mental anxiety.

Today, treatment of varicocele in adolescents seems to be indicated in the following situations:

1. in cases of symptomatic varicocele:
    -scrotal pain or discomfort (Audry et al., 1986; El-Gohary, 1984; Lyon et al., 1982; Reitelman et al., 1987; Sawczuk et al., 1985; Schurholz and Weissbach, 1978);
    -perineal discomfort (Levitt et al., 1987);
    -unspecified subjective symptoms (Janneck, 1979);
    -secondary varicocele (Berger, 1980);
2. in cases of varicocele of marked degree (Audry et al., 1986; Borgmann and Hasse, 1978; El-Gohary, 1984; Kass et al., 1987; Kleinteich and Schickedanz, 1983a; Lyon et al., 1982; Mothes, 1986; Schurholz and Weissbach, 1978);
3. when varicocele is detected at the onset of puberty (Mothes, 1986, 1988);
4. when varicocele progresses in time during observation (Levitt et al., 1987; Mothes, 1986);
5. in cases of scrotal temperature increase, clinically or thermographically noticeable (Janneck, 1979; Mothes, 1986);
6. in cases of right or bilaterally sided varicocele (Hadziselimovic, 1983; Jenny and Hadziselimovic, 1982);
7. in boys with arrested testicular growth or manifest hypo- or atrophy or simple diminished testicular volume and/or consistency (El-Gohary, 1984; Hadziselimovic, 1983; Jenny and Hadziselimovic, 1982, 1987; Kass et al., 1987; Levitt et al., 1987; Lyon et al., 1982; Mothes, 1986, 1988; Reitelman et al., 1987; Sawczuk et al., 1985).
8. in the presence of hormonal pathology: pathological gonadotropin levels in plasma, especially FSH (Hadziselimovic, 1983), pathological LH-RH test (Hadziselimovic, 1983; Kass, 1988; Kass et al., 1988), low testosterone value in plasma (Jenny and Hadziselimovic, 1987);

9. when associated with sperm disturbances: subnormal numbers of mature, motile sperm (Clarke, 1966), oligozoospermia (Johnson et al., 1970); reduction in sperm motility (Johnson et al., 1970), abnormal semen analysis in general (Kass, 1988);

10. according to the grading of varicocele, many authors consider the clinical presence of a varicocele of grade II or III as indication for treatment (Borgmann and Hasse, 1978; Hennebert et al., 1985; Kass et al., 1987, 1988; Kleinteich and Schickedanz, 1981, 1983a; Lyon et al., 1982; Schickedanz and Kleinteich, 1982, 1987; Steeno and Adimoelja, 1975; Steeno et al., 1975, 1976).

Most authors agree that in cases with varicocele grade I an annual follow-up examination and reassessment are required (Kass et al., 1988; Schickedanz and Kleinteich 1982, 1987; Steeno and Adimoelja, 1975; Steeno et al., 1975, 1976).

There is a general agreement on the following statements:

1. Small varicoceles require as much attention as large ones (Johnson et al., 1970);
2. The earlier the varicocele develops, the greater its effect on testicular growth (Lyon et al., 1982);
3. Subjective symptoms or already demonstrated fertility disorder may not be considered as criteria for treatment (Mothes, 1986);
4. grading, clinical symptoms or phlebographic criteria do not allow for prediction of the risk for fertility (Mothes, 1986).

## METHODS OF TREATMENT: TECHNOLOGICAL ASPECTS

Methods of treatment which attend to symptoms, i.e., douches or suspensorium, are rejected (Borgmann and Hasse, 1978; Kiszka and Cowart, 1960; Lindner, 1982). A suspensorium may even increase local temperature (Kiszka and Cowart, 1960).

Operative treatment was performed in the 19th century by a subcutaneous wire loop, which cuts the enlarged veins within 13 to 15 days (Barwell, 1885).

Local resection of the pampiniform plexus or intrascrotal varicosities following Kocher frequently provoked complications, such as hematoma, hydrocele, testicular atrophy or recurrence of the varicosities (Allemann and Jenny, 1980; Clarke, 1966; Lindner, 1982).

Simultaneous resection of the pampiniform plexus and ligation of the internal spermatic vein also involve risk of testicular swelling with complete atrophy (Borgmann and Hasse, 1978).

All actual surgical methods aim at reducing the static pressure in the pampiniform plexus.

The actual surgical approach may be supra-inguinal (Bernardi's or Palomo's method) or high-inguinal (Ivanissevich's method). The methods of Bernardi and Ivanissevich consist in ligating the internal spermatic vein; the method of Palomo in ligating the artery and the vein.

The inguinal method of Ivanissevich is preferred by Allemann and Jenny (1980) (to detect associated inguinal pathology), Audry et al. (1986, 1987), Berezhnoi et al. (1985), Cesaroli et al. (1985), Dias et al. (1985, 1987a,b), El-Gohary (1984, 1986a,b), Isakov and Arbuliev (1969), Isakov et al. (1977) (before 1960), Jenny and Hadziselimovic (1982, 1987), Kiszka and Cowart (1960), Kleinteich and Schickedanz (1983b), Lewis (1950), and by Mothes (1986) (below the age of 16 years in cases of recurrence after initial Bernardi's operation method).

The suprainguinal (high retroperitoneal) approach following Bernardi (ligation of the internal spermatic vein) is preferentially performed by Günther et al. (1974), Hösli (1988), Holschneider et al. (1978), Janneck (1979), Kraeft et al. (1981), Lindner (1982),

Mothes (1986) (below the age of 16 yr), Rost et al. (1975), Schickedanz and Kleinteich (1982) and by Schürholz and Weissbach (1978).

The supra-inguinal (high retroperitoneal) approach following Palomo (ligation of the artery and vein) is preferred by Boeckx and Vereecken (1985), Erokhin (1979a), Gorenstein et al. (1988), Okuyama et al. (1988) and by Vereecken and Boeckx (1986).

Some authors who started with the Ivanissevich's method, changed to Palomo's method because of diminished recurrence (Erokhin, 1979a; Isakov et al., 1977).

Despite the simultaneous ligation of the internal spermatic artery and vein, blood supply of the testicle is maintained since anastomoses exist between the three arteries (the internal spermatic artery, the deferential artery and the cremasteric artery) (Palomo, 1949). Most authors agree with this statement (Günther et al., 1974; Janneck, 1979). Nevertheless, others believe that the arterial blood supply may be diminished at a time of increased vascularity (Holschneider et al., 1978; Kleinteich and Schickedanz, 1983b; Lindner, 1982; Mothes, 1986). For this reason Mothes (1986) only performs Palomo's operation after the age of 16 years.

The tunneling method according to Giuliani or Manteuffel (without ligation) is only seldom practiced (Hecker, 1973; Holschneider et al., 1978; Ribeiro, 1956).

In recent years phlebography with embolization of the internal spermatic vein (with a Gianturco coil, sclerotherapy or balloon occlusion) has become the first choice of treatment since no hospitalization and no general anesthesia is required, no serious effects are noted and this kind of treatment seems to be more efficient and convenient with a success-rate of more than 80% (Braedel et al., 1988; Thon et al., 1989; Vorontsov et al., 1985; Wilms et al., 1988). Nevertheless, in adolescents more failures are noted because of spasms of the spermatic vein and the presence of more collaterals at the pelvic level (Dias et al., 1985, 1987a,b; Lenk et al., 1988) and more danger of rupture of the renal and/or spermatic vein (Audry et al., 1986, 1987). When embolization is impossible, operation can always be performed at a second time.

Mothes (1986) uses a combination of all available methods and techniques. He starts with Bernardi's technique. In case of failure, a phlebography is performed. Occlusion therapy, if possible, is then the first choice of second treatment. If not possible, Ivanissevich's method below the age of 16 years, and Palomo's method after the age of 16 years for reasons already explained.

OUTCOME OF TREATMENT

The results of treatment depend on the technique or method applied. Disappearance of the varicocele was obtained in 48% by Audry et al. (1986, 1987) (Ivanissevich), in 85% by Gorenstein et al. (1988) (Palomo), in 95% by Hösli (1988) (Palomo and Bernardi), in 75% by Holschneider et al. (1978) (Palomo and Bernardi), in 89% by Kiszka and Cowart (1960) (Ivanissevich), in 75% by Holschneider et al. (1978) (high ligation); in 98% by Isakov et al. (1977) (Palomo) against 62% (Ivanissevich); in 62.5% by Kraeft et al. (1981) (Bernardi), in 63% by Lenk et al. (1988) (balloon occlusion) and in 90% by Mothes (1986) (Palomo). Immediate and long term results after embolization of the left internal spermatic vein are also very satisfactory but no definite results are already known (Vorontsov et al., 1985).

Catch-up growth was noted by Audry et al. (1986, 1987), Bennett (1889), Hösli (1988), Jenny and Hadziselimovic (1987), Kass and Belman (1987), Levitt et al. (1987), Lyon et al. (1983), Reitelman et al. (1987) and by Vereecken and Boeckx (1986).

Improvement did not correlate with the degree of varicocele and age at intervention (Vereecken and Boeckx, 1986).

Persistence or recurrence of the varicocele is mostly due to non-ligated or not occluded supernumerary branches of the internal spermatic vein (Lenk et al., 1988; Levitt et al., 1987; Schickedanz and Kleinteich, 1985). Gorenstein et al. (1988) found recurrent varicocele only in patients with increased left renal vein pressure. Follow-up of treated and untreated cases of the same age group revealed that in the treated patients testicular condition and sperm findings are better (Hösli, 1988; Okuyama et al., 1988). Post-operative results were unsatisfactory with the Manteuffel-Giuliani method: only in 20% were postoperative conditions normal (Holschneider et al., 1978); and after local resection of the pampiniform plexus: only normal findings in 36.8% (Holschneider et al., 1978).

## COMPLICATIONS

Operative methods may lead to ligation of lymphatic vessels in connection with the venous system and provoke an hydrocele or non-infectious "orchiepididymitis". For this reason, Bergman and Wahlqvist (1974) stated that early operation to prevent subfertility may be considered only when hydrocele can be avoided. In a follow-up of 21 boys operated on during adolescence and adult age, Hösli (1988) found hydrocele as a complication in three of them. In a literature study of 504 patients operated following Bernardi, Holschneider et al. (1978) noted postoperative complications in 26 (=5.1%). Hydrocele was present in 17 of them; in three of these cases a hydrocele-operation was necessary. Other minor complications were: wound infection (4), inguinal hematoma (2), epididymitis (1) and atelectasis (1). Ligation of the lymph-draining system is more frequently (4 times more) encountered with the technique of Palomo's than of Ivanissevich (Isakov and Erokhin, 1979).

The number of complications has been markedly reduced since the first choice of treatment became occlusion techniques (4 to 5% according to Lenk et al., 1988) or sclerotherapy (Thon et al., 1989). A specific complication of sclerotherapy is sterile thrombophlebitis of the pampiniform plexus, which may last for weeks but finally resolves (Thon et al., 1989).

Steeno et al. (1985, 1987) asked for careful attention and psychological care and follow-up. The fact that school boys and college students are informed about this disease of which they were previously unaware (93.9% of the school boys and 78.3% of the college students), brings psychological problems which cannot be ignored. Depressive reactions, study problems, loss of medical confidence (because of earlier negative findings), and anxiety about their fertility and sexual potency are noted. In not yet operated cases, these "negative" psychological problems were expressed in 28.6% (school boys) to 36.4% (college students). In already operated cases these figures were lower (respectively 22.7 and 27.4 %).

## ASSOCIATIONS WITH OTHER ABNORMAL CONDITIONS

Bennett (1889) already noted that, when occurring on both sides, there is always more or less varicosity of the veins of both lower extremities. Appleby (1955) found varicocele in 5% of the cases associated with varicose veins of the lower extremities.

In 10%, Appleby (1955) noted varicocele to be associated with inguinal hernia, either clinically or at the time of surgery. Allemann and Jenny (1980) recommend Ivanissevich's method since in 1/3 of operated cases associated conditions were found.

In the case of an history of ipsilateral inguinal surgery more collaterals are present (Levitt et al., 1987).

## UNANSWERED OR REMAINING QUESTIONS

Only two authors point to unanswered or remaining questions:

Kass et al. (1987) pose as questions:

1. Do the histologic changes improve following varicocele ligation?
2. Does early varicocele ligation prevent future problems?
3. Which child with a varicocele requires intervention?

Reitelman et al. (1987) have two questions:

1. Does testicular volume correlate with spermatogenesis?
2. Does abnormal testicular histology correlate with impaired fertility?

Further clinical observations and follow-up studies are required to give a satisfactory answer to these questions.

## SUMMARY

Varicocele in the adolescent is a clinical diagnosis and may no longer be considered as an innocuous condition.

Multifactorial causes lie at the basis of the appearance of a left varicocele from the start of puberty on in about 15% of boys (moderate to severe grades in 8%). A right-sided varicocele is very rare and is almost exclusively present in symptomatic varicocele or in patients with situs inversus.

Adolescents are not aware of the presence of a scrotal varicosity, which is seldom painful.

Moderate forms of varicocele have a clinical influence on the ipsilateral or both testicles (volume and/or consistency) in 1/3; severe forms in more than 80% of adolescents. The influence on the testicle(s) increases with the duration of the condition. Adolescents with a pronounced degree of varicocele form a "risk group" for later fertility disturbances.

The pathogenesis or mechanism of injury is also multifactorial. Increase in scrotal temperature, venous stasis and toxic damage have an impact on Sertoli and Leydig cell function, which results in impaired spermatogenesis.

The histological changes are more pronounced in the left testis, are identical as in adult males with varicocele, and worsen with passing of time.

Moderate and severe forms of varicocele, especially when associated with loss of testicular mass or testicular growth failure, are indications for early treatment in order to prevent later fertility problems.

In recent years the first choice of treatment has become occlusion or the application of sclerotherapy methods. If not possible, high ligation of the internal spermatic vein can still be performed.

In adolescents, catch-up growth of the testicles can be obtained with early treatment.

The ultimate goal of treatment is to reduce subfertility or infertility in adult age.

School health centers may play a role in the prevention of male sub- or infertility.

## REFERENCES

Alcalay, J. and Sayfan, J. 1984. Prevalence of varicocele in young Israeli men. Isr. J. Med. Sci., 20: 1099-1100.

Alcalay, J. and Sayfan, J. 1985. Letter to the editor. J. Adolesc. Health Care, 6: 240.

Alcalay, J., Kedem, R., Kornbrot, B. and Sayfan, J. 1986. The ethnic distribution of varicocele. Milit. Med., 151: 327-328.

Allemann, F. and Jenny, P. 1980. Die idiopathische Varicocele mit spezieller Berücksichtigung des Kindes- und Jugendalters, Z. Kinderchir. Grenzgeb., 29: 339-342.

Appleby, G.S. 1955. Varicoceles, a problem in military personnel. W. Virginia Med. J., 51: 76-78.

Arbuliev, M.G. 1967. [Treatment of dilatation of the veins of the spermatic cord in children] (in Russian). Vop. Okhr. Materim. Dets, 12: 52-54.

Arbuliev, M.G. 1968. [On the pathogenesis of varicocele in children] (In Russian with English summary). Urol. Nefrol. (Moskva), 33: 41-44.

Audry, G., Tazi, M. and Brueziėre, J. 1986. Varicocėle chez l'enfant et l'adolescent. Ann. Urol. (Paris), 20: 355-359.

Audry, G., Tazi, M. and Brueziėre, J. 1987. Varicocėle chez l'enfant et l'adolescent. Ann. Pédiatr. (Paris), 34: 625-628.

Barwell, R. 1885. One hundred cases of varicocele treated by the subcutaneous wire loop. Lancet, 1: 978.

Bennett, W.H. 1889. Varicocele particularly with reference to its radical cure. Lancet, 1: 261-268.

Bento, L., Villanueva, A., Zunzunegui, C., Sanchez-Valverde, F., Garcia Bragado, F. and Puras, A. 1984. Varicocele idiopatico en el niño. An. Esp. Pediatr., 21: 842-846.

Berezhnoi, V.I., Malko-Skroz, V.V. and Tritiak, Ia.V. 1985. [The treatment of varicocele in children] (In Russian), Klin. Chir., 6: 70.

Berger, O.G. 1980. Varicocele in adolescence. Clin. Pediatr. (Phila), 19: 810-811.

Bergman, B. and Wahlqvist, L. 1974. Synpunkter på varikocele hos unga män. Lakartidningen, 71: 2929-2930.

Boeckx, G. and Vereecken, R.L. 1985. Faut-il traiter le varicocėle dans le jeune âge? Acta Urol. Belg., 53: 660-671.

Borgmann, V. and Hasse, W. 1978. Die Varicocele testis im Kindesalter. Z. Kinderchir., 25: 236-241.

Braedel, H.U., Steffens, J. and Ziegler, M. 1988. Ergebnisse der ambulanten perkutanen transfemoralen Sklerotherapie der idiopathischen Varicocele. Saarl. Ärztebl., 10: 597-601.

Buch, J.P. and Cromie, W.J. 1985. Evaluation and treatment of the pre-adolescent varicocele. Urol. Clin. North Am., 12: 3-12.

Campbell, M.F. 1944. Varicocele due to anomalous renal vessel: an instance in a thirteen-year-old boy. J. Urology, 52: 502-504.

Cesaroli, G., Beseghi, U., Del Rossi, C., Vanelli, M., Bernasconi, S. and Ghinelli, C. 1985. Il varicocele in eta pediatrica. Pediatr. Med. Chir., 7: 443-444.

Clarke, B.G. 1966. Incidence of varicocele in normal men and among men of different ages. JAMA, 198: 1121-1122.

Diamond, A. and Ravitz, G. 1975. Venographic demonstration of a varicocele in a boy. J. Urol., 114: 640-641.

Dias, A., Hurard, T., Delcour, C. and Schulman, C.C. 1985. La varicocėle chez l'adolescent. Acta Urol. Belg., 53: 629-633.

Dias, A., Hurard, T., Delcour, C. and Schulman, C.C. 1987a. La varicocèle chez l'adolescent. Medical Trends, (4): 47-48.

Dias, A., Hurard, T., Delcour, C. and Schulman, C.C. 1987b. Het varicocele bij de adolescent. Medical Trends, (4): 55-57.

D'Ottavio, G., Lombardo, D., Pozza, D., Provenzano, F. and Zappavigna, D. 1981. Il varicocele idiopatico: considerazioni epidemiologiche e patogenetiche. In "Andrologica Chirurgica". (G. D'Ottavio and D. Pozza) Borla Ed., p. 77-99.

D'Ottavio, D. Pozza, D., Coia, L., De Leoni, M.T., De Benedetto, E. and Spera, G. 1982. Left varicocele at the pubertal age: clinical and histological evaluation for the prevention of infertility. J. Endocrinol. Invest., 5 (Suppl. 1), Abstract #122.

D'Ottavio, G., De Cadilhac, C., Ferrara, B., Pozza, D. and Zappavigna, D. 1983. Affezioni dell'apparato genitale maschile e pubertà. Att. Androl. Chir., p. 9-21.

D'Ottavio, G., Lagana, A., Pozza, D., Toscana, C. and Zappavigna, D. 1987. Il varicocele idiopatico alla puberta: aspetti epidemiologici e clinici. Atti V Congr. Naz. Soc. Ital. Androl., Bologna, Acta Medica Edizioni e Congressi, p. 233-238.

Dubin, l. and Amelar, R.D. 1971. Etiologic factors in 1294 consecutive cases of male infertility. Fertil. Steril., 22: 469-474.

El-Gohary, M.A. 1984. Boyhood varicocele: an overlooked disorder. Ann. R. Coll. Surg. Engl., 66: 36-38.

El-Gohary, M.A. 1986a. Boyhood varicocele. Br. J. Hosp. Med., 35: 183-185.

El-Gohary, M.A. 1986b. Preadolescent varicocele: a survey among schoolboys in Abu Dhabi. Emir. Med. J., 4: 39-43.

Erokhin, A.P. 1979a. [Intraoperative arterio- and venography of the testis in varicocele in children] (in Russian with English summary). Khirurgiia (Mosk.), 11: 55-58.

Erokhin, A.P. 1979b. [On the classification and incidence of varicocele in children] (in Russian with English summary). Klin. Khir., (6): 45-46.

Erokhin, A.P., Vorontsov, Y.P., Vodolazov, Y.A. and Roochkin, A.A. 1980. The pathogenesis of the left-side varicocele in children. Chir. Pédiatr., 21: 391-392.

Garnier, S. and Szatkovnik-Plot, J. 1978. Les varicocèles testiculaires en médecine scolaire. Resultats d'une courte enquête. Hyg. et Med. Scolaires, 31: 243-245.

Gattuccio, F., D'Alia, O., Pirronello, S., Di Trapani, D., Romano, C., Latteri, M.A., Amodeo, G., Alaimo, R. and Chiodo, M. 1988. Varicocele and puberty. A transversal and longitudinal survey. Acta Eur. Fertil., 19: 189-199.

Gorenshtein, A.I., Levin, I.R. and Poplavskii, K.E. 1978. [Regional venous circulatory characteristics in varicocele in children] (in Russian). Klin Khir., (8): 85-86.

Gorenstein, A., Katz, S. and Schiller, M. 1986. Varicocele in children: "to treat or not to treat" - Venographic and manometric studies. J. Pediatr. Surg., 21: 1046-1050.

Gorenstein, A., Katz, S. and Schiller, M. 1988. Surgical treatment of varicocele in children. Isr. J. Med., 24: 172-174.

Gunther, J., May, P. and Braedel, H.U. 1974. Diagnostik und Behandlung der Varikozele, 1974. Deutsche Arzteblatt, 34: 2475-2478.

Hadziselimovic, F. 1983. Im Kindesalter erkennbare Ursachen der männlichen Sterilität. Pathogenese und Behandlung. Praxis, 72: 316-323.

Harrison, R.G. 1966. Male infertility. The anatomy of varicocele. Prog. Roy. Soc. Med., 59: 763-765.

Hecker, W. Ch. 1973. Varikozele bei Jugendlichen. Dtsch. Med. Wochenschr., 98: 1684.

Hennebert, P., Wese, F. and Van Cangh, P. 1985. Le varicocèle de l'adolescent. Acta Urol. Belg., 53: 634-635.

Hienz, H.A., Voggenthaler, J. and Weissbach, L. 1980. Histological findings in testes with varicocele during childhood and their therapeutic consequences. Eur. J. Pediatr., 133: 139-146.

Hösli, P.O. 1988. Varikozele - Resultate nach Frühbehandlung bei Kindern und Jugendlichen. Z. Kinderchir., 43: 213-215.

Holschneider, A.M., Butenandt, O., Schuster, L. Schaupp, D. Tewes, G., Mengel, W. and Hamberger, J. 1978. Operative Therapie der Varikozele im Kindesalter. Z. Kinderchir., 24: 252-265.

Horner, J.S. 1960. The varicocele. A survey amongst secondary schoolboys. Medical Officer, 104: 377-381.

Hotchkiss, R.S. 1944. Fertility in man. J.B. Lippincott Co., Philadelphia, p. 194.

Isakov, Yu.F. and Arbuliev, M.G. 1969. [Pathogenic substantiation of the surgical therapy of varicocele in children] (in Russian with English summary). Urol. Nefrol. (Moskva), 34: 45-47.

Isakov, Yu.F., Erokhin, A.P., Geraskin, V.I. and Vorontsov, Yu.P. 1977. [Concerning the varicocele problem in children] (in Russian with English summary). Urol. Nefrol. (Mosk.), (5): 51-56.

Isakov, Yu.F. and Erokhin, A.P. 1979. [Comparative characteristics of the results of Ivanissevich's and Palomo's operations in children] (in Russian with English summary). Urol. Nefrol. (Mosk.), 5: 30-34.

Isakov, Yu.F., Vodolazov, Yu.A., Polyaev, Yu.A., Vorontsov, Yu.P., Safranov, V.V. and Konstantinov, K.V. 1987. [Endovascular occlusion of vessels in the treatment of some surgical diseases in children] (in Russian with English summary). Khirurgiia (Mosk.), (8): 3-5.

Ivanissevich, O. 1960. Left varicocele due to reflux. J. Int. Coll. Surg., 34: 742-755.

Janneck, C. 1979. Varikozele im Kindesalter. Chir. Praxis, 25: 323-328.

Jenny, P. and Hadziselimovic, F. 1982. Die Bedeutung der Varikozele im Kindesalter. Z. Kinderchir., 35: 90-92.

Jenny, P. and Hadziselimovic, F. 1987. Indikation zur Varicocelen Operation beim Kind und Jugendlichen. In: "Kinderchirurgie. Kongressberichte 1986". B. Thomasson and A.M. Holschneider (eds.), Hippokrates Verlag, Stuttgart, p. 141-143.

Johnson, D.E., Pohl, D.R. and Rivera-Correa, H. 1970. Varicocele: an innocuous condition? South Med. J., 63: 34-36.

Jones, M.A., Sharp, G.H. and Trainer, T.D. 1988. The adolescent varicocele. A histologic study of 13 testicular biopsies. Am. J. Clin. Path., 89: 312-328.

Kass, E.J. and Belman, A.B. 1987. Reversal of testicular growth failure by varicocele ligation. J. Urol., 137: 475-476.

Kass, E.J., Chandra, R.S. and Belman, A.B. 1987. Testicular histology in the adolescent with a varicocele. Pediatrics, 79: 996-998.

Kass, E.J. 1988. Adolescent varicocele: current concepts. Semin. Urol., 6: 140-145.

Kass, E.J., Freitas, J.E. and Bour, J.B. 1988. Pituitary-gonadal function in adolescents with a varicocele. J. Urol., 139: 207A (abstract #180).

Kawamura, K., Sumiya, H., Kataumi, Z., Fuse, H. Miyauchi, T., Ito, H. and Shimazaki, J. 1987. [Clinical study of varicocele in schoolboys] (in Japanese with English summary). Nippon Hingokika Gakkai Zasshi, 78: 113-116.

Kiszka, E.F. and Cowart, T. 1960. Treatment of varicocele by high ligation. J. Urol., 83: 713-715.

Kleinteich, B. and Schickedanz, H. 1981. Hodenvarikozelen - korrigierbare Ursache männlicher Fertilitätsstörungen. Med. Akt., 7: 290-291.

Kleinteich, B. and Schickedanz, H. 1983a. Beitrag zur Varicocele testis bei Kindern und Jugendlichen. Padiatr. Grenzgeb., 22: 383-387.

Kleinteich, B. and Schickedanz, H. 1983b. Kinderchirurgische und andrologische Aspekte der Varicocele testis. Z. ärtzl. Fortbild., (Jena), 77: 253-257.

Kraeft, H., Kriz-Klimek, H. and Holschneider, A.M. 1981. Erfahrungen in der Therapie der Varikozele des Kindersalters. Z. Kinderchir., 34: 272-274.

Lenk, S., Schönberger, B., Schöpke, D., Fahlenkamp, D., Engel, S. and Stösslein, F. 1988. Zur Behandlung der idiopathischen Varikozele bei Kindern und Erwachsenen durch Okklusion der Vena testicularis mit ablösbaren Ballons. Z. Urol. Nephrol., 81: 79-87.

Levitt, S., Gill, B., Katlowitz, N., Kogan, S.J. and Reda, E. 1987. Routine intraoperative post-ligation venography in the treatment of the pediatric varicocele. J. Urol., 137: 716-718.

Lewis, E.L. 1950. The Ivanissevich operation. J. Urol., 63: 165-167.

Lindner, H. 1982. Die Varicocele testis im Kindesalter. Kinderärztl. Prax., 50: 607-612.

Lipshultz, L.I. and Corriere, J.N. 1977. Progressive testicular atrophy in the varicocele patient. J. Urol., 117: 175-176.

Lyon, R.P., Marshall, S. and Scott, M.P. 1982. Varicocele in childhood and adolescence: implication in adulthood infertility? Urology, 19: 641-644.

Lyon, R.P., Marshall, S. and Scott, M.P. 1983. Varicocele in youth. West J. Med., 138: 832-834.

Mothes, W. 1986. Die Varikozele bei Kindern und Jugendlichen. Zentralbl. Chir., 111: 457-460.

Mothes, W. 1988. Die Varikozele bei Kindern und Jugendlichen: Nur ein therapeutisches Problem? Pädiatr. Grenzgeb., 27: 75-78.

Okuyama, A., Koida, T., Itatani, H., Sonoda, T., Aono, T., Mitsubayashi, S., Yoshida, T. and Miyagawa, M. 1981. Pituitary-gonadal function in schoolboys with varicocele and indication of varicocelectomy. Eur. Urol., 7: 92-96.

Okuyama, A., Nakamura, M., Namiki, M., Takeyama, M., Utsunomiya, M., Fugioka, H., Itatani, H., Matsuda, M., Matsumoto, K. and Sonoda, T. 1988. Surgical repair of varicocele at puberty: Preventive treatment for fertility improvement. J. Urol., 139: 562-564.

Oster, J. 1970. Varicocele in children and adolescents. Acta Paediatr. Scand. (Suppl.), 206: 81.

Oster, J. 1971. Varicocele in children and adolescents. Scand. J. Urol. Nephrol., 5: 27-32.

Palomo, A. 1949. Radical cure of varicocele by a new technique: Preliminary report. J. Urol., 61: 604-607.

Petrenko, Iu.I., Sushko, V.I., Khitrik, A.L. and Serdiuk, N.Ia. 1988. [Results of the surgical treatment of varicocele in children] (in Russian with English summary). Klin Khir., (6): 53-55.

Ponchietti, R., Grechi, G. and Dini, G. 1986. Varicocele in adolescents: Ultrastructural aspects. Acta Eur. Fertil., 17: 47-50.

Ponchietti, R., Rangel, A., Grechi, G. and Dini, G. 1987. Ultrastructural changes of Leydig cells in prepubertal varicocele. Acta Eur. Fertil., 18: 347-348.

Pozza, D., D'Ottavio, G., Masci, P., Coia, L. and Zappavigna, D. 1983. Left varicocele at puberty. Urology, 22: 271-274.

Reitelman, C., Burbige, K.A., Sawczuk, I.S. and Hensle, T.W. 1987. Diagnosis and surgical correction of the pediatric varicocele. J. Urol., 138: 1038-1040.

Ribeiro, E.B. 1956. Present conceptions of varicocele and its treatment. J. Int. Coll. Surg., 25: 179-186.

Riebel, Th. 1980. Phlebographie bei Kindern mit Varikozele. Monatsschr. Kinderheilkd., 128: 438-440.

Risser, W.L. and Lipshultz, L.I. 1984. Frequency of varicocele in black adolescents. J. Adolesc. Health Care, 5: 28-29.

Ristic, D., Bonert, D., Knezevic, Z., Selakovic, D. and Marusic, G. 1986. [Ucestalost varikokele u decaka skolskog uzrasta] (with English summary). Med. Pregl., 39: 13-15.

Rost, A., Richter-Reichhelm, M., Kaden, R. and Pust, R. 1975. Die Varicocele als Ursache von Fertilitätsstörungen. Urologe, A14: 282-286.

Russell, J.K. 1954. Varicocele in groups of fertile and subfertile males. Brit. med. J., 1: 1231-1233.

Sawczuk, I.S., Burbige, K.A. and Hensle, T.W. 1985. Asymptomatic varicocele in an infant. Clin. Pediatr. (Phila), 24: 285-286.

Sayfan, J., Soffer, Y., Manor, H., Witz, E. and Orda, R. 1988. Varicocele in youth. A therapeutic dilemma. Ann. Surg., 207: 223-227.

Saypol, D.C. 1981. Varicocele. J. Androl., 2: 61-71.

Schickedanz, H. and Kleinteich, B. 1982. Varicocele testis bei Kindern. Häufigkeit und Operationsindikation. Kinderärtzl. Prax., 50: 496-499.

Schickedanz, H. and Kleinteich, B. 1985. Zur präoperativen Phlebographie der Vena Spermatica Interna als Rezidivprophylaxe von Varikozelen bei Kindern und Jugendlichen. Radiol. Diagn. (Berl.), 26: 265-270.

Schickedanz, H. and Kleinteich, B. 1987. Varicocele testis - eine Kinderkrankheit? Z. Urol. Nephrol., 80: 93-96.

Schürholz, K.H. and Weissbach, L. 1978. Ist die kindliche Varikozele behandlungsbedürftig? Therapiewoche, 28: 1894-1897.

Steeno, O. 1972. Varicocele en infertiliteit. Tijdsch. Geneesk., 20: 1313-1315.

Steeno, O. and Adimoelja, A. 1975. Preventie van fertiliteitsstoornissen door opsporing en behandeling van varicokele op schoolleeftijd. Ann. Ver. Fertiliteitsstudie, 1: 7-8.

Steeno, O., Knops, J., De Clerck, L., Adimoelja, A. and Van de Voorde, H. 1975. Prevention of fertility disorders in male: Detection, incidence and treatment of varicocele at school and college age. Proc. VII Int. Congress School & Univ. Health and Medicine, Mexico, Nov. 24-28: p. 270-273.

Steeno, O., Knops, J., De Clerck, L., Adimoelja, A. and Van de Voorde, H. 1976. Prevention of fertility disorders by detection and treatment of varicocele at school and college age. Andrologia, 8: 47-53.

Thomason, A.M. and Fariss, B.L. 1979. The prevalence of varicoceles in a group of healthy young men. Milit. Med., 144: 181-182.

Thon, W.F., Gall, H., Danz, W. and Sigmund, G. 1989. Percutaneous sclerotherapy of idiopathic varicocele in childhood: A preliminary report. J. Urol., 141: 913-915.

Tulloch, W.S. 1952. A consideration of sterility factors in the light of subsequent pregnancies: Subfertility in the male. Trans. Edinburgh Obstet. Soc., 59: 29-34.

Vercecken, R.L. and Boeckx, G. 1986. Does fertility improvement after varicocele treatment justify preventive treatment at puberty? Urology, 28: 122-126.

Verstoppen, G.R. and Steeno, O.P. 1977a. Varicocele and the pathogenesis of the associated subfertility. A review of the various theories. I. Varicocelogenesis. Andrologia, 9: 133-140.

Verstoppen, G.R. and Steeno, O.P. 1977b. Varicocele and the pathogenesis of the associated subfertility. A review of the various theories. II. Results of surgery. Andrologia, 9: 293-305.

Verstoppen, G.R. and Steeno, O.P. 1978. Varicocele and the pathogenesis of the associated subfertility. A review of the various theories. III. Theories concerning the deleterious effects of varicocele on fertility. Andrologia, 10: 85-102.

Vorontsov, Yu.P., Vodolazov, Yu.A., Erokhin, A.P., Dmitrienkov, B.N., and Ruchkin, A.A. 1979. [Functional estimation of reduction of venous return from the testicle in left-side varicocele in children] (in Russian with English summary). Vestn. Khir., 122: 96-101.

Vorontsov, Yu.P., Vodolazov, Yu.A., Rushanov, I.I., Poliayev, Yu.A., Korznikova, I.N. and Zharov, B.V. 1985. [Endovascular occlusion of the internal seminal vein in varicocele in children and adolescents] (in Russian with English summary). Klin. Khir., (6): 37-38.

Wilms, G., Oyen, R., Casselman, T., Peene, P., Steeno, O. and Baert, A.L. 1988. Solitary or predominantly right-sided varicocele: A possible sign of situs inversus. Urol. Radiol., 9: 243-246.

Wutz, J. 1977. Ueber die Häufigkeit von Varikozelen und Hodendystopien bei 19 jährigen Männern. Klinikarzt, 6: 319-320.

Wyllie, G.G. 1985. Varicocele and puberty - the critical factor? Br. J. Urol., 57: 194-196.

APPENDIX B:

NORMAL AND ELEVATED TESTIS/INTRASCROTAL TEMPERATURE

VALUES - °C

| METHOD | n | | | MEAN | SD | RANGE |
|--------|---|---|---|------|-----|-------|
| 1 | 20 NORMAL | R | | 33.4 | 0.5 | (32.1 - 34.5) |
| MERCURY | | L | | 33.5 | 0.5 | (32.2 - 34.8) |
| THERMOMETER | 20 ELEVATED | R | | 34.7 | 0.6 | (33.3 - 35.7) |
| | | L | | 34.8 | 0.7 | (32.6 - 35.9) |
| 2 | 11 NORMAL | R | | 31.8 | 1.0 | (30.7 - 33.1) |
| INFRARED | | L | | 32.2 | 0.7 | (31.0 - 33.6) |
| | 20 ELEVATED | R | | 33.1 | 1.4 | (30.5 - 34.3) |
| | | L | | 33.2 | 1.4 | (30.7 - 34.3) |
| 3 | 6 NORMAL | R | | 33.0 | 0.5 | (32.2 - 33.5) |
| DBT | | L | | 32.7 | 0.8 | (32.7 - 34.7) |
| | 20 ELEVATED | R | | 34.1 | 0.8 | (32.2 - 35.3) |
| | | L | | 34.1 | 0.8 | (32.6 - 35.6) |

All readings were taken over testis; readings at raphe will be different. Ranges overlap.
A reading above or below mean value probably represents a correct estimate of whether
or not temperature is normal or elevated. For clinical purposes mercury thermometer
reading of >34.0 indicates elevated temperature.
1 Mercury thermometer of the immersion type (graduations 0.05°C).
2 Non-contact infrared thermometry performed with Barnes Thermal Master IT 4M.
3 DBT (Deep Body Thermometer) Terumo CTM204 with PD7 probes.

PARTICIPANTS

Richard D. Amelar, M.D.
New York University School of
   Medicine

Stanley Becker, M.D.
Johns Hopkins University School of
   Public Health

J. Michael Bedford, Vet., M.B., Ph.D.
Cornell University Medical College

Anders R.J. Bergh, M.D.
University of Umeå

André Clavert, M.D.
Centre d'Etude et de Conservation
   des Oeufs et du Sperme Humain
   Alsace

Abraham T.K. Cockett, M.D.
University of Rochester School of
   Medicine and Dentistry

Frank H. Comhaire, M.D., Ph.D.
State University Hospital, Ghent

François Eid, M.D.
Cornell University School of
Medicine

Remy Hsiung, M.D.
Faculté de Medecine Strasbourg

Fabrizio Iacono, M.D.
University of Naples

Fouad Kandeel, M.D.
Harbor Hospital, Torrance CA

Richard Levine, M.D.
Chemical Industry Institute of
   Toxicology

Kevin R. Loughlin, M.D.
Brigham Women's Hospital

Roger Mieusset, M.D.
Centre de Stérilité Masculine,
   Toulouse

Jeffrey A. Miron, Ph.D.
Boston University

Vincenzo Mirone, M.D.
University of Catanzaro

Pablo A. Morales, M.D., M.S.
New York University Medical Center

Michael Oppenheimer, Ph.D.
Environmental Defense Fund

Andrew I. Sealfon, B.S.E.E.
Repro-Med Systems, Inc.

Ahmed Shafik, M.D.
Faculty of Medicine, Cairo
   University

Joseph A. Salisz, M.D.
William Beaumont Hospital

Alfred Spira, M.D., Ph.D.
Hôpital de Bicêtre

Omer P. Steeno, M.D., Ph.D.
Universitaire Ziekenhuisen Louvain

Anna Steinberger, Ph.D.
University of Texas

Hiroshi Takihara, M.D., Ph.D.
Yamaguchi School of Medicine

Giuseppe Tritto, M.D.
Clinique Hartmann
Paris, France

Arthur C. Upton, B.A., M.D.
New York University School of
   Medicine

Veljko Vlaisavljevič, M.D., Ph.D.
Hospital Maribor
Ljubljana, Yugoslavia

Geoffrey M.H. Waites, M.D.
Special Programme of Research,
    Development and Research
    Training in Human Reproduction
World Health Organization

Adrian W. Zorgniotti, M.D.
New York University School of Medicine

$CO_2$ accumulation in testicular
tissue, 281
$^{14}C$-leucine, 34,35
$^{14}C$-serine, 34
$^{14}C$-uridine, 34,35
catecholamines, 282,289,308
cathepsin $D$, 36
cauda epididymidis, 19,20-25,
28,29
cauda fluid, 22,23,25,29
cell membrane injury, 36
cell-to-cell interactions,
33,34,43
$C57BL$ inbred mouse testis, 34
cholesterol, 179
cholesteryl ester hydrolase
($CEH$) activity, 36,179
chronic elevation of temperature, 223
chronic scrotal hypothermia,
227-230
cirsocele, 1
classification of varicocele,
304-306
climate rooms, 9,10
climatic chamber, 187
clinical diagnosis, 33,272
clinical examination, 276,278
clinically manifest varicocele,
271,273
clinically not manifest reflux,
273
closed loop model, 125
clothing, 223
cold receptors, 11,13
collagen deposition, 309
collagen fibers, 145,155,157,
158
communicating veins, 153,158,
162,163,168
computer-graphic procedure,
138,139
computer model,
*see* thermoregulation
conception, 60,62,63,66,67,69
concurrent heat exchange, 130
conductance, 124,126
conduction, 99,100,103,107,
108,124,221
congenital malformation, 54
conical venous configuration,
139
contact thermography, 241,
267-269,271,273,277-279
diagnostic value of, 261-
264
contralateral testis, 35,37,42,
161,162,284,306,307
convection, 99,100,103,108,
124,221
convective heat, 10

cooling by cerebral cortex, 14,
16
copper:constantan thermocouples,
21,25
core body temperature, 33,100,
106-109,123-126,174,175,
177,188,190,202
core testis temperature, 221,222
Coretemp $CTM$ 204, 118,253
corporeal temperature, 217
corrosion resin casting
technique, 138
cortisol, 281-283
counter-current heat exchange
($CCHE$), 123,130,202,221,
222,248,249
mechanism of testis, 137-150
multi-level compartmentation
of, 137-150
simulation models, 137,138,
141,146,150
$CCHE$ - PAMP I and II, 138
simulation experiments, 139
$CCHE$ - PAMP-FEM, 138
simulation experiments, 139
countercurrent heat exchanger,
103,104,109,123,125,126,
129-134
countercurrent thermoregulation
system, 217
cremasteric muscle, 125,129,134,
217,269
cremasterico-darto-venous
complex, 160
cremasteric plexus, 153-161,168
cryptepididymal state, 21-24,27,29
cryptic epididymis, 23
cryptorchidism, 2,33-39,41-43,
49-51,53,203,211,212,217,
218,248
bilateral, 35,37-40,42,43
experimental primary,
in rats, 179-181
experimental secondary,
179-181
unilateral, 40-43
cryptorchid rats, 35-37,39-42
cryptorchids, 20,21
ipsi-lateral testis
temperature, 21
cubital vein, 275
cutaneous temperature receptors,
9,10,13,193,213
cyclic $AMP$, 34,39
cylindrical venous configuration,
139-142,145-147

dartos, 125,129,157,159-161,
169,217
deep body intrascrotal
thermometer, 115-118

varicocele (cont.)
  ligation of, 245,246,248,
    249,296,309,312-315
  primary, 161
  repair, 296
  secondary, 297,302,311
  subclinical, 228,264,267,
    269,271,275,276,278,
    279,281,283,284
  surgery, 245
  symptomatic, 300-301,304
  venous tension in, 161
varicocelectomy, 1,2,3,229,231,
  253-260
  effect on fertility, 289-292
  failed, 201,202,228
varicocele/reflux, 272,275-279
varicocelogenesis, 161
Varicoscreen, 268,269
varicosity, 168
vasal plexus, 153-155,158,165,168
vas deferens, 20,23,24,29
vascular compartment,
  changes in, 310
vascular hemodynamics, 214
vascular microarchitecture of
  testis, 138,146
vasodilatation, 214,244
vasomotor changes, 9

vasomotor reflex, 194
vena cava inferior, 297
venography, 261,271,275-279,
  302,303
venous dilatation, 297
venous Doppler, 241
venous gradient excursion
  difference (VGED), 139-145,
  147
venous outflow, 123,129,131-134
venous plexus, 138-140,153-156,
  158,168,171
venous pooling, 283,284
venous port, 147,148
venous renal hypertension, 297
venous return, 123,124,126
  blood, 99,100,108
  flow, 133
  heat, 100
ventrobasal complex of thalamus, 14
volume-dependency, 142

wall temperature, 10,187-189
warm receptors, 11,13
water bath thermometer, 217
Womersley's theory, 145
wool keratin, 13

zero-heat-flow method, 115,116